Best Regards

Phil Fields

Introduction to
Industrial Engineering
and Management Science

**McGRAW–HILL SERIES IN INDUSTRIAL ENGINEERING
AND MANAGEMENT SCIENCE**

Consulting Editor

James L. Riggs, Department of Industrial Engineering, Oregon State University

Gillett: Introduction to Operations Research: A Computer-Oriented
        Algorithmic Approach
Hicks: Introduction to Industrial Engineering and Management Science
Riggs: Engineering Economics
Riggs and Inoue: Introduction to Operations Research and Management Science:
        A General Systems Approach

Introduction to Industrial Engineering and Management Science

PHILIP E. HICKS

Visiting Professor of Industrial Engineering
Oklahoma State University, Stillwater

McGRAW-HILL BOOK COMPANY

New York St. Louis San Francisco Auckland Bogotá Düsseldorf
Johannesburg London Madrid Mexico Montreal New Delhi
Panama Paris São Paulo Singapore Sydney Tokyo Toronto

This book was set in Press Roman
by Hemisphere Publishing Corporation.
The editor was B. J. Clark and
the production supervisor was Milton J. Heiberg.
R. R. Donnelley & Sons Company was printer and binder.

INTRODUCTION TO INDUSTRIAL ENGINEERING AND MANAGEMENT SCIENCE

4567890 DODO 898765

Library of Congress Cataloging in Publication Data

Hicks, Philip E.
    Introduction to industrial engineering and management
science.

    (McGraw-Hill series in industrial engineering and
management science)
    Includes index.
    1. Industrial engineering. I. Title.
T56.H47   658.5   76-43307
ISBN 0-07-028767-8

To VERMONT

Contents

Preface

In the past twenty-five years, there has been considerable change in the field of industrial engineering, and two fields—operations research and management science—have come into existence. The three fields possess a common body of knowledge that, for lack of a name, is referred to throughout this text as IE/OR/MS (industrial engineering/operations research/management science). This is not to imply that the three fields are the same. Indeed, each field has its own relatively exclusive specialties. There is, however, an increasing common body of knowledge concerning "human systems" of considerable interest to all three disciplines. One of the primary objectives of this text is to shed light on both the differences between these fields and the interests they have in common.

In the author's opinion, it is desirable for college students in a discipline to take an introductory course that surveys the discipline early in their program. It is hoped that such a course will convince the students that they either *are* or *are not* interested in the discipline as a life's work. If they are, serious learning can begin; if not, the search for the right discipline can continue. Thus, one primary objective of this text is to give students a sufficient insight into the nature of IE/OR/MS that they can either make a commitment to the field or be convinced to continue the search.

Within industrial engineering alone the scope of activity has greatly increased. Modern industrial engineers are educated far differently from their predecessors of twenty years ago. The lingering tendency to consider industrial engineers as "stopwatch engineers" is no longer appropriate. They must still be technically trained in and able to manage a methods engineering activity, for example, but their education today goes far beyond methods engineering. This text attempts to describe the training of a modern industrial engineer as a combination of the best from traditional industrial engineering and the best of all that has been added in recent years. This author also does not agree with those "modernists" who believe that traditional industrial engineering should be de-emphasized. Indeed, it needs to be revitalized. It is a forgotten and yet very necessary and potentially research-rich aspect of industrial engineering.

The area common to IE/OR/MS is a bewildering one today. There has been so much development, in such a short time, common to the three disciplines (and some others to a lesser degree) that describing this field to someone unfamiliar with it, such as a first-year college student, is indeed a challenge. This text is an attempt to meet that challenge.

One development of considerable interest to a broad spectrum of disciplines is what is generally referred to as the "systems approach." The common bond in IE/OR/MS is primarily a joint interest in the analysis and design of productive systems. In fact, the considerable growth in demand for industrial engineers, and operations researchers and management scientists as well, can be explained as a renewed interest in the "generalist" approach to solving many problems today which simply do not lend themselves to solution by the "reductionist" approach. It is also the opinion of this author that students should not only be introduced to their discipline early in their studies, they should be provided with a historical perspective on it. The first three chapters of this text, therefore, offer a brief but, it is hoped, relevant history of engineering, industrial engineering, and some related disciplines.

Chapters 4 and 5 survey traditional industrial engineering or production management in two parts. First, consideration is given to production systems design. This is followed by a discussion of techniques for production systems control.

Chapter 6, on operations research, gives a sampling of the type of quantification that has invaded both industrial engineering and management over the past twenty years. The techniques employed have been carefully selected in the hope of not exceeding the expected mathematical capability of a first-year college student. Because of this limitation in their selection, the techniques employed throughout the text may not be truly typical of the fields they are intended to represent. In fact, both statistics and engineering economy, so much a part of IE/OR/MS today, are almost totally unrepresented here for this reason.

Chapter 7 is intended to be a gathering together of representative "bits and pieces" related to systems. Systems seem to be in the forefront of IE/OR/MS today. In the opinion of the author, systems analysis and design will receive considerable attention in the years to come. Our knowledge of how to analyze general systems is minuscule compared to the need for analysis in this mushrooming and important technical area. This chapter includes a number of articles reprinted from the technical literature to give the reader a closer sense of what systems is all about. Each instructor should decide how all or some portion of them will best serve the educational goals of his or her course.

Chapter 8 is an attempt to summarize the present status of IE/OR/MS and to forecast future prospects in these fields.

This book was planned to support a three-semester-hour course. It can be employed for a two-semester-hour course by assigning the first three chapters as additional outside reading without lecture or by completely omitting them. Or it can be lengthened to a four-semester-hour course by including topics of special interest to the instructor, such as a detailed explanation of the specific departmental curriculum, or by assigning special report topics to the students.

At various places throughout the text, attention is directed specifically to productivity. After all, this is what IE/OR/MS is all about. Americans have enjoyed a standard of living and a sense of national security that is unequaled anywhere in the world. Our productive ability, to a significant extent, underlies this enviable position. Any reasonable means to increase productivity should be sought, regardless of whether it is an extension of the traditional methods or an entirely new creation. It is the blending of the old and new that represents the future of IE/OR/MS.

Thanks is due to Ginny Miller and Olivia Garcia for their early help in typing the manuscript. Special thanks is given to Mary Ann Westbrook for her most able assistance in completing the manuscript typing.

Philip E. Hicks

Introduction to
Industrial Engineering
and Management Science

Part One

The Past

Chapter 1

A Brief History of Engineering

Human history becomes more and more a race between education and catastrophe.

H. G. Wells

One of the first engineers in this world may well have been a fellow by the name of Joe Ogg. He is the lead in a cartoon film (4) produced for the American Institute of Industrial Engineers depicting primitive attempts at industrial organization. The film introduces concepts such as specialization of labor, methods study, material handling, and quality control, with respect to the production of arrows and hides. Ogg's son, Junior, emerges as a hero because his radical concepts are readily accepted by his peers and the fruits of their efforts immediately validate his approach. This could only happen in the movies.

In contrast, Sprague de Camp states (6, p. 13), "The story of civilization is, in a sense, the story of engineering—that long and arduous struggle to make the forces of nature work for man's good." In this sense, it is obvious that engineering is as old as civilization itself. What may not be readily apparent is that although industrial engineering is the newest of the major fields of engineering (i.e., civil, mechanical, electrical, chemical, and industrial engineering), fundamental principles of industrial engineering were employed in the age of Ogg. This age lies somewhere between 1 million and 10,000 years ago.

At this time humans were entering an "agricultural revolution," a revolution as important as the "industrial revolution" of more recent times. The change from a nomadic existence to one predicated on a relatively fixed location for raising crops and edible animals represented a necessary precondition for industrial development. Some historians believe these changes first occurred in Syria and Iran at about 8000 B.C.

Three to four thousand years later, villages developed in the river valleys along the Nile, Euphrates, and Indus Rivers, and with this centralization of population came the beginning of civilization in the form of writings, local government, and eventually science. The first engineers were architects, irrigation specialists, and military engineers. The villages depended on irrigation for raising crops, but the best location for farming was often a poor location for military defense; consequently, the building of walls for defending a city became one of the earliest assignments of engineers of antiquity. Shelter is one of the basic human needs, and it is fair to concede that architects preceded engineers in meeting this need. In the design and construction of public buildings, however, a point was reached at which engineering skills became essential to success.

Innovation in the form of inventions was extremely slow during these times. The time required to survive, with respect to fulfilling both military and agricultural needs, left little time for experimentation. Because of the limited forms of communication between the disparate populations of the day, many of the earlier inventions had to be reinvented at a later date before they became permanent features of civilization. Villages along the trade routes between China and Spain progressed more rapidly because the knowledge of innovations made at other locations reached them. It is a well-known fact of highway siting today, as it was in the expansion of the railroad system in the United States, that development follows routes of transportation. In any event, those who lived away from the trade routes typically progressed somewhere between "not at all" and "partially," as measured against the progress of trade route populations.

In considering this period with respect to the development of technology, it is academic to attempt to distinguish between invention and engineering. Invention in those times constituted the beginning of engineering, which was then and still is the process of employing scientific knowledge for the betterment of humanity. Engineering has always progressed because it embodies an ever-increasing store of knowledge concerned with meeting human needs. Compared to the impact that engineering made on human development during those times, there is relatively little in recorded history about engineering. Historical records of those times were kept primarily by priests, who were more inclined to record detailed accounts of the poets and kings of their day than to describe physical things. Today, as then, engineering is one of those "taken for granted" professions. If you were to take a sidewalk survey and ask, "Who put American astronauts on the moon?", the likely response would be "our scientists." A careful examination of personnel records in the space program, however, would uncover a considerable predominance of engineers and technicians over scientists. Sprague de Camp states (6, p. 25), "Everybody has heard of Julius Caesar—but who knows about his contemporary Sergius Orata, the Roman building contractor who invented central indirect house heating?" The material improvements brought about by engineering have had as much effect on human progress as any of the political, economic, or social developments to date, yet the public takes much of this development for granted.

Science and engineering have made enormous strides in the past three centuries, whereas before the eighteenth century they had highly variable rates of progress. At times they got off on "benders" such as astrology and alchemy, and at other times they were judged theologically unacceptable. Throughout history, despite barriers along the way, science and engineering have always moved closer to ultimate truth.

EGYPTIAN ENGINEERING

Some of the greatest engineering works of all time were performed by the Egyptians, one of the earliest on record being the wall of the city of Memphis. This old capital was located approximately twelve miles north of where Cairo is today. Sometime after the building of the wall, Kanofer, the royal architect of Memphis, had a son named Imhotep, whom historians consider to be the first engineer to be known by name. He was known more as an architect than as an engineer, but his accomplishments included fundamental elements of engineering.

A number of factors made Imhotep's accomplishments possible: (1) a religious belief of the time that to enjoy the hereafter it was necessary to preserve one's remains intact, (2) an almost unlimited supply of slave labor, and (3) a patient attitude on the part of those who controlled the resources of the day. During the reign of King Joser, the time was ripe for Imhotep's invention of the pyramid. The technical skills demanded in designing, organizing, and controlling a project of this magnitude distinguishes it as one of the earliest of the major engineering feats of all time.

Of all the pyramids, King Khufu's was the greatest. The Great Pyramid, as it is now called, was 756 feet square at the base and originally reached a height of 480 feet. It contained approximately 2,300,000 blocks of stone, each weighing $2\frac{1}{2}$ tons on the average. The accuracy with which the base was oriented with respect to north-south, east-west alignment was approximately 6 minutes of arc maximum in error, and the base was less than 7 inches from being a perfect square. These were remarkable feats considering the limited knowledge of geometry and lack of instruments at that time. Imhotep was honored for his accomplishments by being added to the list of Egyptian gods of the day after his death. It is interesting that pyramid building, which started around 3000 B.C., lasted for only about one hundred years. Yet these massive engineering structures are second in magnitude only to the Great Wall of China among the structures of antiquity.

Pyramid building was truly a feat when we consider that neither the screw nor the block and tackle were known. There was no mechanical advantage attained other than by means of the lever. The inclined plane was used extensively, however, and one of the leading theories about how the pyramids were built is that inclined planes or ramps were built around the pyramid, eventually burying it. When the top was reached there followed the task of "unburying" the pyramid. This explains how simple methods supported by unlimited labor produced "hard to believe" results.

The Egyptians were some of the earliest draftsmen of all time. Drawings were essential to successful pyramid building and were recorded on papyrus, stone slabs, and even wood. Some of the problems they had to overcome, for example, are suggested by records which indicate that they believed π to equal exactly 3.16. Consider also that the horse was not used as a work animal until 1300 years later. Although the Egyptians built impressive structures, they produced few truly significant innovations in building with stone—brute force and size were their forte.

The Egyptians were also builders of dikes and canals and had elaborate systems of irrigation. When the land to be irrigated was above the level of the river, a device called a "swape" or "shaduf" was used to lift water to a level from which it would drain down to the land. The swape consists of a bucket attached by rope to the long end of a pivoted pole counterweighted at the short end. The operator adds force to the counterweight to lift the bucket and to rotate the pole about the pivot. What seems so amazing today is that many of these earliest devices are still in common use in Egypt.

As early as 3200 B.C. the art and science of irrigation had become a ceremony conducted by the king and called "cutting of the dikes." King Menes, the first dynastic pharaoh, had conquered lower Egypt, yet he personally concerned himself with the administration and ceremonies associated with the irrigation system. A principal responsibility of governors of nomes in Egypt was the continuous surveillance and maintenance of canals. The lack of materials to provide holding power, other than reeds from the marshes, made early detection of leaks in the mud banks essential to prevent disaster.

MESOPOTAMIAN ENGINEERING

Another great watershed culture developed in northern Iran between the Tigris and Euphrates Rivers. The Greeks named this land Mesopotamia—"the land between the rivers." Although Egyptians excelled in the art of building with stone, much of our science, engineering, religion, and commerce are derived as much from Iran as Egypt. At the beginning of recorded history a people of unknown origin called the Sumerians built city walls and temples and dug irrigation canals that may have been the first engineering accomplishments of this world. The Sumerians were gradually overtaken by a considerable influx of Arabian nomads interested in becoming farmers and city dwellers. The city of Babylon which resulted housed a number of short-lived empires and was later conquered by the Assyrians.

As in Egypt, surveillance of the mud banks of canals was an important activity. For four thousand years these canals served a population more dense than the one there today. When the Mesopotamians had learned to irrigate their land and wall their cities, their attention turned to the construction of temples. Historians indicate that the tradition of having a politician break ground for a public building by turning a spade of earth started in Mesopotamia.

The Assyrians were a militaristic people and in those times, as now, war seems to have been a catalyst to invention. The Assyrians were the first to employ weapons made of iron. The manufacture of iron had been known for seven or eight centuries, having been discovered by a tribe called the Chalybes in Asia Minor. The Assyrians also invented the siege tower, and this device became a standard piece of military equipment for the next two thousand years, until the invention of the cannon made it obsolete. At different times during this period the device was also called a "belfry" or "helepolis." A further Assyrian improvement was to add the battering ram to the siege tower.

At around 2000 B.C. the Assyrians made a significant advance in transportation. They learned that a horse could be tamed and ridden. This innovation resulted in a considerable military advantage; they had invented cavalry.

GREEK ENGINEERING

Around 1400 B.C. the center of learning shifted, first to the island of Crete, and then to the ancient city of Mycenae, Greece. Their water and irrigation systems were modeled after earlier Egyptian systems, but demonstrated an improved use of materials and labor. The engineers of this period, like Japanese engineers after World War II, were better known for their use and development of borrowed ideas than for creativity and invention.

Most Greek history begins at about 700 B.C., and the period from approximately 500 to 400 B.C. is often called the "Golden Age of Greece." An amazing number of significant accomplishments in the areas of art, philosophy, science, literature, and government were the reason for naming this narrow slice of time in human history.

At about 440 B.C. Pericles retained architects to build temples on the Acropolis, which was a large, flat-topped rock overlooking the city of Athens. A path up the western slope led through a huge gateway at the top known as the Propylaia. The marble ceiling beams of this structure were reinforced by wrought iron, which is the first known use of metal as a structural component in a building design.

The Parthenon, another classic structure of ancient Greece, has steps leading to it that are not horizontal. The steps curve upward to the center to give the appearance, as the result of an optical illusion, of being horizontal. In bridge building today it is generally recognized that bridges that curve upward to the center appear comforting, if not horizontal, whereas horizontal bridges give the appearance of sagging at the center.

Those who supervised the building of these ancient structures did not have a title that translates to "engineer." Such a person was called an "architekton," meaning one who had served an apprenticeship in the standard methods of constructing public buildings. Architects received about a third more pay than masons. There was no classroom training, and what they learned they learned at work. It was totally "on-the-job" training, as we call this type of learning process today.

There is little question that Aristoteles of Stagyra, commonly referred to as Aristotle today, was one of the great geniuses in human history. His contributions have been some of the most significant in the history of science. There is some question among historians as to the authorship of a short article entitled "Mechanika" (Mechanics); although most historians credit Straton of Lampsakos with the authorship, some credit Aristotle. This uncertainty about authorship is unfortunate in light of the fact that Mechanika is generally believed to be the first known engineering text. Such fundamental engineering concepts as the theory of the lever were discussed in this article. It also contains a diagram illustrating a gear train of three gears, shown as circles, which is the first known description of gearing. These gears more than likely did not have teeth. A great deal of slippage had to take place before the advantage of teeth and the means to produce them became known.

Some measure of the difficulties under which engineers labored because of their technical ignorance is suggested by an underlying assumption in the design of the water clock of Ktesibios, in Alexandria, at about 270 B.C. It was assumed that the time between sunrise and sunset was 12 hours; therefore an hour of time was variable as a function of the time of year, being longest in midsummer and shortest in midwinter. The clock compensated accordingly for the time of year.

The greatest contribution to engineering by the Greeks was the discovery of science itself. Plato and his pupil Aristotle are probably the best known of the Greeks for teaching the belief that there is a consistent order to nature which can become known. A belief in a consistent, repeatable order in nature in the form of natural laws is essential to the existence of science. Aristotle was probably the greatest physical scientist of this period in history; his works have represented the foundation of science for the past 2000 years. The combined abstract thought of Plato, Aristotle, and Archimides has probably not been equaled since their time.

A distinction must be drawn, however, between their contributions to the philosophy of science and to innovation in engineering. Whereas they excelled in abstract thought, their accomplishments in engineering can only be judged as modest. Their philosophical search for truth, particularly that of Plato and Aristotle, included a snobbish disdain for experimentation or invention, which by its very nature involved manual work. Aristotle believed this type of work should be performed by slaves or "base mechanics," and that they should not be permitted citizenship. This snobbery often appears to exist in mathematics departments, as judged by some engineering professors at many universities in this country. In fact, however, they simply have different goals, which we cannot afford to do without. Mathematicians are continually reproving old truths and searching for new truths, whereas engineers are anxious to learn what mathematics exists so that they may apply it in our present world. This dual role of science and engineering first becomes distinguishable in Greece. The Greeks, specifically Dionysius, were the first known to hire people to invent machines of war. This practice has been passed down through the ages until today, in this country, a large fraction of our federal budget is annually allocated to the defense of our way of life. It has not yet become apparent, since the days of Dionysius, that a nation can logically discontinue expenditures for defense and transfer a like amount to the enjoyment of life.

Another reason for the failure of Greece to produce engineering structures comparable in magnitude to those of earlier watershed societies was a decline in the use of slave labor for accomplishing such feats. The Greeks developed a study called "hubris," which was a belief in the need to exercise moral and physical laws of restraint in the application of a mastered technique. They came to believe that straining humans and beasts to the limit and beyond in quarrying and transporting stone monoliths of several tons was both inhumane and unnecessary. These dehumanizing exercises had reached a peak in Egypt and appear at various times later in history, for example, Stonehenge in Britain a thousand years later. What the Greeks may have lacked in engineering accomplishments, however, they more than made up in the fields of art, literature, philosophy, logic, and politics. It may be interesting to note that surveying, as developed by the Greeks and later by the Romans, is considered the first applied science in engineering, and was to be practically singular in existence as an applied science for the twenty centuries to follow.

The Greeks attempted to employ disciplined order in military engagements. Their armies marched into battle with dressed ranks and files keeping step to flutes. They were convinced that a solid front of spears and shields was superior to the onrush of a mob. It is difficult to judge now whether disciplined order or the equipping of their soldiers with steel for the first time made them superior in battle. Steel weapons, compared to the then conventional soft wrought iron or bronze weapons, obviously offered a sizable advantage.

In 305 B.C. Demetrius had produced the most fearsome war machine up to that time. The belfry, designed by the engineer Epimachos, had nine stories, was 50 to 75 feet square at the base, and stood between 100 and 150 feet tall. The entire rig weighed approximately 180 tons, had eight huge iron-rimmed wheels, and was pushed and pulled by 3,400 soldiers (belfry pushers). Each of the nine stories had a water tank and buckets for putting out fires on the belfry. One of the belfry defenses that developed seems now to have been quite imaginative. This method of defense involved determining the path of a belfry, collecting sewage and wash water, and even adding scarce drinking water if necessary, and then dumping all of this at night in the path of the belfry. A belfry was a sufficiently unmaneuverable monster that if enough liquid was added to the right soil and given enough time to soak, a bogged belfry was the inevitable outcome. This is an early example of the commonly expressed belief in military weaponry circles today that for every offensive weapon there is at least one potentially effective defensive weapon. The belfry was a fairly standard offensive weapon for years until the invention of the cannon made walls ineffective as a line of defense.

Archimedes, although best known for what we now call "Archimedes' principle," was a versatile mathematician and engineer. He made many important discoveries in the areas of plane and solid geometry, such as a more accurate estimate of π and laws for finding centers of gravity of plane figures. He also determined the law of leverage and proved it mathematically. While in Egypt he invented what has been called "Archimedes' screw," which is a helix enclosed within a tube and turned to lift water. This device was used extensively in water systems and mining centuries later. Archimedes was also a shipbuilder and an astronomer. Typical of his inventiveness was a crane he had installed on one of his larger ships, which was used with a grappling hook to lift the bow of a small attacking ship until its contents were emptied, and then release it aft first into the water. Archimedes was one of the great minds of all time.

ROMAN ENGINEERING

The Roman engineers had more in common with their counterparts in the earlier watershed societies of Egypt and Mesopotamia than the Greek engineers who had just preceded them. The Romans used simple principles, slave labor, and time to produce far-reaching practical improvements for the benefit of the Roman Empire. Roman contributions to science, compared to those of the Greeks, were limited; however, they did produce a number of outstanding soldiers, leaders, administrators, and jurists. The Romans extensively applied much of what had preceded them, and they may well have been the best engineers of antiquity. What they lacked in originality they made up in extensive application throughout an expanding empire.

Roman engineering for the most part was civil engineering, particularly the design and construction of permanent structures, such as aqueducts, roads, bridges, and public buildings. One exception was military engineering, and another minor exception, for example, was electroplating. The profession of "architectus" was both respected and popular; in fact, Emperor Tiberius' son Drusus was an architect.

One interesting innovation of architects of the day was the reinvention of central indirect house heating, which had originally been used around 1200 B.C. in Beycesultan, Turkey. The original invention occurred at a time when, because of the lack of communication and patent protection, significant inventions sometimes had to be reinvented before they became a permanent part of technology. Strangely enough, after the fall of the Roman Empire, central indirect house heating did not appear again until modern times.

One of the great triumphs of public construction during this period was the Colosseum, which was the largest public gathering place of its kind until the construction of the Yale Bowl in 1914.

Roman engineers made significant improvements in road building. One reason for this was a belief that communication was essential in maintaining an expanding empire, and another was a belief that a road properly constructed would last for a long time with a minimum of maintenance. Roman roads were known to last for as long as one hundred years before needing a major overhaul. Only in very recent times has road construction returned to a "high initial cost–low maintenance" basis.

Probably the best-known triumph in road building of antiquity is the Appian Way. This road was begun in 312 B.C. and was the first major surfaced road in Europe. The road originally went 162 miles from Rome to Capera. In 244 B.C. it was extended to Brindisi, and it was such a prestigious accomplishment in its day that both sides of the road leaving Capera were lined with funeral monuments of aristocrats of the day.

Heavy traffic was a reality in Rome during this time. Julius Caesar once ordered all wheeled vehicles off the city streets during the day in the hope of providing a partial solution to the traffic problems. At the peak of the Roman Empire, the road system contained approximately 18,000 miles of roads spread between the Euphrates Valley and Great Britain.

Roman aqueducts, as compared to earlier ones, were larger and more numerous. Most of what is known today about the Roman water system is derived from the book *De Aquis Urbis Romae* by Sextus Julius Frontinus, who was Curator Aquarium of Rome from A.D. 97 to 104. Frontinus kept records on water usage that indicate that 17 percent was used by the emperor, 39 percent was privately used, and 44 percent was publicly used. It is estimated today that between 100 and 300 million gallons of water were used each day in Rome. The 44 percent for public use was further subdivided into 3 percent for military barracks, 24 percent for public buildings, which included eleven public baths, 4 percent for theaters, and 13 percent for fountains. There were 856 private baths at the time of his report. In any case, the administration of water in Rome was a sizable and important task. Much of the water that was supposed to enter the city never did because of hidden taps to private users. Even in Roman times water gates were a problem.

Roman aqueducts were all built employing essentially the same design, which involved semicircular arches of stone set on a row of piers. When an aqueduct crossed a gorge it often required multiple levels of arches. One of the best preserved today is the Pont du Gard at Nimes, France, which has three levels, as shown in Fig. 1-1. The lowest level also included a road.

Romans used lead pipes and later suspected that they were unhealthy. However, lead poisoning was not specifically diagnosed until Benjamin Franklin wrote a letter in 1768 concerning its use.

Emperor Claudius in A.D. 40 had his engineers attempt the drainage of Lake Fucinus by use of a tunnel. The outflow was to be used for irrigation. On the second attempt to tap the lake the outflow was far greater than was expected, and a few picnic tables and guests were lost to the occasion. This angered the emperor's wife, and she lost her temper. Later, reflecting on the fact that the

Figure 1-1 Pont du Gard aqueduct.

emperor might retaliate for her outburst, she decided to poison him with toad-stools.

One of the early alchemists of this time, a woman referred to as Mary the Jewess, is believed to have invented the still. If not, she offered the first recorded description of a still.

A book by Athenaios, called *Mechanikos*, discusses siege engines, flying bridges, rams, tortoises, belfries, and other similar devices. They were improvements in the stock of military weaponry of the day. At about A.D. 100 one of the best technical writers of all time, Hero of Alexandria, produced engineering manuscripts entitled *Mechanics, Pneumatics, Siegecraft, Automation-making, The Surveyor's Transit,* and *Measurement and Mirrors.* He was a prolific technical writer. He also developed a steam engine, or "aeolipile," which worked on the reaction principle similar to that of a rotary lawn sprinkler.

At about A.D. 200 a battering ram called an "ingenium" was invented for attacking walls. Many years later the operator of an ingenium was referred to as an "ingeniator," which many historians believe to be the origin of the word engineer.

Roman engineering declined after A.D. 100 and advances were modest. One factor believed to have contributed to the eventual fall of the Roman Empire, at about A.D. 476, was that while Roman science and engineering had stagnated during this period, the same was not true with respect to the barbarians to the north. Another factor that retarded growth in science and engineering was laws enacted about A.D. 301 that were intended as reforms by Diocletian in the form of price and wage controls, and laws that forced every man in the empire to pursue his father's trade. This was done, at least in part, in the hope of providing economic stability.

One innovation during this period was the invention of public street lighting in the city of Antioch at about A.D. 350.

The fall of Rome is synonymous with the end of ancient times. In the times that followed, the medieval period, caste legislation and religious influence greatly retarded engineering development. The period A.D. 600 to 1000 is referred to as the "Dark Ages" by many historians. Engineering and architecture ceased to exist as professions during this period.

It was in the thirteenth century that St. Thomas Aquinas argued at length that science and religion were compatible. Ghazzali, who was a scholar of Greek science and philosophy, concluded that science drew people away from God and therefore was bad. Europeans followed St. Thomas while Islam followed Ghazzali. This difference in philosophy, to a large extent, underlies the highly divergent rates of technical development in these two different cultures. It is not universally accepted today that either of these great scholars was correct. It is apparent, however, that Europe has enjoyed technical superiority in the world and its rewards for centuries, whereas technical development in the Islam culture has been limited.

In the years immediately following the fall of the Roman Empire, technical leadership did shift to the Byzantine capital of Istanbul. Massive walls, as high as 40 feet in some places, held off barbarians for ten centuries to follow.

ORIENTAL ENGINEERING

After the fall of the Roman Empire engineering development shifted to India and China. The ancient Indians were skilled in ironwork, and they possessed the secret of good steel before Roman times. Austria and India were the two main iron centers during the rise of the Roman Empire. At a later date, ingots of Indian steel were used in Damascus by Syrian smiths to make Damascene sword blades. They represented one of the few truly superior steels available at that time. For approximately two centuries Jundishapur, India, was the world's science capital.

At about A.D. 700 a Mesopotamian monk by the name of Severos Sebokht informed Western civilization of the Indian system of numbers, which we have since called Arabic numerals.

One of the greatest feats of all time was the Great Wall of China. The distance from one end of the wall to the other is approximately 1400 miles; however, there are more than 2550 miles of wall in total. Most of the wall is about 30 feet high, 25 feet thick at the base, and narrows to about 15 feet thick at the top. There is a paved road running along the top.

The wall has 25,000 towers on its main part with another 15,000 towers separated from the main wall. The wall was not high enough to keep invaders from scaling it, but they had to leave their horses at the wall. Without horses they were often in trouble with local garrisons who were on horseback, and the invaders were often content to make it back over the wall to the waiting horses.

Canals have flourished in China for thousands of years. Most Chinese canals are of a size suitable for irrigation but limited with respect to navigation, and the canal lock was not known during this time. They did use sluices, but they were of limited value. After 3000 years of canal digging, the Chinese canal system has in excess of 200,000 miles of canals. The largest canal in China, the Yun-ho or Grand Canal, extends 1200 miles from Tientsin to Hangchow. It was constructed over a span of a thousand years. It is one of those examples of Oriental timeless determination and patience.

The Chinese were early and rather unique bridge builders. Some of their earliest bridges were suspension bridges employing cables made from bamboo fiber. One of the most important inventions of all time, paper, was made by the Chinese. Tsai Lun, at about A.D. 105, wrote a report to the emperor of the day on a process for making paper and was acclaimed for his development. Block printing was used later in the tenth century in the kingdom of Shu for producing the first paper money in the world.

Gunpowder is believed to have been invented by the Chinese. It is somewhat ironic that this Chinese invention, with the cannon, should eliminate walls as a viable defense.

The Chinese developed geared machinery at a very early date. Some historians believe geared machinery in China goes back as far as 400 B.C. The Chinese were the first to invent escapement mechanisms for clocks. Later, Peter Henlein of Nuremberg, Germany, around A.D. 1500, invented the spring-driven watch. Maximilian I of Bavaria in 1800 used to quip (6, p. 304), "If you want troubles, buy a watch." The first watches were about the size of a present-day alarm clock, were hung from a chain, and had only one hand.

Another important discovery of the Chinese was the magnetic needle. It rapidly spread into common use around A.D. 1200.

The Arabs later learned from the Chinese the method for making paper, and produced paper in large quantities. Communication of ideas significantly increased after this time. Chemistry greatly increased as a science in Arabia, and the process for making gunpowder was also learned and spread quickly. Gutenberg's invention of movable type in Germany was another giant step toward improved communication. It was now possible to disseminate knowledge without hand copying. The breadth of dissemination of knowledge made possible by printing was a precondition to the extensive developments that have followed.

EUROPEAN ENGINEERING

The Middle Ages, sometimes referred to as the medieval period, covered a span of years from approximately A.D. 500 to 1500, whereas the Dark Ages are usually associated with the period A.D. 600 to 1000. The professions of engineer and architect were not recognized during the Dark Ages, and this work was continued by artisans, such as master masons. The literature of the Dark Ages was predominantly religious in nature, and science and engineering were de-emphasized by those in power. The feudal ruling powers were conservative and in many ways attempted to maintain a status quo. Legislation, for example, required a son to follow his father's trade. In the 1500s, however, a number of important scientific discoveries took place in engineering and mathematics, which suggests that although science had been de-emphasized, a revolution in thinking concerning the nature and behavior of matter was taking place. Motion, force, and gravity received considerable attention in the high Middle Ages and beyond.

One invention that contributed to the demise of the way of life involving castles surrounded by walls was the cannon. The cannon was invented in Germany in the fourteenth century, and by the fifteenth century castles were no longer defensible.

The Renaissance, which literally means "rebirth," started in Italy during the fifteenth century. The rediscovery of the classics and a revival in learning lead to a reevaluation of the scientific concepts of antiquity.

One of the obvious limits to the development of engineering has been the ease with which comparative thoughts could be communicated from one person to another. The invention of eyeglasses in 1286, and considerable growth in printing in Europe in the fifteenth century, were two critical events related to the expansion of engineering thought. Of course, another major factor at any time is the attitude of a society toward a profession. During the Renaissance, engineers were again members of a respected profession and some were even well paid. Filippo Brunelleschi was a well-known engineer of the early 1400s and, like most well-known Renaissance engineers, was both a military and a civil engineer, as well as an architect and artist. One of his developments was perspective drawing.

In 1474 the Republic of Venice developed the first formal patent law, and in

1594 Galileo was issued a patent on a device for lifting water. Although the early patent law developed in Venice was in need of considerable improvement before it could offer effective protection, it did represent a first attempt at encouraging innovation by protecting the commercialization of an invention. However, the cost today for acquiring a patent and the delays associated with the patent system certainly limit its effectiveness as an inducement for the average citizen.

In 1514 Pope Paul III was faced with the problem of replacing the architect Bramante after his death during the rebuilding of St. Peter's Cathedral. An artist and engineer by the name of Michelagniolo Buonarroti, known to us as Michelangelo, was selected to see the project through to conclusion. His work in completing St. Peter's Cathedral is well known. It is less known, however, that in Florence, and again in Rome, he was called on to design fortifications for the city. After building them, he was convinced the fortifications would not hold because of the incompetence of the defenders, and slipped through the lines of the attacking enemy. Michelangelo was a stubborn individualist, and his face was disfigured by a broken nose he received in a fight with a fellow sculptor.

Another of many enemies of Michelangelo was Leonardo da Vinci. Like Michelangelo, da Vinci is best known for his artistic endeavors. However, he was an active, almost continuously absorbed scholar. He tried to master astronomy, anatomy, aeronautics, botany, geology, geography, genetics, and physics. His studies of physics represented a broad coverage of what was known at that time. He had a scientific curiosity that got him into trouble on occasion. He was dismissed by Pope Leo X when the Pope was informed that da Vinci was learning human anatomy by dissecting the real thing. From a purely scientific point of view, what better way is there to learn human anatomy?

In 1483 da Vinci moved to Milan. He submitted the following résumé to Duke Lodovico Sforza in the hope of gaining employment (6, pp. 363–364):[1]

Having, My Most Illustrious Lord, seen and now sufficiently considered the proofs of those who consider themselves masters and designers of instruments of war and that the design and operation of said instruments is not different from those in common use, I will endeavor without injury to anyone to make myself understood by your Excellency, making known my own secrets and offering thereafter at your pleasure, and at the proper time, to put into effect all those things which for brevity are in part noted below—and many more, according to the exigencies of the different cases.

I can construct bridges very light and strong, and capable of easy transportation, and with them pursue or on occasion flee from the enemy, and still others safe and capable of resisting fire and attack, and easy and convenient to place and remove; and have methods of burning and destroying those of the enemy.

I know how, in a place under siege, to remove the water from the moats and make infinite bridges, trellis work, ladders, and other instruments suitable to the said purposes.

[1] Excerpt from *The Ancient Engineers* by L. Sprague de Camp. Copyright © 1963 by L. Sprague de Camp. Reprinted by permission of Doubleday & Company, Inc., and Barthold Fles, Literary Agent.

Also, if on account of the height of the ditches, or of the strength of the position and the situation, it is impossible in the siege to make use of bombardment, I have means of destroying every fortress or other fortification if it be not built of stone.

I have also means of making cannon easy and convenient to carry, and with them throw out stones similar to a tempest; and with the smoke from them cause great fear to the enemy, to his grave damage and confusion.

And if it should happen at sea, I have the means of constructing many instruments capable of offense and defense and vessels which will offer resistance to the attack of the largest cannon, powder, and fumes.

Also, I have means by tunnels and secret and tortuous passages, made without any noise, to reach a certain and designated point, even if it be necessary to pass under ditches or some river.

Also, I will make covered wagons, secure and indestructible, which entering with their artillery among the enemy, will break up the largest body of armed men. And behind these can follow infantry unharmed and without any opposition.

Also, if the necessity occurs, I will make cannon, mortars, and field pieces of beautiful and useful shapes, different from those in common use.

Where cannon cannot be used, I will contrive mangonels, dart throwers and machines for throwing fire, and other instruments of admirable efficiency not in common use; and in short, according as the case may be, I will contrive various and infinite apparatus for offense and defense.

In times of peace I believe that I can give satisfaction equal to any other in architecture, in designing public and private edifices, and in conducting water from one place to another.

Also, I can undertake sculpture in marble, in bronze, or in terra cotta; similarly in painting, that which it is possible to do I can do as well as any other, whoever he may be.

Furthermore, it will be possible to start work on the bronze horse, which will be to the immortal glory and eternal honor of the happy memory of your father, My Lord, and of the illustrious House of Sforza.

And if to anyone the above-mentioned things seem impossible or impracticable, I offer myself in readiness to make a trial of them in your park or in such a place as may please your Excellency; to whom as humbly as I possibly can, I commend myself.

Duke Lodovico Sforza evidently was not impressed and did not hire da Vinci after reading his résumé. Da Vinci was later commissioned by the duke as the result of an association da Vinci had with another artist. The duke had a habit of paying late, if at all, however, which resulted in da Vinci quitting once, but reconsidering later.

Leonardo da Vinci was one of the great geniuses of all time. He anticipated many engineering developments that were to follow; to name a few: the steam engine, machine gun, camera, submarine, and helicopter. It is likely, however, that he had little influence on engineering thought at the time. His research was an unpublished mish-mash of thoughts and sketches. He was an impulsive researcher and never summarized his research for the benefit of others through publication. His research was recorded from right to left in his notebooks, possibly for ease of writing because he was left-handed.

Another great genius of the time was Galileo. At the age of 25 he was appointed professor of mathematics at the University of Pisa. He studied mechanics, discovered the fundamental law of falling bodies, and studied the harmonic motion behavior of the pendulum. He lectured on astronomy in Padua and Florence, and was later called before the Inquisition in 1633 because of his belief that the sun, not the earth, was the center of our universe. In 1638 he published probably his greatest mathematical works which were shortly thereafter placed on the Index Expurgatorius, prohibiting their being read in all Catholic countries. In his later life, under house arrest, he concentrated on the less controversial topic of mechanics.

Trusses to support roofs were employed in the medieval period, but they were cumbersome and often included members that added to the weight of the truss but did not contribute to its strength. Truss design was not understood at the time. Because of the use of trial and error methods in the design of structural members, public buildings, particularly churches, were known to collapse on unsuspecting believers. The roof of the Beauvais Cathedral collapsed twice in the thirteenth century, and in the sixteenth century a spire was added which shortly thereafter fell to the ground. Of course, cathedrals were and still are monumental structures with large open spans, which have always challenged the ingenuity of architects and engineers. Adrea Palladio is believed to have been the first engineer to truly understand the forces in trusses and offered bridge truss designs in 1570, in Venice, in which all of the bridge members served a useful purpose in the design.

In 1560 Giovanni Battista della Porta started a society in Naples called the Academy of the Secrets of Nature. It was similar to such predecessors as Plato's Academy, the Lyceum of Aristotle, and the Museum at Alexandria. Considerable communication was occurring during this time between a number of European scientists. The academy was closed, however, because of clerical suspicion. In 1603 the Academy of the Lynx was established and it still exists today. Galileo was one of its members. Its membership hoped to put nonmonastic monasteries at various locations around the world. The Royal Society of London was publicly chartered in 1662 following a series of secret meetings. Boyle, Hooke, and Newton were members. In 1666 the French Academy was formed, and in 1700 the Berlin Academy came into being.

In 1540 Biringuccio wrote an outstanding treatise on metallurgy, and in 1912, it was translated into English by Herbert and Lou Henry Hoover. Herbert Hoover was a young engineer at the time; he is the only engineer in American history to have become president of the United States.

One of the most important discoveries in the history of engineering mechanics was made by Simon Stevin in the Netherlands in the late 1500s. By considering a "triangle of forces" he made it possible for engineers to deal with resulting forces acting on structural members. Stevin wrote a treatise on fractions and also did work that later led to the development of the metric system.

A number of significant mathematical developments were taking place at this time. Around 1640 Fermat and Descartes independently discovered analytical geometry. An English clergyman by the name of William Oughtred, at about 1622, devised the first slide rule that used the sum of logarithms to determine a product.

One major change had occurred in the approach of science since the Middle Ages and earlier. The concept of testing a hypothesis, and rejecting or accepting it on the basis of the outcome of an experiment, proved to be a vast improvement. The "scientific method" had come into use. We know now that progress is slow without it.

Descartes and Leibnitz both independently discovered differential calculus. Newton discovered integral calculus and later described the reciprocal relationship between differential and integral calculus. His discoveries were made at Woolsthorpe at about 1665 because Cambridge was closed as a result of the plague.

Jean Baptiste Colbert was a minister under Louis XIV and established the first formal engineering school in 1675. The Corps du Génie, as they were called, were military engineers trained by Sebastien le Prestre de Vauban, a well-known French military engineer.

In 1771 a small group of engineers, frequently called on to give testimony on harbor and canal projects, formed the Society of Engineers. John Smeaton was a leader of this group and was the first to call himself a "civil" engineer to distinguish himself as having interests apart from military engineering. This society became the Institution of Civil Engineering in 1828, initiating specialization within engineering.

In 1795 Napoleon authorized the establishment of the École Polytechnique, which was the first of a number of such schools that appeared in Europe during the nineteenth century. Others followed, such as the Eidgenossisches Polytechnicum in Zurich in 1855, the Polytechnic schools at Delft in 1864, and others in Chemnitz, Turin, and Karlsruhe. In 1865 the Massachusetts Institute of Technology was founded, the first such school in the United States.

During the medieval period, the main sources of power were water, wind, and animals. It was not until the eighteenth century that the fantail was invented, which by means of gearing kept the main wind propellers turned into the wind. The fantail is one of the first known self-regulating devices in engineering history.

Although a number of those who preceded him were responsible for minor developments, Thomas Savery is given much of the credit for development of the steam engine. In 1698 he received a patent for a steam-powered device for draining mines, which he advertised in a book he later wrote entitled *The Miner's Friend*. Thomas Newcomen developed a much improved steam engine in 1712, shown in Fig. 1-2, which was also used for pumping water from a mine. These early engines were very inefficient but did represent the initial development of power from heat engines. It is difficult to imagine at what point our civilization would be today without the development of heat engines.

A number of scientific developments in the seventeenth century preceded the development of the steam engine. Robert Boyle studied the elasticity of air and discovered the law relating temperature, pressure, and volume which carries his name today. Robert Hooke experimented with the elasticity of metals, and discovered the law of elasticity which bears his name. Christian Huygens determined the relationship for centripetal force, and Sir Isaac Newton developed the three basic laws of motion.

James Watt, following Newcomen, made such significant improvements in the steam engine that he is often given partial credit for its origination, along with Savery

Figure 1-2 A Newcomen engine. [*From Kirby et al. (2, p. 163).*]

and Newcomen. During an experiment in 1782, he found that a brewery horse could expend 32,400 foot-pounds of energy per minute. A year later, he standardized on 33,000 foot-pounds per minute as equaling 1 horse-power, and this equivalent is still in use today.

Richard Trevithick, in 1804, was the first to run a steam locomotive on tracks. He later proved that smooth wheels could run on smooth tracks if grades were not excessive. One of Trevithick's locomotives was exhibited on a circular track in London in 1808 until the locomotive ran off the track and overturned. So few shillings had been spent to see it that it was not replaced on the track.

George Stephenson, after being employed as a cowhand, became a steam engine fireman and later a pump engine tender. At the age of thirty-two he constructed his first steam locomotive, and later was instrumental in amending an act passed in 1821, to employ steam locomotion rather than horses for a railway to run from Stockton to the Willow Park colliery. He used the 4-foot, 8½-inch gauge rail that had been used previously for horse-drawn wagons. This railroad gauge is still the most commonly used throughout the world.

As we are all aware, following the development of the railroad systems in Europe and America, engineering developments came at an ever-increasing rate. The first half of the twentieth century produced an almost unbelievable number of engineering developments, and there is little doubt that two world wars were catalysts to much of this development.

The invention and development of automobiles and airplanes in the United States were significant factors in twentieth century engineering development. Inventions by Thomas Edison which initiated the power industry and Lee De Forest's invention of the "electronic valve," which gave considerable impetus to the communications industry, were also highly significant events.

Until 1880 engineering was either civil or military and for all but the last one hundred years was both. In 1880 the American Society of Mechanical Engineers was founded, followed by the American Society of Electrical Engineers in 1884 and the American Institute of Chemical Engineers in 1908. The American Institute of Industrial Engineers, the last major field of engineering to become organized, was incorporated in 1948.

SUMMARY

The first engineers often had a dedication to purpose born of necessity. The first engineering discipline—military engineering—developed to help meet a basic need for survival. Every period in history has had different social and economic climates and pressures that have greatly influenced both the path and progress of science and engineering. We need to be reminded occasionally that what we are raised to believe is normal may well be a fleeting social or economic fancy representing a point in time. Thomas Malthus, at the turn of the nineteenth century, was the leading economic theorist of his day. It is interesting to compare his views, as expressed in the following quotation from "An Essay on the Principle of Population" (3, p. 80), with those of present-day welfare economists:

All the children born, beyond what would be required to keep up the population to this level, must necessarily perish, unless room be made for them by the deaths of grown persons ... To act consistently therefore, we should facilitate, instead of foolishly and vainly endeavouring to impede, the operations of nature in producing this mortality; and if we dread the too frequent visitation of the horrid form of famine, we should sedulously encourage the other forms of destruction, which we compel nature to use. Instead of recommending cleanliness to the poor, we should encourage contrary habits. In our towns we should make the streets narrower, crowd more people into the houses, and court the return of the plague. In the country, we should build our villages near stagnant pools, and particularly encourage settlements in all marshy and unwholesome situations. But above all, we should reprobate specific remedies for ravaging diseases; and those benevolent, but much mistaken men, who have thought they were doing a service to mankind by projecting schemes for the total extirpation of particular disorders.

It is difficult to accept today that people could have believed that not so long ago.

The history of engineering is also replete with examples of how virtually nothing of value has been gained without a struggle. Galileo's house arrest for insisting that the sun is the center of the universe, Edison's dogged tests of thousands of materials in the hope of finding a filament material, and the individual who later insisted that the Boston Edison company control the speed of its generators so that he could invent and market the electric clock—all are examples of the human struggle to learn. The need to be of a persevering nature to be successful in engineering is as true today as when Leonardo da Vinci was trying to sell his services as an engineer.

It is also interesting to note, in considering the history of engineering, that engineering is in its infancy. Prior to 1900 only two closely allied fields, civil and military engineering, represented the bulk of engineering progress in the United States. The remaining disciplines have all had their most significant development since 1900.

REFERENCES

1   Beakley, George C., and H. W. Leach: *Careers in Engineering and Technology,* The Macmillan Company, Toronto, 1969.
2   Kirby, Richard S., S. Withington, A. B. Darling, and F. G. Kilgour: *Engineering in History,* McGraw-Hill Book Company, New York, 1956.
3   Oser, Jacob: *The Evolution of Economic Thought,* Harcourt, Brace and World, Inc., New York, 1963.
4   "The Story of Joe Ogg," a 16-millimeter sound film produced at Kansas State University for the American Institute of Industrial Engineers, Norcross, Ga., 1969.
5   Sandström, Gösta: *Man the Builder,* McGraw-Hill Book Company, New York, 1970.
6   Sprague de Camp, L.: *The Ancient Engineers,* Doubleday & Company, Inc., Garden City, N.Y., 1963.

REVIEW QUESTIONS AND PROBLEMS

1   Differentiate between the agricultural revolution and the industrial revolution.
2   How do routes of transportation affect economic and technical development?
3   What is the name of the first engineer known by name, and what engineering project did he construct?
4   What two developments gave the Assyrians an advantage in battle?
5   Who authored the first text in engineering?
6   Were the Greeks or the Romans more innovative with respect to engineering developments?
7   Are traffic problems an outcome of developments of the twentieth century?
8   What is the origin of the word "engineer"?
9   What is the origin of the number system in common use today in the United States?
10 What invention significantly reduced the effectiveness of walls as a line of defense?
11 What was the relative status of engineers during the Dark Ages?

12 Briefly describe some of the personality traits of Michelangelo and Leonardo da Vinci.
13 Compare the closeness of engineering and art during da Vinci's time as compared to the present.
14 At what approximate time in history did formal engineering education begin?
15 What does history tell us about the effect a time of war has on engineering developments?

Chapter 2

A Brief History of Industrial Engineering

The aspiring man needs to recognize the merits of his older contemporaries without letting himself be hindered by their shortcomings.

Goethe

It is difficult to say when industrial engineering began. Certainly in the age of Ogg there were production problems associated with making arrows which have their parallel today. If the individual in a toy factory today most concerned with how to make arrows is an industrial engineer, does that mean that when Ogg was deciding how to make arrows he was doing industrial engineering? The basic "what, how, where, and when" of production analysis has been a questioning approach for centuries.

Adam Smith's *Wealth of Nations* (18), in 1776, was one of the first works promoting "specialization of labor" to improve productivity. He observed in pin making that division of the task into four separate operations increased output by a factor of almost five. Whereas one worker performing all the operations produced 1,000 pins per day, ten workers employed on four more specialized tasks could produce 48,000 pins per day. The concept of designing a process to efficiently use the work force had arrived.

Around 1800 Matthew Boulton and James Watt, Jr., sons of prominent steam engine developers, attempted organizational improvements in their Soho, England, foundry that were well ahead of their time. Their efforts were pioneering prototypes for industrial engineering techniques to follow. At about this time, an increasing number of mechanical improvements, such as Arkwright's spinning jenny, were making a considerable influence on productivity. The industrial revolution of this period was freeing humans and beasts as sources of power in industry. The development of water and steam power and other mechanical devices is the usual primary connotation given to the term "industrial revolution."

In 1832 Charles Babbage, a self-made mathematician, again suggested division of labor for improved productivity in his book *On the Economy of Machinery and Manufacturers.* In fact, his "difference engine," the prototype of the modern mechanical calculator, was conceived after he heard about French attempts to produce handbook tables by dividing the calculation task into small steps requiring simple operations. Later, his "analytical engine" was a mechanical prototype of our modern computers. His analytical engine was never completed; the British government abandoned the project after he had spent £17,000 in development. Babbage, somewhat reminiscent of Leonardo da Vinci, was a tireless researcher who had little patience in completing what he had already conceived. Babbage was also aware of the need for improved organization in industry; he toured a number of plants in England and the Continent in the hope of improving his knowledge of the "mechanical art."

After the American Revolution, there was a considerable demand in the United States for muskets, and independence made it possible to produce manufactured goods. Eli Whitney found backers to support the concept of manufacturing interchangeable parts in producing muskets. However, his backers became quite impatient when, after a considerable time had elapsed and much money had been spent, they learned he was still making tools to make parts. Eventually, however, his efforts did produce cheap, interchangeable parts in large quantities. The concept, which is readily accepted today, of producing an expensive set of dies to produce a million parts cheaply was simply not understood at the time. Whitney's invention of the cotton gin typifies many highly significant mechanical improvements of the day, but there is little question that his concept of "tooling-up" for interchangeable parts was the major innovation of this period.

Around the turn of this century, Henry Ford, on observing carcasses on a moving conveyor in a slaughterhouse, got the idea for progressive assembly of automobiles by use of conveyors. Conveyors have become so much a part of our industrial heritage today that it becomes necessary in an industrial engineering course dealing with materials handling to offer a job lot problem for which the use of conveyors is a poor choice of approach. This shock seems necessary to convince students that conveyors help most of the time but not all of the time. There is little question that the mass production of Fords gave considerable impetus to the mass production concept in the United States.

In 1886 Henry Towne of the Yale and Towne Company, in "The Engineer as Economist," a paper in the *Transactions of the American Society of Mechanical Engineers,* stressed the need for engineers to be concerned with the profitability effects of their decisions. Until this time engineers were primarily battling the elements, and costs were assumed to be a necessary and relatively uncontrollable price for winning the battle against nature. Another member of the American Society of Mechanical Engineers (ASME), much impressed by the concepts offered by Towne, was Frederick W. Taylor. Taylor is often referred to today as the "father of industrial engineering." Based on the accomplishments he made, in light of the times in which they were made, the title seems most appropriate. Whereas the industrial revolution brought new sources of power that made widespread industrialization possible, Taylor offered the concept that it was an engineering responsibility to design, measure, plan, and schedule work.

TAYLOR

Frederick Winslow Taylor was from a well-to-do Philadelphia family. Following attendance in French and German schools, he entered Phillips Exeter Academy in Exeter, New Hampshire, to prepare for entrance into Harvard Law School. Two things happened at Phillips Exeter Academy which affected his life and subsequently the development of industrial engineering. First, he was influenced by George Wentworth, a professor of mathematics at the academy, who determined the time to solve math problems for homework by timing problems in class. And second, although Taylor passed his Harvard entrance exams his eyesight became impaired, and rather than enter law school he became a machinist's apprentice in the Enterprise Hydraulic Works in 1874. Nine years later he was married, had received a mechanical engineering degree from Stevens Institute a year earlier, and had just been promoted to chief engineer at the Midvale Steel Company. His efforts at Midvale Steel led the way to what came to be known as "scientific management."

To appreciate Taylor's accomplishments, it is necessary to understand the prevailing work environment of the times. The owner-manager, along with the sales and office staff, typically had little direct contact with the production activity. In most cases, a superintendent was given full responsibility for producing the products demanded by the sales staff. All planning and staff functions were informally performed by the superintendent, who had to deal with journeymen mechanics in attempting to get work done. There were no recognized staff functions, and work methods were determined by the individual mechanics on the basis of personal experience, preferences, and what tools were readily available. Taylor, influenced by both Towne and Wentworth, developed the concept that work design, work measurement, production scheduling, and other staff functions were engineering responsibilities. His attempts to implement his concept revolutionized industrial productivity.

In 1881 Taylor began a study of metal cutting that went on for twenty-five years, ending with publication in 1907 of the longest paper (more than 200 pages) ever published in the *Transactions of the American Society of Mechanical Engineers* (19). Before this study the geometry of metal-cutting tools and speeds and feeds for metal cutting were determined by experience or rules of thumb. Taylor, along with other experimenters who assisted him, turned metal cutting into a science. Phenomenal increases in the rate of metal cutting have resulted from these initial experiments.

Later, at Bethlehem Steel, Taylor made an analysis of shoveling tasks. He noted that although there was a considerable variety of shoveling tasks performed at the mill, the same type of shovel was used for all tasks. A shovelful of rice coal weighed only $3\frac{1}{2}$ pounds, whereas a shovelful of iron ore weighed 38 pounds. Taylor reasoned that with this degree of variability in shovel-load weights, the same type of shovel could not be ideal for all tasks. After some experimentation he found that $21\frac{1}{2}$ pounds seemed to represent an ideal weight for a shovel load, and then designed shovels of different sizes for different tasks such that in all cases a shovel load would weigh approximately the ideal amount. Shoveling productivity in the mill improved dramatically. In a period of three and one-half years the number of workers performing shoveling tasks was reduced from 500 to 140. Even more important than the improvement in shoveling productivity was the concept of applying engineering analysis to human work situations. Taylor initiated the practice, which is very common today, of performing an engineering analysis of work requirements, specifying the exact method, tools, and equipment to be employed, and then training the worker to perform the operations as specified.

Another classic work situation in which Taylor performed a pioneering study was the handling of pig iron. He eventually determined an optimum method of handling, optimum pace, and optimum work and rest periods. He then carefully selected workers to perform this task and carefully trained them to perform the task exactly as he specified. As a result of his analysis, dramatic changes in pig-iron handling productivity resulted.

Analysis of the work requirements and specifications for a method to perform an operation is now called "work design" or "methods study." The shoveling and pig-iron studies were primarily concerned with the design of work. However, Taylor also pioneered the activity we now generally call "work measurement." This activity is concerned with determining the amount of time an operator should be allowed for performing an operation. The reciprocal of such a time value permits determination of the amount of production that is expected of an employee in a given time period, such as a day. Taylor very carefully timed employees performing a task exactly as he specified, which established a basis for the amount they should produce in a day. Taylor invented stopwatch timestudy, which is still used extensively to determine the time to perform an operation, commonly referred to today as a "time standard."

Before Taylor, labor control was attempted through direct supervision. His development of timestudy led to time standards, which are the underlying basis for control of labor costs and are a necessary input to scheduling and pricing activities in industry.

In June, 1895, Taylor presented his first significant paper, entitled "A Piece Rate System," at a meeting of the ASME. His paper was not well received, primarily because it was assumed to be another attempt at devising another wage payment plan, as the title suggests. A number of wage payment plans had been attempted, often embodying unsound and unethical rate-cutting practices. Rate cutting is the practice of reducing a time standard once it has been reached in the hope of increasing output at the same pay level, when such a reduction in time is not justified. Apparently, the general distaste for wage payment plans in the engineering community drew attention away from the significant breakthroughs in management thought embodied, but for the most part missed, within the paper. Taylor was disappointed at the reception his paper received, but concluded that he had not properly presented his findings and thoughts and decided to present an improved paper at a later date.

In June, 1903, at the Saratoga, New York, meeting of the ASME, he presented a second paper, entitled "Shop Management" (20). Again, the reaction was less than enthusiastic; in fact, it was generally regarded to be an insignificant work by the engineering community. The paper was later well received by many plant management personnel, who in everyday practice work closer to the innovations suggested by Taylor. Henry Towne, a past president of the ASME, drew attention to what he considered to be interesting concepts in the paper, and a short time thereafter it was the center of controversy of management thought of the day. When one considers the content of the educational experience of the mechanical engineer during this period, it is easier to understand why plant personnel perceived the innovations suggested, whereas most of the mechanical engineers of the day seemed to miss their significance. The mechanical engineers then typically received little education in production management. In fact, the need for engineers trained in production management was the basic justification for initiating the first options in industrial engineering within mechanical engineering departments at universities, and later separate departments of industrial engineering. When one reviews "Shop Management" today, it is interesting to note the number and diversity of concepts embodied within one paper. To name a few:

1 Methods study
2 Time study
3 Standardization of tools
4 A planning department
5 The "exception" principle of management
6 Instruction cards for workers
7 Slide rules for metal cutting
8 Mnemonic classification systems for parts and products
9 A routing system
10 Costing methods
11 Employee selection in relation to the job
12 Task idea permitting a bonus if the job is completed in the specified time.

To understand the significance of the above concepts it is essential that one consider the environment of scientific thought of the times. In the latter half of the nineteenth century, mechanical engineering had become established in Europe. In the United States, however, during this same period Copley (5) states, "The very idea that there could be a true science of mechanical devices continued to be generally scorned." Copley (5) offers the following quotation from a statement by Taylor which typifies the state of the mechanical art in the United States during this period:

I can remember distinctly the time when an educated scientific engineer was looked upon with profound suspicion by practically the whole manufacturing community.

The successful engineers of my boyhood were mostly men who were endowed with a fine sense of proportion—men who had the faculty of carrying in their minds the size and general shape of parts of machinery, for instance, which had proved themselves successful, and who through their intuitive judgment were able to make a shrewd guess at the proper size and strength of the parts required for a new machine.

It was my pleasure and honor to know intimately one of the greatest and one of the last of this school of empirical engineers—Mr. John Fritz—who had such an important part in the development of the Bessemer process, as well as almost all of the early elements of the steel industry of this country.

When I was a boy and first saw Mr. Fritz, most of the drawings which he made for his new machinery were done with a stick on the floor of the blacksmith shop, and in many cases the verbal description of the parts of the machines which he wished to have made were more important than his drawings. Time and again he himself did not know just what he wanted until after the pattern or model was made and he had an opportunity of seeing the shape of the piece which he was designing. One of his favorite sayings whenever a new machine was finished was, "Now, boys, we have got her done, let's start her up and see why she doesn't work."

The engineer of his day confidently expected that the first machine produced would fail to work, but that by studying its defects he would be able to make a success of his second machine.

In 1909 Taylor attempted again to shed light on his concepts by offering another paper to the ASME, entitled "Principles of Scientific Management" (20). By this time his approach had received considerable notoriety, particularly as the result of railroad rate case hearings before the Interstate Commerce Commission in which his concepts represented the center of controversy. Louis Brandeis, representing Eastern industrialists, contended that the rate increase was unjustified because the railroads had failed to take advantage of the Taylor system. Because of the controversy "Scientific Management" had created, eighteen months had passed without action on his paper submitted to ASME. Taylor, feeling obliged to clarify much of the confusion surrounding his techniques, withdrew the paper from ASME, published it privately, and sent it to all ASME members at his own expense. "Principles of Scientific Management," in contrast to his earlier papers, was philosophical in nature. For the most part, it attempted to describe and justify his approach.

In 1911 and 1912 Taylor was questioned at length by a special committee of the U.S. House of Representatives concerning the Taylor system (20). In looking back over Taylor's efforts, it is not difficult to understand why his approach was met with considerable criticism at the time. First, his approach was very different and successful. He increased output concurrently with a reduction in overall labor cost, while paying higher wages. He taught workers how to work, and then expected them to work almost to capacity for somewhat higher wages. In some applications, Taylor had produced a fourfold increase in production. He had no shortage of workers wanting to receive the increased wages and was never "struck" by labor. It is doubtful that a union today, however, would permit the above distribution of increased productivity.

Another source of criticism against Taylor was his frankness with respect to human behavior, as indicated in the following quotation from "Scientific Management" (20, p. 59):

> Now one of the very first requirements for a man who is fit to handle pig iron as a regular occupation is that he shall be so stupid and so phlegmatic that he more nearly resembles in his mental make-up the ox than any other type. The man who is mentally alert and intelligent is for this very reason entirely unsuited to what would, for him, be the grinding monotony of work of this character. Therefore the workman who is best suited to handling pig iron is unable to understand the real science of doing this class of work. He is so stupid that the word "percentage" has no meaning to him, and he must consequently be trained by a man more intelligent than himself into the habit of working in accordance with the laws of this science before he can be successful.

It is difficult to clearly identify the first course taught in industrial engineering in the United States. An elective course was taught in 1904 by Professor Kimball at Cornell University concerning the economics of production, and included concepts offered by both Towne and Taylor.

In 1913, concerned about the effects of the Taylor system, Congress added an amendment to the government appropriation bill stipulating that no part of the appropriation should be made available for the pay of any person engaged in timestudy work. Later, a law was passed making it illegal to use a stopwatch in a post office; it contained both a possible fine and imprisonment for violation. As late as 1947, the Military Establishment Appropriation Act and the Navy Department Appopriation Act specified that wages for performing a stopwatch timestudy could not be paid, nor incentive wages for employees, from these funds. In July, 1947, a bill was passed in the House of Representatives allowing the War Department to use timestudy; and in 1949 all federal restrictions against the use of stopwatches were discontinued.

In reflecting on Taylor's accomplishments, it is interesting to note that he joined Bethlehem Steel in 1898 and, because of the confusion he created in the twelve years that followed, Robert P. Lindeman, then president of Bethlehem Steel, summarily dismissed him in 1901. Not until around 1910 had he received any visible acceptance of his concepts, in the midst of considerable heated controversy. Following his dismissal from Bethelehem Steel he divided his time between consulting and lecturing in the hope of explaining his concepts. He died of pneumonia in 1915.

GILBRETH

Frank B. Gilbreth was born on July 7, 1968, in Fairfield, Maine. At college age, after moving to Boston with his family, he would have liked to enter the Massachusetts Institute of Technology; however, the family budget was already strained by the education of an older sister, and so he took a job as a bricklayer's helper instead.

He had a very questioning attitude about his work, but it would seem that he received too few satisfying answers. Bricklaying was an evolved art, and under detailed analysis there was considerable room for improvement. To make a long story short, before he was thirty years of age he was the owner of a profitable construction firm with offices throughout the world.

Where, in the past, 120 bricks laid per worker per hour was normal, Gilbreth's innovations resulted in an average production rate of 350 bricks per worker per hour. These rate increases were not gained by making bricklayers work faster, but by a more effective method. In analyzing the standard method for laying exterior brickwork, Gilbreth reduced the number of motions from eighteen to five. Instead of having a bricklayer bend over and pick up a brick from a pile of bricks on a relatively unadjustable scaffold, rotate the brick to find the best side, and then lay the brick by tapping with mortar of often poor consistency, he did otherwise. First, he had bricks sorted at delivery by low-cost laborers, who made "packets" of bricks preoriented for laying. He then provided adjustable scaffolds, the proper location of bricks and mortar, and mortar of proper consistency. The result was a vast improvement in productivity with less fatigue. Gilbreth was always in search of the "one best way."

Frank Gilbreth married Lillian Moller Gilbreth, a Phi Beta Kappa psychology graduate of the University of California, who later received a Ph.D. from Brown University. The Gilbreths worked very closely together and, six boys and six girls later, the combination of engineer and psychologist had made significant inroads into the analysis of human work behavior.

Of particular interest to Frank Gilbreth was the analysis of fundamental motions of human activity. He classified the basic motions into what he called "therbligs," which is almost Gilbreth spelled backwards, such as: search, find, transport empty, pre-position, grasp, and so forth. In an attempt to analyze motions in more detail, he employed industrial motion picture cameras in a technique he labeled "micromotion study" at an ASME meeting in 1912. Because cameras were generally hand-cranked and not of constant speed, he included a clock called a microchronometer, graduated to 1/2000 of a minute, to provide a time dimension to filmed activity. Today, micromotion filming is done at 1000 frames per minute, which automatically spaces successive frames of activity 0.001 minute apart. Time is determined by counting frames, with a counter integral to the projector.

Gilbreth also studied the motions of parts of the human body, typically the hands. Using an open lens, he filmed a light attached to a point on the body in a darkened room. The photo produced by this technique is called a cyclegraph. By adding "blips" in the light path at fixed time intervals he was able to add a time dimension to the motion path photograph. The photograph thus produced was called a chronocyclegraph. These techniques and others were used to study fundamental motions in human activity to determine average times for such motions under varying conditions.

Gilbreth was influenced greatly by Taylor, but whereas Taylor applied his methods almost exclusively to the industrial shop, Gilbreth brought out the generality of these techniques by applying them to fields such as construction, canal building, education, medicine, and the military. His wife added the "human factor" dimension to their work, which led eventually to organization theory and analysis of management practice.

By 1924 Frank Gilbreth had become internationally famous for his contributions. He died three days before he was to leave to give invited papers at the World Power Conference in England and the International Management Conference in Czechoslovakia. Lillian Gilbreth continued their work after his death and became the most distinguished woman engineer in the United States to date. In her later years, she possessed an enthusiasm and charm that served as an inspiration to many. Two of the Gilbreths' children described their family in the book *Cheaper by the Dozen*, which was made into a very enjoyable movie of the same name.

LATER TRADITIONALISTS

One of Taylor's associates at Midvale Steel was Carl Barth, a mathematician, who had started his career in the drafting room at Midvale Steel. Barth became deeply involved in the metal-cutting experiments of Taylor and developed slide rules, such as the one illustrated in Fig. 2-1, to be used by workers to quickly calculate feed and speed parameters for a particular operation. Barth also performed some early fatigue studies in an effort to establish appropriate fatigue allowances in timestudy. Taylor had more than a dozen associates helping him with different aspects of the study. Proponents of operations research today suggest that one unique feature of operations research as compared to earlier approaches is the use of the "team approach." This differs little in concept from the approach in Taylor's metal-cutting study, as indicated in the following quotation (19, p. 35):

> Mr. White [Mansel White] is undoubtedly a much more accomplished metallurgist than any of the rest of us; Mr. Gantt [H. L. Gantt] is a better all around manager, and the writer of this paper has perhaps the faculty of holding on tighter with his teeth . . . Mr. Barth [Carl G. Barth], who is a very much better mathematician than any of the rest of us, has devoted a large part of his time . . . to carrying on the mathematical work.

Another associate of Taylor was Henry Laurence Gantt, who is best known today for a type of chart, used for scheduling production equipment, which bears his name.

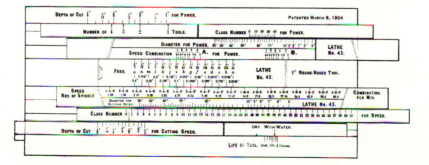

Figure 2-1 A Barth slide rule. (*From Charles D. Flagle, W. H. Huggins, and R. H. Roy,* Operations Research and Systems Engineering, *John Hopkins Press, Baltimore, 1960, p. 18.*)

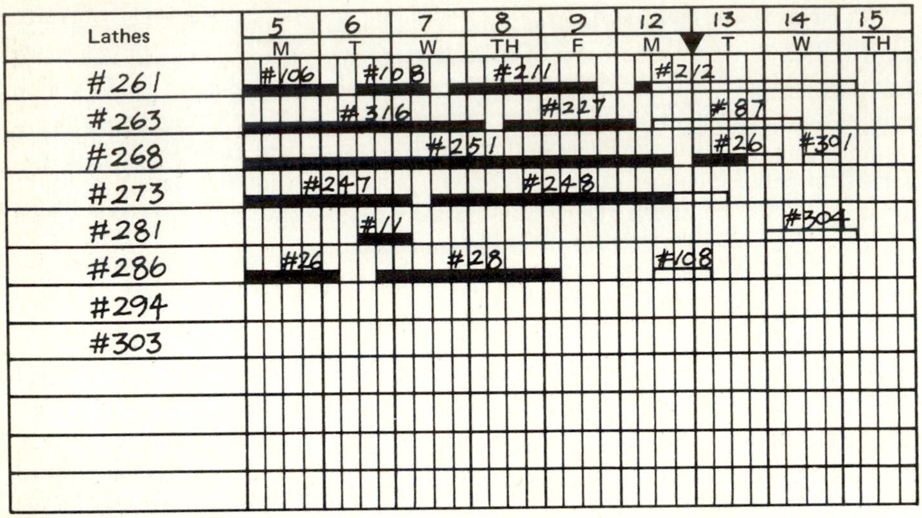

Lathes	5 M	6 T	7 W	8 TH	9 F	12 M	13 T	14 W	15 TH
#261	#106	#108		#211		#212			
#263		#316			#227		#87		
#268				#251				#26	#301
#273		#247			#248				
#281		#11							#304
#286	#26		#28			#108			
#294									
#303									

Figure 2-2 A Gantt chart.

The Gantt chart, illustrated in Fig. 2-2, is used to graphically show the work that has been scheduled ahead for each machine and the progress of jobs to date. Gantt charts are, therefore, a means of planning production and observing and planning the utilization of equipment. Variations of the Gantt chart are in common use in many of the smaller "job shop" production plants today.

Gantt also developed a wage incentive plan in 1901, which paid workers a bonus if they worked above the standard rate of activity. The plan encouraged cooperation between management and labor and softened the tougher approach used by Taylor. Taylor did not believe in retaining a person who could not work above the standard rate; he preferred keeping only a "first-class" person and paying a bonus for work above standard.

Harrington Emerson attempted to use Taylor's approach and some ideas of his own in analyzing labor efforts in the Santa Fe Railroad system. Emerson reorganized the management of the company, and employed improved shop practices, standard costing, and tabulating machines for accounting purposes. His improvements resulted in reported annual savings well in excess of $1 million per year for the line. He later wrote a book entitled *Twelve Principles of Efficiency* (6), in which he attempted to elucidate his approach. Emerson's success precipitated modernization efforts in a wide range of industrial firms, because his approach had applicability across the commercial and industrial fields.

Morris L. Cooke attempted to employ scientific management in city governments. Later, he and Philip Murray, president of the Congress of Industrial Organizations, published "Organized Labor and Production," a pamphlet that brought out the desirability of a common goal of "optimum productivity" for both labor and management.

Dwight V. Merrick, following the timestudy work of Taylor, performed a study of elemental times, which was published in the journal *American Machinist*. Merrick, like Gantt, also developed a wage incentive plan, as a "hoped for" improvement over those offered by Taylor and Gantt.

Times of war have always provided a stimulus to improved technology. Franklin D. Roosevelt, through the Department of Labor, recommended the use of time standards during World War II, and a significant improvement resulted from their use. In fact, Regional War Labor Board III in the middle East Coast area encouraged wage incentives and issued guidelines for their use.

A rather interesting study of human performance was made at the Hawthorne Works of the Western Electric Company, the manufacturing subsidiary of the Bell Telephone System, starting in 1927. One part of the study was concerned with the relationship between illumination and productivity. One area in the plant was provided with increased illumination, and a considerable increase in productivity was noted. Later it was realized that the increased production was related to management interest in the study (i.e., frequent visits to the study area by top-level management) and, to a much lesser degree, to the increased illumination. It became a classic example of the necessity to include a "control group" in a study of this type. The control group should have experienced all effects, except the increased illumination, experienced by the other group (i.e., including management attention). It would have been noted that the productivity of the control group had increased also. This erroneous assumption of the causal relationship of an effect, when a control group has not been employed, has acquired the name "the Hawthorne effect."

EARLY MODERNISTS

For better or for worse, in considering the history of industrial engineering, one can classify post-Taylor practitioners in the first half of this century as either "later traditionalists" or "early modernists." The later traditionalists are those who extended and developed the philosophical concepts of Taylor and later developed graphical approaches (such as the Gantt chart, the flow process chart, and the multiple activity chart) to assist in implementing the underlying methodology of scientific management. By early modernists is meant those pioneers of modern industrial engineering techniques who employ mathematical description and analysis as a means to further extend the preceding philosophy.

F. W. Harris was one of the earliest to convert a graphical description of the simplest of inventory models into mathematical terms. Unfortunately, another gentleman made extensive application of the Harris formula, and it became known after him as the "Wilson formula." In 1931 F. E. Raymond wrote the first book on inventory control, in which he endeavored to detail the utility of inventory control in manufacturing.

In 1924 W. A. Shewhart of the Bell Telephone Laboratories offered the first description of a "control chart," and in 1931 published the first text on statistical quality control (16). Professor Eugene Grant of Stanford in 1946 published a text on quality control (8), which has been a mainstay and is presently in its fourth edition. Earlier, in 1930, Grant and W. G. Ireson published the text *Principles of Engineering*

Economy, which was a pioneering effort in its area. For years, at least until recently, a course in engineering economy has been required for most engineering students in the United States. Grant and Ireson's text, which is now being revised for its fifth edition, has been a mainstay in this field.

Texts by Barnes (2) and later Niebel (14) and Mundel (12) extended the methods and timestudy efforts of Gilbreth and Taylor. All of these books contain techniques that bridge the definitions of later traditionalists and early modernists. Charting methods are still an effective methodology for performing production analysis and often defy useful mathematical description. Such a technique as work sampling, however, is applied statistical sampling and depends on mathematical description for its solution.

The "objective rating method" of Mundel, in which the inherent difficulty of a method is a factor to be considered, at least suggests, as compared to earlier rating methods, that this is indeed a complex area. Harold Smalley, who had been a student of Mundel, in a lengthly article (17) on work measurement in the *Journal of Industrial Engineering,* highlighted the controversial nature of many of the underlying concepts in this field. What has always seemed ironic to this author is the lack of attention being given to this field in light of its importance with respect to productivity nationally. There are generally recognized inherent inconsistencies between competing approaches in this field, which suggests that research is needed, and yet research support has been minimal over the years. Krick's text (9) admirably identifies many of these theoretical inconsistencies, as uncomfortable as they may seem to the typical sophomore industrial engineering student, and draws attention to the need for continued research in this area.

Mallick and Gaudreau (10), Muther (13), and Apple (1) offered early texts in the area of plant layout. Plant layout has been, and for the most part in application still is, dictated by graphical and conceptual techniques. Other texts, such as those by Reed (15) and Moore (11), have attempted to include more of an early modernist point of view. However, in light of the number of variables to be considered, it is likely that much of the basic workload of plant design will continue to be done by the traditional techniques developed to date. That is not to say that modern techniques will not be employed. Digital computer simulation and queueing theory have been successfully and extensively employed in this field and will continue to be employed. Other quantitative techniques are making inroads into the field.

ORGANIZATIONS

There is little question that industrial engineering developed as an offshoot of mechanical engineering, and the American Society of Mechanical Engineers was the first organization to represent industrial engineering interests. Symbols for flow process charting commonly employed in production analysis are still defined by ASME standards. Also, a number of industrial engineering programs are still options within mechanical engineering departments at universities in the United States.

In 1911, when the controversy over scientific management was at its height, the Taylor system was being discussed in the railroad rate case study before the Interstate

Commerce Commission, and Taylor himself was giving testimony before a committee of the U.S. House of Representatives, Morris Cooke and Harlow Persons organized a conference on scientific management at the Amos Tuck School of Dartmouth College.

A year later, the Efficiency Society was formed in New York City, and the Society to Promote the Science of Management was initiated, which three years later, in 1915, became the Taylor Society. In 1917 the Society of Industrial Engineers (SIE) was formed to specifically represent the interests of production specialists and managers, as compared to the focus on general management philosophy that had developed within the Taylor Society.

A number of individuals who wished to develop corporate training programs for managerial staff formed the American Management Association (AMA) in 1922. This is still the major organization in the United States representing the art and science of management.

In 1936 the Taylor Society and the Society of Industrial Engineers merged to form the Society for the Advancement of Management (SAM). This society combined the interests of production specialists, production managers, and those interested in general management philosophy.

It seems obvious today that Wyllys G. Stanton, a professor of industrial engineering at Ohio State University, was a man of vision. In 1948 he called a meeting of a number of his associates to consider the formation of a new professional organization to represent industrial engineers. In August, 1948, twelve members formed the Columbus Chapter of the American Institute of Industrial Engineers (AIIE), with Eldon Raney as their president. A month later, AIIE was incorporated under the laws of the state of Ohio. One year later there were student chapters at the University of Alabama, Columbia University, Georgia Tech, Northeastern University, Ohio State University, Oklahoma State University, the University of Pittsburgh, Syracuse University, Texas Tech, and Washington University. AIIE's first headquarters was Stanton's office at Ohio State University, and shortly thereafter it was moved to Columbus, Ohio. In 1960 the office was moved to the United Engineering Center in New York City, and only recently was moved again to a suburb of Atlanta, Georgia.

The first official journal of the AIIE was the *Journal of Industrial Engineering*. Colonel Frank Groseclose, professor emeritus and former director of the School of Industrial Engineering at Georgia Tech, was its first editor. He produced the first issue of the journal in June, 1949. By 1969 a dichotomy of interests had developed within the AIIE, which led in that year to the offering of two journals—*Industrial Engineering* for the practitioners and *AIIE Transactions* for the academicians.

In recent years, AIIE has fostered the development of divisions within the institute. Typical of these is the Facilities Planning and Design Division of AIIE. Divisions are assuming a greater role each year in representing special interests. The division newsletters, for example, have been useful in allowing special interests to communicate more effectively within the institute. It is interesting to note that the development of the division concept within the ASME (e.g., the Production Management Division) may have been one of the factors that led to the identification of industrial engineering as a separate engineering discipline.

Since World War II, operations research, management science, and systems engineering have had a considerable impact on industrial engineering practice. The history of these related fields is the topic of the next chapter. Modern industrial engineering draws from all of these disciplines, as will be discussed at length later in this text.

SUMMARY

It is doubtful that anyone has contributed more to the industrial engineering profession than Frederick W. Taylor. How tragic it seems that a man who contributed so much should have been arguing for acceptance of his ideas three years before his death. His efforts, ultimately, did not go unnoticed. In 1918 Georges Clemenceau of the French Ministry of War referred to Taylor's methods as (3, p. 14) "the employment in every kind of work of the minimum of labor through scientific research into the most advantageous methods of procedure in each particular case," and encouraged their use. Lenin wrote in *Pravda* (3, pp. 14–15):

> We should try out every scientific and progressive suggestion of the Taylor System.... To learn how to work—this problem the Soviet authority should present to the people in all its comprehensiveness. The last word of capitalism in this respect, the Taylor System, as well as all progressive measures of capitalism, combined the refined cruelty of bourgeois exploitation and a number of most valuable scientific attainments in the analysis of mechanical motions during work, in dismissing superfluous and useless motions, in determining the most correct methods of work, the best systems of accounting and control, etc. The Soviet Republic must adopt valuable and scientific technical advance in this field. The possibility of socialism will be determined by our success in combining the Soviet rule and the Soviet organization of management with the latest progressive measures of capitalism.

The recent period of high inflation has resulted in a renewed concern for productivity in the United States. It is interesting to note that in the last thirty years there has been a trend away from incentives in this country. During the same period a minimal amount of research has been encouraged to improve techniques in this area. This lack of research is evident when one considers that AIIE has been in existence for more than twenty-five years and has not yet developed rating films, used in timestudy, to support the education of future industrial engineers or to assist in training timestudy technicians in industry. In fact, there are at least four major methods of performing "rating" in timestudy, all of which are theoretically inconsistent with one another and produce different results, and yet there has never been any definitive research study under AIIE sponsorship to determine which method is preferable.

It is no wonder, however, that AIIE has not accomplished all that it might have. This chapter covers a period of history from about the turn of the century to the beginning of World War II. It has been possible to describe the scope of industrial engineering during this period in fairly concise terms. For the most part, industrial engineers were concerned with the design of manufacturing plants and controls for operating them. Since World War II, with the emergence of operations research, management science, systems engineering, and computer science, it is no longer possible to describe the domain of industrial engineering in simple terms. It is hoped that the following chapters will offer insight into the kinds of work industrial engineers perform. Industrial engineers receive a rather unique education today, by comparison not only with other engineers but with students in other colleges as well, which uniquely prepares them to analyze a broad spectrum of productive activity. The key to understanding the breadth of this activity is the economist's definition of the word "production." In economics, production refers to either a product or a service. The present-day technology of industrial engineering is sufficiently general with respect to application to analyze production in such diverse areas as manufacturing, banking, hospitals, defense systems, distribution, retailing, shipbuilding, construction, the chemical industry, insurance, goodwill industries, dental offices, and mail-order houses. The work of an industrial engineer today can be so diversified that the following definition, adopted by AIIE, is about as specific as possible:

> Industrial engineering is concerned with the design, improvement, and installation of integrated systems of people, materials, and equipment; drawing upon specialized knowledge and skill in the mathematical, physical, and social sciences together with the principles and methods of engineering analysis and design, to specify, predict, and evalute the results to be obtained from such systems.

This text endeavors not to define industrial engineering but to describe the activities of industrial engineers, so that the reader may acquire an understanding of the capability and consequently the role of industrial engineering.

REFERENCES

1   Apple, James M.: *Plant Layout and Materials Handling,* 2d ed., The Ronald Press Company, New York, 1963.
2   Barnes, Ralph M.: *Motion and Time Study,* 6th ed., John Wiley and Sons, Inc., New York, 1968.
3   Blair, Raymond N., and C. Wilson Whitston: *Elements of Industrial Systems Engineering,* Prentice-Hall, Inc., Englewood Cliffs, N.J., 1971.
4   Buffa, Elwood S.: *Production-Inventory Systems,* Richard D. Irwin, Inc., Homewood, Ill., 1968.
5   Copley, Frank Barkley: *Frederick W. Taylor,* Harper & Row Publishers, Inc., New York, 1923.
6   Emerson, Harrington: "Twelve Principles of Efficiency," *Eng. Mag.,* 1912.
7   Grant, Eugene L., and W. Grant Ireson: *Principles of Engineering Economy,* 4th ed., The Ronald Press Company, New York, 1960.
8   Grant, Eugene L., and Richard S. Leavenworth: *Statistical Quality Control,* 4th ed., McGraw-Hill Book Company, New York, 1972.
9   Krick, Edward V.: *Methods Engineering,* John Wiley & Sons, Inc., New York, 1962.
10 Mallick, Randolph W., and A. T. Gaudreau: *Plant Layout Planning and Practice,* John Wiley & Sons, Inc., New York, 1951.

11 Moore, James M.: *Plant Layout and Design,* The Macmillan Company, New York, 1962.

12 Mundel, Marvin E.: *Motion and Time Study,* 3d ed., Prentice-Hall, Inc., Englewood Cliffs, N.J., 1960.

13 Muther, Richard: *Practical Plant Layout,* McGraw-Hill Book Company, New York, 1955.

14 Niebel, Benjamin W.: *Motion and Time Study,* 5th ed., Richard D. Irwin, Inc., Homewood, Ill., 1972.

15 Reed, Ruddell, Jr.: *Plant Layout,* Richard D. Irwin, Inc., Homewood, Ill., 1961.

16 Shewhart, Walter A.: *Economic Control of Quality of Manufactured Product,* D. Van Nostrand Co., New York, 1931.

17 Smalley, Harold E.: "Another Look at Work Measurement," *J. Ind. Eng.,* vol. 18, no. 3, March, 1967.

18 Smith, Adam: *An Inquiry into the Nature and Causes of the Wealth of Nations,* Edwin Cannan (ed.), Random House, Inc., New York, 1937 (originally published in 1776).

19 Taylor, Frederick W.: "On the Art of Cutting Metals," *Trans. ASME,* vol. 28, pp. 31–350, 1907.

20 Taylor Frederick W.: *Scientific Management,* Harper & Row Publishers, Inc., New York, 1947 (comprises "Shop Management," "The Principles of Scientific Management," and "Testimony Before the Special House Committee").

21 Vaughn, Richard C.: *Introduction to Industrial Engineering,* Iowa State University Press, Ames, 1967.

REVIEW QUESTIONS AND PROBLEMS

1 What was Eli Whitney's main contribution to engineering development?

2 What new role did Towne believe an engineer should accept as part of his responsibility?

3 Who initiated an approach that came to be known as scientific management?

4 Differentiate between methods study and work measurement.

5 What is rate cutting, and why was it employed?

6 What was the attitude of government toward stopwatch timestudy in the first half of this century?

7 What is micromotion study?

8 In what industry did Harrington Emerson make extensive use of the Taylor system?

9 What is meant by the Hawthorne effect?

10 Does most of the plant layout activity performed today come closer to fitting the later traditionalist or the early modernist definition?

11 From what major engineering discipline did industrial engineering come?

12 Who was primarily responsible for initiating the American Institute of Industrial Engineers?

13 Was interest in the Taylor system confined solely to the United States?

14 In what way does the economist's definition of production suggest an expanding role for industrial engineering?

Chapter 3

An Introduction to Some Related Disciplines

Those who cannot remember the past are condemned to repeat it.

Santayana

Today, more than ever, industrial engineering means different things to different people. In fact, one of the best ways to develop an understanding of modern industrial engineering is through an understanding of how it relates to other fields. It would be convenient, for purposes of explanation, if there were clearly defined boundaries between subdisciplines of industrial engineering as well as fields related to industrial engineering; unfortunately, that is not the case. The fields most commonly referred to today as subdisciplines of or related to industrial engineering are: management, computer science, statistics, operations research, management science, human engineering, and systems engineering. There are those in each of these disciplines who believe their field is separate and distinct from industrial engineering.

Management and statistics are well-developed disciplines that preceded the development of industrial engineering and are of a broader scope than is generally envisioned for industrial engineering. It seems likely that these two fields will continue to be related to industrial engineering for some time. The other disciplines mentioned above, however, cannot be as easily related to industrial engineering at present, and are even more difficult to categorize with respect to the future. The education of the modern industrial engineer involves some combination of most of the above disciplines. In any particular instance, what that combination is depends on the industrial engineering department and then the company where the individual was trained. What may or may not be apparent at this point is the diversity of the typical degree program in industrial engineering. Whereas depth in a single discipline is the primary strength of an electrical, mechanical, or civil engineering program, breadth of understanding across disciplines both within and outside the college of engineering, as well as depth in industrial engineering, is the primary strength of an industrial engineering program.

The following introduction to each of these sub and related disciplines is intended to offer both relevant history and a limited comparative understanding of the present nature of each discipline. In the final chapter of this text, following a more detailed description of some of these fields, an effort will be made to consider the likely future role and prospects of these disciplines.

MANAGEMENT

Of all the disciplines mentioned above, management was one of the earliest to emerge in human history. If management is the art and science of directing human effort, then it must have begun when one person attempted to get another to work. There is considerably less than a unanimity of opinion today as to how best to do that.

The recognition of the need for planning, organizing, and controlling human effort can be traced back at least as far as early Egyptian times. The execution of these functions is essential if one is to build a pyramid, for example, in a reasonable period of time.

The establishment of a minimum wage, as another example, goes back to the Code of Hammurabi, 1800 B.C., as the following excerpt indicates: "If a man hire a field laborer, he shall pay him 8 gus of grain per year" (15). It may be of interest to note that we have vacillated between minimum wage laws and maximum wage laws, finally returning to the minimum wage laws of today. In the seventeenth and eighteenth centuries it was common practice for a local justice of the peace in England to establish maximum wage laws to "protect" the entrepreneur. In the 1760s the labor supply became plentiful again and this practice was discontinued.

The development of basic principles of management can be traced from the days of the Egyptians to the present through such diverse writings as those of Nebuchadnezzar, Plato, Socrates, Alexander the Great, the disciples of Christ, Ghazzali, Sir Thomas More, Machiavelli, Adam Smith, Thomas Jefferson, and James Mill (12). What must be obvious from this list is the diversity of application of management—from the management of war to the planning, organizing, and controlling of human effort in a search for lasting peace.

With the possible exception of an introductory statement or paragraph about pre-twentieth century management thought, most modern texts in management begin their development with a discussion of the scientific concepts of Frederick W. Taylor. Many authors refer to Taylor as the "father of scientific management." In the second chapter of this book, it was mentioned that Taylor is commonly referred to as the "father of industrial engineering" by industrial engineers. The question obviously arises whether Taylor's scientific management concepts were an academic extension of engineering or management.

There is little question that the subdivision of management commonly referred to as production management has a great deal in common with industrial engineering. In most business colleges, production management is a sequence of two to three courses at the undergraduate level, which attempt to familiarize management students with concepts and techniques specific to the analysis and management of a production activity. Industrial engineering, on the other hand, is an engineering degree concerned with the analysis, design, and control of *productive* systems. By a productive system is meant any system that produces either a product or a service. Production management courses are often primarily concerned with teaching management students how to manage (i.e., direct human efforts) in a production environment, with less attention paid to the analysis and design of productive systems.

Industrial engineering students, on the other hand, are primarily taught how to analyze and design productive systems and the control procedures for efficiently operating them. Except for a possible course or two concerned with the fundamental understanding of management concepts for directing the human effort associated with such systems, it is generally assumed that industrial engineers will not operate the systems they design. The training of a race car driver is analogous to management education, whereas the designing of the car is somewhat analogous to industrial engineering education. The race car driver wants to know first and foremost how to run the car and is less concerned with a detailed understanding of how it runs. The industrial engineer designs the car with a driver in mind but with no intention of getting behind the wheel on the day of the race. The engineer does intend to be there, however, to observe the performance of the car and assist with appropriate adjustments. The engineer's concern after the initial design is with design improvements or the continued development of procedures that result in optimum performance.

There are those in industrial engineering today, the author included, who believe that when industrial engineers become part of the human effort solely concerned with operating a system (i.e., without concern for improvement) they are not making full use of their expertise. That is not to say that industrial engineers should not become managers. They do, in large numbers; but they should recognize that when they do they are practicing management, not industrial engineering. It might be noted, at this point, that the technical education of industrial engineers provides them with a quantitative analytical capability that is of considerable value for managing a technically oriented company. The business graduate is far better trained to manage a small business, such as a small loan company office or a shoe store, but managing an electronics manufacturing company is another matter entirely. Industrial engineering education is commonly considered one of the best, if not *the* best, available today at the college level for anyone aspiring to a high-level management position in a technical company.

OPERATIONS RESEARCH

When World War II began, there was a small group of military researchers, headed by A. P. Rowe, who were interested in the military use of a technique called radiolocation, which was developed by civilian scientists. Some historians consider this research to be the identifiable starting point for operations research. Others believe that studies that have the characteristics of operations research work can be traced further back in time. Archimedes' analysis and solution of the naval blockage of Syracuse for the King of Syracuse in the third century is considered by some to be the beginning. F. W. Lanchester, in England, just before World War I, developed mathematical relationships representing the firepower of opposing forces, which when solved through time could determine the outcome of a military engagement. Thomas Edison also made studies of antisubmarine warfare. Neither the studies of Lanchester nor those of Edison had any immediate impact; along with Archimedes, they were early examples of the employment of scientists for determining the optimum conduct of war.

Not long after the outbreak of World War II, the Bawdsey Research Station, under Rowe, became involved in devising optimum-use policies for a new early-warning detection system called radar. Shortly thereafter, this effort developed into an analysis of all phases of night operations, and the study became a model for operational research studies to follow.

In August 1940, a research group was organized under the direction of P. M. S. Blackett of the University of Manchester to study the use of a new radar-controlled antiaircraft system. The research group came to be known as "Blackett's circus." The name does not seem unreasonable in light of their diverse backgrounds. The group was composed of three physiologists, two mathematical physicists, one astrophysicist, one Army officer, one surveyor, one general physicist, and two mathematicians. The formation of this group seems to be commonly accepted as the beginning of operations research.

In 1941 Blackett and part of his group became involved in the problems associated with the detection of ships and submarines by airborne radar. This study led to Blackett's becoming director of Naval Operational Research of the British Admiralty. The remaining part of his group later became the Operational Research Group of the Air Defense Research and Development Establishment, and then splintered again, forming the Army Operational Research Group. Within two years after the beginning of the war, all three major services had operational research groups.

As an example of these earlier studies, the Coastal Command was having difficulty sinking enemy submarines with a newly developed antisubmarine bomb. The bombs were triggered to explode at depths of not less than 100 feet. After detailed study, a Professor Williams concluded that maximum kill likelihood would occur at a depth setting of 20 to 25 feet. The bombs were then set for the minimum possible depth setting of 35 feet, and kill rate increases from different estimates ranged from 400 to 700 percent. Research was immediately undertaken to develop a firing mechanism that could be set to the optimum depth of 20 to 25 feet.

Another problem considered by the Admiralty was the relative merits of large versus small convoys. The results indicated that large convoys fared much better.

Within a few months after the United States entered the war, operations research activities were initiated in the Army Air Force and the Navy. By D day (the Allied Invasion of France) twenty-six operations research groups, with approximately ten scientists per group, had developed in the Air Force. A similar development had also occurred in the Navy. Philip M. Morse of the Massachusetts Institute of Technology headed a group in 1942 to analyze the sea and air attack data against German U-boats. Another study was undertaken later to determine the best maneuvering policy for ships in convoy attempting to evade enemy planes, including the effects of antiaircraft accuracy. The results of the study showed that large ships should attempt severe changes in direction, whereas small ships should change direction gradually.

One final example of early military operations research was a study headed by Ellis A. Johnson. The effort involved computer-simulation war gaming to determine optimum policies for the deployment of sea mines. The efforts of this group ultimately culminated in Operation Starvation. The study showed that B-29s could be extremely effective in mining if employed at night at an altitude of 5,000 feet. This approach to mining reduced loss rates to one-tenth of the previous rate. Operation Starvation cost the Japanese war effort 1,200,000 tons of shipping with B-29 loss rates of less than 1 percent.

What is surely obvious by now is that operations research developed to meet the need for analysis of operational systems during World War II. The first chapter of the text by McCloskey and Trefethen (19) is an excellent account of this early development of operations research.

At first, operations research dealt with existing weapons systems and, through analysis, typically mathematical, searched for optimum system policies for their use. Today, operations research still performs this function in the military sphere; however, and far more important, operational system needs are now analyzed, by employing mathematical models, and an operational system (or systems) is designed to offer optimum capability.

The success of operations research in the military sphere was fairly well documented by the end of World War II. In 1946 General Arnold persuaded Donald Douglas of the Douglas Aircraft Corporation to manage a Research ANd Development project (RAND) for the Air Force, and $10 million of Air Force funds was allocated to the project. The RAND Corporation today plays a major role as part of the Air Force research establishment.

From the beginning of the development of operations research as a discipline, certain characteristics have been commonly identified with it; these are: (1) the systems approach, (2) mathematical modeling, and (3) the team approach. On both sides of the Atlantic, throughout the development of operations research during World War II, these characteristics prevailed.

A systems approach was necessary for maximizing the effectiveness of the military capability available at the time. The days of making high-level decisions concerning the conduct of a war that involved sophisticated systems by comparison to general strategies learned from previous wars or to a game of chess had come to an end.

The digital computer and the systems approach were necessary preludes to mathematical modeling of operational military systems. Applied mathematics had proved to be useful in the analysis of economic systems, and the use of operations research in the analysis of military systems likewise proved to be very fruitful.

For an analysis of a military operational system to be technologically feasible, it was necessary to have adequate technical breadth to represent all relevant sub-components of the system. The team approach, therefore, proved to be both necessary and effective.

Operational research in England and operations research in the United States in post-World War II days had different characteristics and rates of acceptance in the industrial sector of the respective economies. In the United States, private consulting and industrial engineering were familiar improvement activities before the war. Industry experimented with operations research, somewhat as if it were a doubtful competitor to consulting and industrial engineering. Although U.S. firms had become flexible with respect to experimenting with new approaches, they often jealously guarded the obtained results from their competitors, limiting dissemination of the experience gained.

Prewar British industry, by comparison, had a more traditional trade orientation, and had not integrated improvement activities into the industrial structure to the extent that U.S. industry had. Consequently, operational research, particularly in light of its wartime successes and the backlog of opportunities for improvement in British industry, was welcomed as a needed, if not a somewhat all-encompassing, improvement activity. Much of what had been done in the United States by industrial engineers and consultants was being sold as operations research in England, one exception possibly being "work study." Work study is the equivalent terminology in England for what is now called "methods engineering" and was previously called "motion and time study" in the United States.

One of the first courses in operations research was offered in 1948 at the Massachusetts Institute of Technology. The following year a lecture series was presented at University College, London. Case Western Reserve became the first university to offer a degree program, and shortly thereafter universities such as Johns Hopkins and Northwestern were staffing departments of operations research.

The group of scientists who had initiated the operations research lecture series at University College had a year earlier formed the Operational Research Club. The club later initiated the first periodical in operations research, entitled *The Operational Research Quarterly*. The first operations research publication in the United States was *Operations Research with Special Reference to Non-Military Applications,* published by the Committee on Operations Research, which was formed in 1949 by the National Research Council. A year later the Operations Research Society of America was chartered, with Philip M. Morse as its first president. The quarterly *Journal of the Operations Research Society of America* was first published in November 1952.

There seems to be little question today that operations research has been effective in the analysis of operational systems in the military sphere. It is not as obvious that it has been or will be successful as a separate discipline in the industrial sector. The following quotation constitutes the summary of the report "A Limited

Survey of Industrial Progress in Operations Research" prepared by McKinsey and Company, Management Consultants, and reprinted in Lehrer's *The Management of Improvement* (18, p. 114):

> Despite the impression one might gain from the business press that operations research is firmly integrated into the decision-making processes of a great many American business enterprises, the evidence uncovered in our limited survey will not support such a conclusion. A factual evaluation of current realized operations research contributions in a selected sample of well-managed companies indicates that progress in a large proportion of them has been slight so far. Actual accomplishments in terms of new decision-making practices is modest, and management acceptance and understanding of the O.R. approach to problem solving are growing at a slow rate in these companies.
>
> Our survey did not in any way indicate, however, that lack of progress means lack of opportunity. The potential of operations research methods to contribute significantly to better, more profit-oriented decision making is unquestionably still there. The inability of companies to exploit this potential is not attributable to inadequate techniques or methodologies, but rather to the continuing lag in the number of professional people truly skillful in the application of these techniques in a way that gains the confidence and support of management.

This quotation is offered as food for thought. If such opportunity exists in the industrial sector of our economy, why is it that operations research scientists have not been more successful in its application? More will be said about this later.

SYSTEMS ENGINEERING

Two highly significant works were published in 1948; one was Norbert Wiener's *Cybernetics, or Control and Communication in the Animal and the Machine* (29), and the other was Claude Shannon's *The Mathematical Theory of Communication* (28). Wiener derived the word "cybernetics" from a Greek word meaning "steersman," and his subject was the generality of negative feedback in systems spanning the biological and physical world.

The most commonly used example of negative feedback is the thermostat. When the temperature drops sufficiently below some desired value, the thermostat initiates the heating portion of the cycle, and heat is added to cause the temperature difference to be negated until a temperature is reached that is greater than the desired temperature. Heating is then stopped to permit cooling to negate the overheating. "Negative feedback" means that some action is taken to oppose or negate an unacceptable difference.

Figure 3-1 is a conceptual model of negative feedback in management systems. An apparent condition is compared with a goal, and if a sufficient difference (i.e., error) exists, management action is taken to reduce the difference. The action should result in a change in the apparent condition so that later comparison of the apparent condition with the goal will cause the controlling action to cease. Assume that a manufacturer wishes to have 100 units of inventory on hand. After reviewing his inventory status he notes that he has only 80. If 20 is a sufficient difference, he would

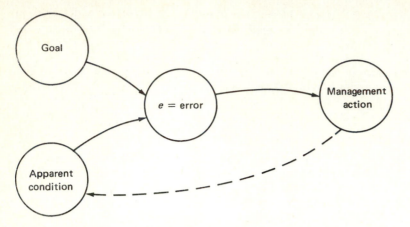

Figure 3-1 Negative feedback in management systems.

likely perform the management action of ordering more material in an attempt to raise his inventory level closer to the desired level. The order, after a purchase delay, should bring the apparent condition closer to the goal, removing the need for additional management action. This concept is consistent with the "exception" principle of managerial control, which says that management attention should be directed to situations in which abnormal values are known to exist. It is management's job to cause an undesirable value to return to a normal or steady-state level.

It is the generality of this concept and other characteristics of systems that constitutes the significance of Wiener's text. "Homeostasis" is a word commonly employed in the biological sciences in connection with the regulatory processes in living organisms. Analogous regulation can be identified in such diverse systems as water flow in irrigation or current flow in electrical networks.

Wiener's work is generally considered to be the starting point of what is now commonly referred to as "general systems theory." Schlager (27) reported in 1956, on the basis of a review of a nationwide survey, that the first known use of the term "systems engineering" was in the Bell Telephone Laboratories in the early 1940s. Considering the problems the Bell System faced at that time in expanding its system, it is understandable that the term might well have been conceived there. The RCA Corporation, in the years just preceding this, had recognized the need for a systems engineering point of view in the development of a television broadcasting system.

In 1946 the newly created RAND Corporation developed a methodology which they labeled "systems analysis." Quade and Boucher, in *Systems Analysis and Policy Planning,* define systems analysis as "a systematic approach to helping a decisionmaker choose a course of action by investigating his full problem, searching out objectives and alternatives, and comparing them in the light of their consequences, using an appropriate framework—in so far as possible analytic—to bring expert judgment and intuition to bear on the problem" (25, p. 2).

It now appears that recognition of cybernetics, which was popularized by Wiener in 1946, was delayed in the Soviet Union because of the unfavorable intellectual climate until 1948, at which time a Soviet writer defined cybernetics as "the new science of purposeful and optimal control over complicated processes and operations which take place in living nature, in human society, and in industry" (7).

Both operations research and systems analysis, as practiced until 1950, were concerned with determining the optimum selection, deployment, and control of existing operational systems. In 1950 the RAND Corporation initiated research concerned with identifying optimal security policy and national strategy. The initial reaction of the Pentagon was that RAND was invading their domain. It was the Pentagon's belief that it could not be assumed that techniques applicable to weapons systems optimization would be effective in the more complex area of strategy and policy planning. It seems now that they were at least partially correct. Techniques had to be extended or conceived to handle additional complexities posed by the problems associated with this increased scope of analysis. Studies to date, however, have been effective in offering improved objectivity and reducing the extent of crystal-ball policy development at the Pentagon. Policy planning is still partly science and partly art and will likely continue to be so for a long time into the future. It is the improved ratio of science to art for which systems analysis deserves some credit.

It has been said that systems analysis is to operations research as strategy is to tactics. Enthoven attempts to clarify this distinction as follows (8)[1]:

> Again, analysis at this level cannot prove the optimality of any national security policy. I don't doubt for a moment that, given a specified set of ships and aircraft and equipment, and a particular task such as tracking down and killing submarines in a given area, operations analysis can indicate the optimal way to go about doing it. There, only one value judgment enters in. That is, that it is desirable to kill enemy submarines. You cannot do that at the national policy level. Rather, at that level, analysis can only trace out implications of alternative policies.

Lying somewhere between the operations research study to determine an optimal tactic and the systems analysis study to support the formation of national policy is the "cost-effectiveness study," which is concerned with identifying the best of available alternatives in light of associated costs and system effects.

In a paper published in 1951, Bertalanffy (3) indicated the need for identifying the underlying general principles of organismic systems that exist across disciplines, and the need for developing a universal language common to all disciplines. The considerable increase in attempts to perform interdisciplinary research in recent years has drawn attention to the problems of language between disciplines. The social barriers between disciplines and the fears associated with becoming involved in areas beyond the comfortable and secure domain of one's specialty may be equally significant barriers to effective interdisciplinary research. The proper conduct of interdisciplinary research is a newly emerging science across disciplines.

In 1956 Boulding referred to general systems theory as "a level of theoretical model-building which lies somewhere between the highly generalized constructions of

[1] Reprinted by permission from Alain C. Enthoven, "Operations Research and the Design of the Defense Program," *Proceedings of the 3rd International Conference on Operations Research*, Drenod, Paris, p. 534, Copyright 1964, ORSA. No further reproduction permitted without consent of the copyright owner.

pure mathematics and the specific theories of the specialized disciplines" (5, p. 197).[1a]
A number of analogous systems are commonly referred to in engineering. The flow of
water in pipes has its parallel in electrical circuits, whereas entropy in thermodynamic
systems is a highly useful concept for describing certain aspects of information systems
(1, p. 347). Even the fundamental laws of motion have their parallels in political
science, and the shock absorbers and springs of a front-end suspension have analogous
electrical circuit representations. It is no longer necessary to build a front-end
suspension to determine its dynamic characteristics; analog simulation of the
analogous electrical circuit can provide the results. In industrial engineering, a digital
simulation of an entire computer-controlled physical distribution center can be
employed to test the effectiveness of the system before the first piece of equipment to
be used in the system is installed.

Johnson stated that "Models can be developed which have applicability to many
fields of study. An ultimate but distant goal will be a framework (or system of systems
of systems) which could tie all disciplines together in a meaningful relationship" (17,
p. 6). To reiterate, however, the greatest problem at present may be the lack of a
common technical language, as indicated, possibly to the point of cynicism, in the
following quotation from Boulding: "Hence physicists only talk to physicists,
economists to economists—worse still, nuclear physicists talk only to nuclear physicists
and econometricians to econometricians. One wonders sometimes if science will not
grind to a stop in an assemblage of walled-in hermits, each mumbling to himself words
in a private language that only he can understand" (5, p. 198).[1a]

In 1956, both Boulding and Bertalanffy played a part in forming the Society for
the Advancement of General Systems Theory. A year later the name was changed to
the Society for General Systems Research. These societies were the culmination of
efforts that began at a 1954 meeting of the American Association for the
Advancement of Science (AAAS) in Berkeley, California.

At about the same time that the commonality of disparate systems was becoming
fully appreciated, a more quantitative understanding of the processes of decision-
making was being reached. Von Neumann and Morgenstern's *Theory of Games and
Economic Behavior* (22) initiated game theory. First, working with two-person
zero-sum games (i.e., if one player wins, the other must lose a like amount), they were
able to analytically model and determine the outcome of a competitive game,
assuming a fixed playing strategy for each person. This work was part of the
foundation of a major area of study commonly referred to as decision theory; the
book by Raiffa and Schlaifer (26) has been a classic in this area for some time.

Some fairly clear differences have emerged over the years between operations
research and systems engineering. Although the early philosophers of operations
research believed it to be the beginning of an analytical attack, via mathematics, on
large-scale problems, a review of the operations research literature shows that most
problems must be stated with a limitation on the number and complexity of
representations if one hopes to reach analytically sound solutions. Some operations
research problems involve a large number of equations—a linear programming solution,

[1a] Reprinted from "General Systems Theory, the Skeleton of Science," by Kenneth E.
Boulding, *Management Science*, April, 1956, pp. 197, 198, published by The Institute of
Management Sciences.

for example—but the complexities of representation in a single equation may, and often do, make the entire set of equations unsolvable. For many problems today the techniques of operations research offer solutions that were unavailable in the recent past. For a much larger number of problems, however, analytical techniques do not exist for their solution. This is particularly true for problems concerning large-scale systems with multiple objectives, particularly if the system is socioeconomic rather than purely physical.

Systems engineering, however, seems to have developed with less dependence on "hard" mathematical representation of all aspects of a system. Digital simulation is a much more frequently employed technique in systems engineering, particularly if the system cannot be "tightly" represented and solved analytically because there is no appropriate analytical technique or the data are not in the form required for a specific operations research technique.

Another characteristic difference between operations research and systems engineering seems to be one of scope. The following quotations may shed some light on this fundamental difference:

> Systems analysis . . . differs in scope from operations research in the conventional sense, and it is not performed exclusively or even primarily by people who might be identified as operational researchers . . . It is a discipline with a logic of its own, similar in many respects to that of operations research, but also different in some fundamental aspects.
>
> Like operations research, this kind of analysis can and must be honest, in the sense that the quantitative factors are selected without bias, that the calculations are accurate, that alternatives are not arbitrarily suppressed, and the like. But it cannot be "objective" in the sense of being independent of values. Value judgments are an integral part of the analysis; and it is the role of the analyst to bring to light for the policy-maker exactly how and where value judgments enter so that the latter can make his own value judgments in the light of as much relevant information as possible (8).[1b]

> When large-scale and expensive systems problems are encountered, large numbers of people are required for their solution. This inevitably necessitates highly technical decisions and high-level compromises in the organizations involved. These decisions are often made in the face of large uncertainties. The burden of such decisions and compromises has become so great that the systems engineering function has evolved to help organize the factual bases for them (14, p. 6).[2]

> Operations research is usually concerned with the operation of an existing system, including both men and machines. Thus we find operations research looking at military operations, supermarkets, factories, farms, etc., and examining specific functions within these operations such as inventory control, distribution of raw and finished materials, waiting lines, advertising, etc. The object usually is to optimize, or to make better use of materials, energies, people and machines

[1b] Reprinted by permission from Alain C. Enthoven, "Operations Research and the Design of the Defense Program," *Proceedings of the 3rd International Conference on Operations Research*, Drenod, Paris, p. 530, Copyright 1964, ORSA. No further reproduction permitted without consent of the copyright owner.

[2] From *A Methodology for Systems Engineering* by A. Hall. © 1962 by Litton Educational Publishing, Inc. Reprinted by permission of Van Nostrand Reinhold Company.

already in existence and at hand. In contrast, systems engineering emphasizes the planning and design of new systems to better perform existing operations or to implement operations, functions or services never before performed (14, p. 18).[3]

There has been a recent tendency to broaden the definition of operations research so that it is practically synonymous with system design. However, there is a fundamental difference in approach: The operations analyst is primarily interested in making procedural changes, while the system engineer is primarily interested in making equipment changes (13, p. 130).

It would seem, therefore, that operations research attempts to determine how a system can be optimally employed. Systems engineering, however, is more concerned with the design of a system in light of its system objectives.

One of the more recent, and possibly most controversial and significant, contributions to the field of systems engineering is that of Jay W. Forrester of the Massachusetts Institute of Technology. Forrester's *Industrial Dynamics* (9) offers an almost totally new view of systems analysis. As the name of his book indicates, the approach was first applied to industrial management system problems.

Forrester viewed the role of the industrial manager as that of controller of the rates of flow of resources. The development by his staff of a digital computer compiler called DYNAMO (24) provided the means to continuous system representation of a macrosystem by a set of equations relating its significant elements. By testing various management policies (i.e., parameter values and modified representations of variables) Forrester demonstrated that simulation of changes could provide an insight into the behavior of the modified system.

Digital simulation was by no means new at this point. What was new was the representation of large-scale systems, including representation of the management or decision processes as an integral part of the simulation.

Forrester's research probably became most controversial when he dared to model systems such as urban growth (10) and the world (11, 21). Economists argued, for example, that his urban model possessed an inadequate representation of the "Keynesian" view of economics. Throughout human history there have always been times when the leading economic theory of the day may not have provided the answers needed. A review of economic thought, such as that offered by Oser (23), can raise doubts about the validity of any leading economic theory of the day. In any case, conclusions drawn from experiments performed on Forrester's model of urban growth raised interesting, if not embarrassing, questions for those responsible for locating federally subsidized housing projects in areas such as Watts, Los Angeles. Forrester showed that the effect of building federally subsidized housing in Watts was to attract low-income, high-unemployment people into an area that lacked employment opportunities, compounding the unemployment problem. A shift in location policy was immediately effected to encourage dispersion of federally subsidized housing in relation to the availability of employment opportunities.

A possibly valid criticism of much industrial dynamics research that has been

---

[3] From *A Methodology for Systems Engineering* by A. Hall. © 1962 by Litton Educational Publishing, Inc. Reprinted by permission of Van Nostrand Reinhold Company.

published is related to the adequacy of the data to support the models developed. The computer adage "garbage in—garbage out" applies equally well to industrial dynamics modeling.

There seems to be general agreement today that, although industrial dynamics needs improvement, it is one of the best techniques for solving many pressing socioeconomic system problems. The following quotation from a multi-university research report for a study concerned with modeling the Chesapeake Bay states this point of view as follows (2, pp. 23–24):

> Two models of systems developed during the last few years are of particular relevance to this study. The first has been summarized by M. D. Mesarovic, et al., of the Systems Research Center of Case Western Reserve University and concerns itself with hierarchial, multilevel systems. . . . A second model has been proposed by J. W. Forrester. It is a computer model of social systems directed initially at urban systems but expanded to include a global perspective.

At the first lecture of the Bromilow Lecture Series recently initiated at New Mexico State University, Dean Joseph Hogan of Notre Dame reflected on the past, present and future of engineering practice. In his lecture he said (16), "I believe the methodology used in this study [21] [as originally developed by Dr. Jay Forrester of M.I.T.] must continue to be refined to provide us with more accurate guidelines on how to predict the interrelationships of the large scale systems existing in society today."

In summary then, what are systems engineers and what is their training? Goode and Machol describe a systems engineer as follows (13, p. 8):

> We are led to the concept of the system-design team, a small group of engineers or scientists, to lead a large project and organize the system effort. Such men have been variously called *engineering scientists, system engineers, system analysts,* or *large-scale-system designers.* The technique has been variously called the *systems approach* or the *team development method.*
>
> It is toward this man and his teammates that these discussions are directed. With the realization that not enough can be learned in all the required fields to make him a specialist, enough is introduced to make him aware of the language and problems of the specialist. This *generalist* is a new quantity in the engineering world, and his education must be begun.

Quade comments on the state of the field as follows (25, p. 18): "It is not easy to tell someone how to carry out a systems analysis. We lack an adequate theory to guide us. This must be expected, for systems analysis is a fairly new discipline, and history teaches us that good theory usually comes late in the development of any field and after many false starts." Hall seems to indicate that the field is so young that specifically trained "systems engineers" do not really exist as yet. He states (14, p. 16):

> It will be clear that with present-day training methods, "ideal systems engineers" are unavailable. However, by associating people in teams, a good mixture of qualities can be achieved. Achieving a good mixture of available talent is really an *ad hoc* solution to the problem. It is obvious that a larger supply of individuals

with a better blend of qualities is needed for systems engineering. Also needed is
research on the best types and amounts of talent to blend for the different kinds
of work to be done.[4]

It should be noted that since 1962, the year in which Hall's text was first published, a
number of systems engineering degree programs have been initiated, and a number of
industrial engineering departments have changed their names to industrial and
systems engineering departments to reflect this shift in industrial engineering
education.

Blair and Whitston conclude the introductory chapter of their book with a
discussion of general systems theory, stating (4, p. 38): "The practical influence of this
source of thought on human activity systems engineering lies largely in the future.
Probably it will *be* industrial engineering's future."

It is not apparent yet whether industrial engineering should claim the
responsibility for developing a science of systems within engineering, or whether
systems theory is so general that it should be developed across all engineering
disciplines and not identified with any one discipline. Engineers from all disciplines
have been active in developing system theory, as indicated in the following quotation
from Quade and Boucher (25, p. 10):

> The uses of systems analysis are also being explored on other levels of
> government. In California, for example, major problem areas of concern to the
> state have been subjected to systematic analysis by engineers of a number of
> aerospace firms: transportation systems (North American Aviation), criminal
> justice and the prevention of delinquency (Space-General), the flow of
> information needed for the state's operation (Lockheed), the control and
> management of wastes (Aerojet-General), regional land-use information systems
> (TRW Systems), and the state's social welfare operations (Space-General).

The industrial engineering field in particular, however, has grown considerably in the
direction of systems analysis. The following quotation from Conway and Schultz
indicates this trend (6, p. 2):

> Prior to 1950, nearly all industrial engineering was practiced in the manufacturing
> phase of the mechanical goods industries. Today people with this background are
> practicing, often under names other than industrial engineering, in transportation,
> distribution, military logistics, weapons systems analysis, finance, public
> health, and the service industries, as well as in manufacturing and as fre-
> quently in the process industries as with the mechanical manufacturing
> industries.

Chapter 7 of this book offers an additional description, through example, of systems
engineering accomplishments to date.

[4] From *A Methodology of Systems Engineering* by A. Hall. © 1962 by Litton Educational
Publishing, Inc. Reprinted by permission of Van Nostrand Reinhold Company.

OTHER RELATED DISCIPLINES

Very little has been said so far about the disciplines of computer science, statistics, management science, and human engineering. They all play a part in modern industrial engineering, and their respective contributions need to be understood.

Computer science has developed as a field along with the widespread development and use of the digital computer in every facet of our commercial, industrial, and educational world. It is difficult to conceive of any field of science or engineering today that can operate effectively without the aid of a digital computer. This development, however, raises the question in educational circles of the appropriate organizational domain of computer science in a university environment. Who should it belong to if everyone needs it? Engineering? Arts and sciences? If all first-year engineering students need a course in FORTRAN programming, who should teach them, a computer scientist or an engineering professor?

A more basic question raised by the development and use of computers is, To what extent does the development of a piece of equipment constitute the basis for a separate discipline? It is quite common today and will become more common in the future, as more professors are trained in the use of the computer, that the appropriate specific computer aspects of any science or engineering course will be taught along with the underlying technical material. What does that leave for the computer scientist? Basic, commonly employed compilers can be taught either in computer science or in the science or engineering department concerned. Economy of scale is the advantage of having all students take the course in computer science, but the nonhomogeneity of mathematical capability across colleges results in the typical home economics or education major not being able to keep up and the mathematics, chemistry, physics, or engineering student becoming "bored to tears" with the class. The best compromise might be to have all engineering students take the course together in engineering, and likewise in the other colleges.

What then is the role of computer science? Computer science can effectively and efficiently teach students across disciplines advanced computer techniques of a general nature for which technical knowledge in a specific field is not necessary. They can also teach and research improvements in computer technology. It would appear that the future for many computer science programs is to be teaching general programming for their own and other students, doing continued research in computer technology, and giving technical support to the typical campus computer facility. The computer is becoming such a common component of any science or engineering field that it seems that any attempt to teach all computer use exclusively in a computer science department would constitute a considerable deterent to the logical development of computer use in specific disciplines.

Statistics has been and will continue to be a field distinctly separate from industrial engineering. However, the approach of industrial engineering has changed significantly, more so than that of any other engineering discipline, to view the world around us as "stochastic" in nature rather than "deterministic." By deterministic it is meant that all actions under consideration in a particular study situation are assumed to be certain. Stochastic implies that at least one aspect of the study situation has a probability of occurrence associated with it that must be considered.

Assume for the moment that John is trying to decide whether to date Joan or Margaret on Friday night. In a deterministic sense, John would likely compare his estimate of fun with Joan, F_J, at a cost of C_J, with his estimate of fun with Margaret, F_M, at a cost of C_M. This problem can be complicated enough if one adds the often nonlinear effects of marginal utility of increased fun as one approaches bankruptcy. The stochastic view of the situation, in addition, would require estimation of probabilities associated with varying levels of fun with both Joan and Margaret, as well as the probability and associated negative return associated with going to pick up Joan or Margaret and finding out she has decided to date Marvin instead, and forgotten to mention it.

The stochastic view of the world has so pervaded industrial engineering practice and education that a beginning course in probability and statistics has now become the most important prerequisite in a typical industrial engineering degree program. Industrial engineering is well ahead of other engineering disciplines in this development, and it seems likely that the improved insight it offers to problems will ultimately result in all disciplines shifting toward a more stochastic view of the world.

Management science is a field that developed closely allied to operations research in the 1960s. The underlying techniques were identical to those of operations research. What was different was the background of the management scientist and the area of application of the discipline. Management science, for the most part, was an outgrowth of the desire in many business administration or industrial management programs to offer degree options of a quantitative nature employing the discovered techniques of operations research. Because these programs developed in business colleges, the area of application was most often the one they were most familiar with—management. Whereas an operations research study might offer a linear programming solution to a problem involving refueling submarines, a management science study might well involve a linear programming solution to a portfolio selection problem.

In many instances the leaders in the operations research movement were also leaders in the management science movement. In the affluent 1960s, two closely allied societies with independent publications coexisted with little difficulty. Since then weakening of the economy has resulted in at least partial merger of management science with the operations research society, with the initiation of the joint publication *Interfaces*. In even more recent years the two societies have held joint conferences and have a joint business office which suggests that merger into one society at some future time is likely.

Human engineering, sometimes called human factors or ergonomics, is an engineering parallel to a field generally referred to as either industrial psychology or experimental psychology. Industrial engineering systems by nature are often human-machine systems, in contrast to hardware systems in electrical engineering, for example. The design of human-machine systems involves determining the best combination of human and machine elements. A typical course summarizes the considerable research that has been performed to date in industrial psychology and human engineering to familiarize the industrial engineering student with human-machine systems design. The text by McCormick (20) has been used extensively for this purpose. A significant amount of human engineering research is now being performed in industrial engineering departments, complementing the continuing research in industrial psychology.

SUMMARY

What, then, are modern industrial engineers? First and foremost, they are engineers. They take the same science, mathematics, and engineering courses as other engineering students. In addition, they take courses in their own discipline, have an interest in management, and have acquired a capability in computer science, operations research, systems engineering, and human engineering, while developing a stochastic view of the world. Industrial engineering students, more than any other engineering students, try to find a class in another department in another college in various remote corners of a campus. Industrial engineers have a far broader training than students in any other engineering disciplines. It is probably their greatest asset when it comes time to leave campus, as most students do sooner or later, and go to work.

REFERENCES

1 Beer, Stafford: *Decision and Control,* John Wiley and Sons, Inc., New York, 1966.
2 Beers, Roland F., et al.: "The Chesapeake Bay," a proposal prepared by The Johns Hopkins University, the University of Maryland, and the Virginia Institute of Marine Science, 1971.
3 von Bertalanffy, Ludwig: "Problems of General Systems Theory," *Hum. Biol.,* December, 1951.
4 Blair, Raymond N., and C. Wilson Whitston: *Elements of Industrial Systems Engineering,* Prentice-Hall, Inc., Englewood Cliffs, N.J., 1971.
5 Boulding, Kenneth E.: "General Systems Theory, the Skeleton of Science," *Manage. Sci.,* April, 1956.
6 Conway, Richard W., and Andrew Schultz, Jr.: "Industrial Engineering's Decade of Development," *The Cornell Engineer,* January, 1973.
7 Deckert, Charles R. (ed.): *The Social Impact of Cybernetics,* p. 18, Simon & Schuster, Inc., New York, 1967.
8 Enthoven, Alain C.: "Operations Research and the Design of the Defense Program," in *Proceedings of the 3rd International Conference on Operational Research,* Drenod, Paris, 1964.
9 Forrester, Jay. W.: *Industrial Dynamics,* M.I.T. Press, Cambridge, Mass., 1962.
10 Forrester, Jay W.: *Urban Dynamics,* M.I.T. Press, Cambridge, Mass., 1969.
11 Forrester, Jay W.: *World Dynamics,* Wright-Allen Press, Inc., Cambridge, Mass., 1973.
12 George, Claude S.: *The History of Management Thought,* Prentice-Hall, Inc., Englewood Cliffs, N.J., 1968.
13 Goode, Harry H., and Robert E. Machol: *Systems Engineering,* McGraw-Hill Book Company, New York, 1957.

14 Hall, Arthur D.: *A Methodology For Systems Engineering*, D. van Nostrand Co., Princeton, N. J., 1962.
15 Harper, Robert F.: *The Code of Hammurabi, King of Babylon*, p. 157, University of Chicago Press, Chicago, 1904.
16 Hogan, Joseph C.: "The Practice of Engineering," First Annual Dean Frank Bromilow Lecture, New Mexico State University, Las Cruces, February 21, 1975.
17 Johnson, Richard A., et al.: *The Theory and Management of Systems*, McGraw-Hill Book Company, New York, 1963.
18 Lehrer, Robert N.: *The Management of Improvement*, Reinhold Publishing Corp., New York, 1965.
19 McCloskey, Joseph F., and Florence N. Trefethen (eds.): *Operations Research for Management*, The Johns Hopkins Press, Baltimore, Md., 1954.
20 McCormick, Ernest J.: *Human Factors Engineering*, McGraw-Hill Book Company, New York, 1957.
21 Meadows, Dennis L., et al.: *Limits to Growth*, Wright-Allen Press, Inc., Cambridge, Mass., 1972.
22 von Neumann, J., and O. Morgenstern: *Theory of Games and Economic Behavior*, Princeton University Press, Princeton, N.J., 1947.
23 Oser, Jacob: *The Evolution of Economic Thought*, Harcourt Brace Jovanovich, Inc., New York, 1963.
24 Pugh, Alexander, III: *DYNAMO II User's Manual*, M.I.T. Press, Cambridge, Mass., 1970.
25 Quade, E. S., and W. S. Boucher (eds.): *Systems Analysis and Policy Planning*, American Elsevier Publishing Co., New York, 1968.
26 Raiffa, Howard, and Robert Schlaifer: *Applied Statistical Decision Theory*, M.I.T. Press, Cambridge, Mass., 1961.
27 Schlager, K. J.: "Systems Engineering—Key to Modern Development," *IRE Trans. Eng. Manage.*, vol. 3, pp. 64–66, 1956.
28 Shannon, Claude: *The Mathematical Theory of Communication*, University of Illinois Press, Urbana, 1948.
29 Wiener, Norbert: *Cybernetics, or Control and Communication in the Animal and the Machine*, John Wiley & Sons, Inc., New York, 1948.

REVIEW QUESTIONS AND PROBLEMS

1 What two fields are related to but not subdisciplines of industrial engineering?
2 How does production management as an area of study differ from industrial engineering?
3 What is the difference in connotation between operational research and operations research?
4 What was the name of the first operations research periodical?
5 Give an example of negative feedback.
6 What is meant by the phrase "systems analysis is to operations research as strategy is to tactics"?
7 Differentiate between operations research and systems engineering with respect to approach and scope.
8 Forrester observed that providing federally subsidized housing in an area of low employment may increase the unemployment problem. Is this an example of positive feedback, and why?

9 Compare the relative extent of development of the fields of management and systems engineering.

10 How does industrial engineering compare with other engineering disciplines in the extent to which a probabilistic view has replaced the more classical, deterministic view of natural phenomena?

11 How does management science differ basically from operations research?

12 What characteristic of industrial engineering systems often makes human engineering an unavoidable consideration in the design of a system?

Part Two

The Present

Chapter 4

Production Systems Design

Any truly labor-saving device will win out. All that you have to do to find proof of this is to look at the history of the industrial world. And, gentlemen, scientific management is merely the equivalent of a labor-saving device.

F. W. Taylor

Since the days of Taylor, industrial engineers have been concerned with the design of manufacturing plants. At first, attention was centered on activity within an employee's workplace, and this type of analysis came to be known as methods engineering. Later, attention was also given to methods of handling materials between workstations and the relative spatial arrangement of all entities within a plant. These two areas of analysis are commonly referred to as materials handling and plant layout, respectively. The sum of the three activities above—methods engineering, materials handling and plant layout—is commonly referred to as plant design. Obviously, a plant design can be no better than its workstation designs because employees perform their functions at workstations (i.e., that is where the action is).

METHODS ENGINEERING

A typical text on methods engineering covers both methods study and work measurement. Whereas methods study is concerned exclusively with design, work measurement is employed in design but plays its primary role in cost control, which is considered in more detail in the next chapter. Because of the considerable interdependence between methods study and work measurement, however, work measurement is reviewed in this chapter.

Methods Study

The detailed design of workstations, and to a limited extent their interrelationships, is called methods study. In the planning stages, an estimate is made of the time it will take an employee to perform a given task at a workstation. Later, when the employee has learned the task and the conditions affecting the task have stabilized (e.g., tooling is available and working properly) management normally requires a detailed restudy of the job. Through observation and analysis, an industrial engineer or technician determines a "time standard" for performing the task, including allowances. This time becomes management's official opinion of how long it should take a normal employee at a normal rate of activity to perform the necessary work required on one unit of product, including prorated allowances. This time, therefore, serves the purpose of providing management with a basis for determining employee performance by comparing the actual number of units produced by an employee over a defined period of time, with the number the employee would have produced during this same period utilizing the "standard time" per unit. The process of determining the standard time for an operation is called work measurement. The term "methods engineering" connotes both methods study and work measurement, which respectively attempt to answer the questions, How should a task be performed? and How much time should it take to perform the task?

Many of the traditional approaches in methods engineering can be classified as "charting methods." These methods have changed very little in the last forty years, yet they still possess a general utility that has not been replaced by most "modernist" approaches. In general, they offer a graphic picture or dimension to a problem and also channel data collection with respect to it.

The operation process chart has been in use for many years to display in one figure the operations and inspections, and their sequence, for a complete product. Figure 4-1 is a typical operation process chart. If the plant has a volume requirement that dictates a production line layout, the operation process chart offers a possible first hint as to a likely relative arrangement of processing areas. If you took the roof off a plant and looked at the operations from a helicopter, Fig. 4-1 could represent the general relative locations of various product flows in the plant. In fact, it is rare that we have an opportunity to do this, whereas plant personnel can view detailed arrangement within limited areas with little difficulty. This is one of the reasons why areas in a plant are more likely to be poorly located with respect to one another, as compared to the arrangement of equipment within areas.

After the creation of an operation process chart, the next step typically is to

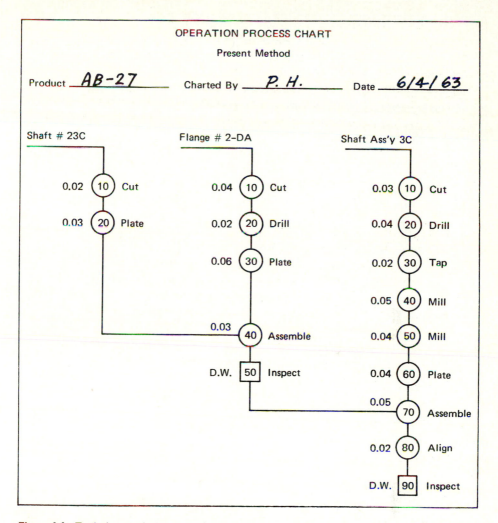

OPERATION PROCESS CHART

Present Method

Product __AB-27__   Charted By ___P. H.___   Date ___6/4/63___

Shaft # 23C

0.02 ⑩ Cut

0.03 ⑳ Plate

Flange # 2-DA

0.04 ⑩ Cut

0.02 ⑳ Drill

0.06 ㉚ Plate

0.03 ㊵ Assemble

D.W. 50 Inspect

Shaft Ass'y 3C

0.03 ⑩ Cut

0.04 ⑳ Drill

0.02 ㉚ Tap

0.05 ㊵ Mill

0.04 ㊿ Mill

0.04 ㉠ Plate

0.05 ⑺ Assemble

0.02 ⑻ Align

D.W. 90 Inspect

Figure 4-1 Typical operation process chart.

perform a more detailed analysis of each part or assembly of the total product. Whereas analysis was restricted only to operations and inspections in using the operation process chart, the flow process chart includes additional consideration of moves, delays, and storages. Figure 4-2 is a typical flow process chart. In effect, analysis has shifted from intraoperation handling to interoperation handling. Methods study is predominantly the study of transportation within a workstation, whereas materials handling is concerned with the transportation of material between workstations.

It may be instructive to reflect on the fact that a typical production process embodies only two main functions: (1) transformations and (2) moves. The transformations cause the product to change in nature and value as combined materials

FLOW PROCESS CHART

Page 1 of 2

PART NAME __Gizmo__

PROCESS DESCRIPTION __Machine base and assemble, and finish__

DEPARTMENT __Machine shop, Assembly, and Finishing__

PLANT __XYZ Products Co.__

RECORDED BY __I. M. Looking__   DATE _____

	SUMMARY	NO.
○	OPERATIONS	
⇨	TRANSPORTATIONS	
☐	INSPECTIONS	
D	DELAYS	
▽	STORAGES	
	TOTAL STEPS	
	DISTANCE TRAVELED	

STEP	Operations Transport Inspect Delay Storage	DESCRIPTION OF PRESENT METHOD					
1	○⇨☐D▽	in storage at receiving					
2	○⇨☐D▽	to position at mach 2	walkie	6'			
3	○⇨☐D▽	at mach. 2					
4	○⇨☐D▽	into mach. 2	hand	4'			
5	○⇨☐D▽	turn					
6	○⇨☐D▽	to table	hand	4'			
7	○⇨☐D▽	on table					
8	○⇨☐D▽	to mach. 3	hand	4'			
9	○⇨☐D▽	drill					
10	○⇨☐D▽	to table	hand	4'			
11	○⇨☐D▽	on table					
12	○⇨☐D▽	into mach. 4	hand	3'			
13	○⇨☐D▽	drill					
14	○⇨☐D▽	to skid	hand	4'			
15	○⇨☐D▽	on skid					
16	○⇨☐D▽	to Assembly Dept.	walkie	10'			
17	○⇨☐D▽	at end of assembly bench	•				
18	○⇨☐D▽	onto bench to assy. position	hand	5'			
19	○⇨☐D▽	assemble					
20	○⇨☐D▽	to inspection position	hand	3'			
21	○⇨☐D▽	inspect					
22	○⇨☐D▽	to skid at end of assy. bench	hand	8'			

Figure 4-2 Typical flow process chart. (*From James M. Apple:* Plant Layout and Materials Handling, *2d ed., p. 165. Copyright © 1963, The Ronald Press Company, New York.*)

offer increased utility (i.e., an assembled bicycle is worth more than the sum of its parts). If you do not believe this, ask the parents who have to assemble a tricycle on Christmas Eve if they would prefer to purchase it already assembled. In a plant, the moves merely represent a necessary evil in terms of cost, which should be minimized. Movement of parts and assemblies within a plant does not add to a product's functional utility. For this reason, the single main criterion in plant layout is minimum material handling cost. The flow process chart is particularly effective in tracing the sometimes almost unbelievable distance a part travels in a plant, particularly if the plant layout has been allowed to evolve over time with little detailed analysis.

In a practical sense, a typical production employee works in a workstation not a plant. Therefore, it is the sum effect of well-designed workstations that results in a productive plant. For this reason, most methods engineering is performed in relation to workstations. Figure 4-3 offers basic biometric data for typical benchwork. The goal in methods study typically is to develop a best work design for a workstation such that a best sequence of movements of the hands of an employee results in a minumum operation time (i.e., the product transformation is accomplished at a minimum cost). Figure 4-4 is an operation sheet that specifies the workplace arrangement and sequence of motions to be employed. An ideal workstation minimizes intraoperation handling as a result of detailed analysis and specification of the exact method to be employed. A number of common principles of good design have evolved over the years and are commonly referred to as "principles of motion economy." Twenty-two motion

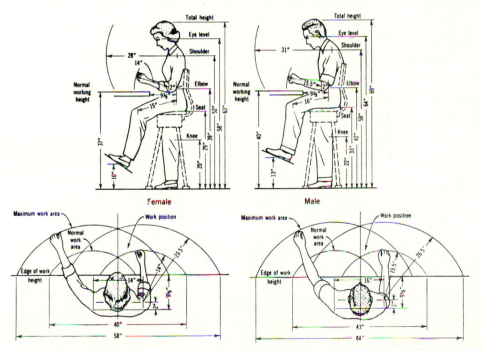

Figure 4-3 Dimensions of normal and maximum work areas. [*From Barnes (4, p. 260).*]

HIXY CLAMP COMPANY

Operation Sheet

Oper. Desc. _BONDING OF HOLDER_   Oper. No. _125_

Drawn By _P. H._   Dept. _4-12_   Date _4/7/66_   Sheet _1_ of _1_

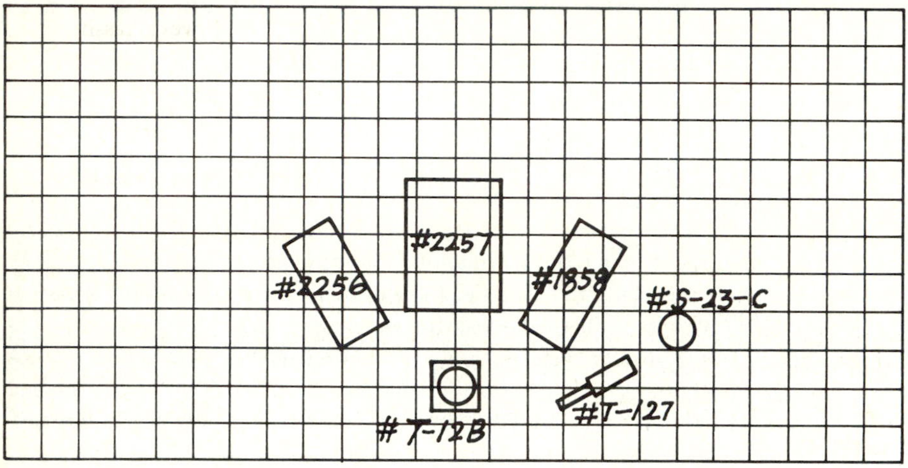

```
                    ┌──────────┐
              #2256 │  #2257   │  #1858
                    │          │        #S-23-C
                    └──────────┘           ○

                      ◉
                   #T-12B        #T-127
```

Parts   Supplies   Tools

#2256   #23-C Epoxy   #12B Clamp
#2257   #127 Spreader
#1858

Operation

Step	Description
1	LOCATE #2257 IN CLAMP.
2	ALIGN #2256 WITH RESPECT TO #2257.
3	SPREAD EPOXY ON #2257 & #2256 PER SPEC. NO. 116-2.
4	DRY FOR 10 MINUTES.
5	SCREW #1858 TO ¼" DEPTH.
6	RELEASE ASSEMBLY FROM CLAMP AND DISPOSE.

Figure 4-4 An operation sheet.

economy principles from Barnes' text, which have been in common use for a number of years, are as follows (4, pp. 221–305):

1   The two hands should begin as well as complete their motions at the same time.

2   The two hands should not be idle at the same time except during rest periods.

3   Motions of the arms should be made in opposite and symmetrical directions and should be made simultaneously.

4   Hand and body motions should be confined to the lowest classification with which it is possible to perform the work satisfactorily.

5   Momentum should be employed to assist the worker wherever possible, and it should be reduced to a minimum if it must be overcome by muscular effort.

6   Smooth continuous motions of the hands are preferable to straight-line motions involving sudden and sharp changes in direction.

7   Ballistic movements are faster, easier, and more accurate than restricted (fixation) or "controlled" movements.

8   Work should be arranged to permit easy and natural rhythm wherever possible.

9   Eye fixations should be as few and as close together as possible.

10   There should be a definite and fixed place for all tools and materials.

11   Tools, materials, and controls should be located close to the point of use.

12   Gravity feed bins and containers should be used to deliver the material close to the point of use.

13   "Drop deliveries" should be used wherever possible.

14   Materials and tools should be located to permit the best sequence of motions.

15   Provisions should be made for adequate conditions for seeing. Good illumination is the first requirement for satisfactory visual perception.

16   The height of the work place and the chair should preferably be arranged so that alternate sitting and standing at work are easily possible.

17   A chair of the type and height to permit good posture should be provided for every worker.

18   The hands should be relieved of all work that can be done more advantageously by a jig, fixture, or foot-operated device.

19   Two or more tools should be combined wherever possible.

20   Tools and materials should be pre-positioned wherever possible.

21   Where each finger performs some specific movement, as in typewriting, the load should be distributed in accordance with the inherent capacities of the fingers.

22   Levers, crossbars, and hand wheels should be located in such positions that the operator can manipulate them with the least change in body position and with the greatest mechanical advantage.

The use of these principles in combination with a questioning attitude has traditionally been very productive in developing efficient workstation designs. In fact, it is this author's opinion that far too little innovative, detailed methods study is done in industry today.

There is a class of charts commonly referred to as "multiple activity charts." These charts can be effective in analyzing situations in which at least two resources are employed within an operation. The object typically is to so arrange the sequence of events with respect to the resources employed that a minimum unit production time is determined.

One such chart, for example, is the "man-machine chart," illustrated in Fig. 4-5. This chart involves the analysis of three resources—an operator and two Jones and Lamson internal thread grinders. The general approach in such an analysis is to attempt

---

MAN & MACHINE PROCESS CHART

Subject Charted Grinding internal threads of housing   Chart No. 912
Drawing No. B-1976   Part No. B-1976-3   Chart of Method Proposed
Chart Begins Pick up part and load machine   Charted By E. B. Watmough
Chart Ends Remove ground part from machine   Date 3-17   Sheet 2 of 3

ELEMENT DESCRIPTION	OPERATOR	JONES & LAMSON INTERNAL THREAD GRINDER NO. 1	JONES & LAMSON INTERNAL THREAD GRINDER NO. 2
Open chuck and remove part machine #2	.0007		
Lay aside part	.0009		
Pick up new part from table	.0009		Idle
Place part in diaphragm chuck machine #2	.0012		
Close diaphragm chuck Mch. #2	.0007		
Start machine (spindle) mch. #2	.0003		
Start grinding wheel mch. #2	.0003		
			Loading .0058
Run grinding wheel forward 6" machine #2	.0028	Idle	
Engage feed	.0005		
Walk to machine #1	.0011		
Run grinding wheel back 1" machine #1	.0008	Machine Handling .0013	
Engage feed machine #1	.0005		
Inspect pieces from machine #1 and machine #2 previous cycle	.0050		Grind (first pass) .0075
		Grind (second pass) .0075	

Figure 4-5 A man-machine chart. [*Reproduced with permission from* Time and Motion Study, *Fifth Edition, by B. W. Niebel (Homewood, Ill.: Richard D. Irwin, Inc., 1972 c.) p. 147.*]

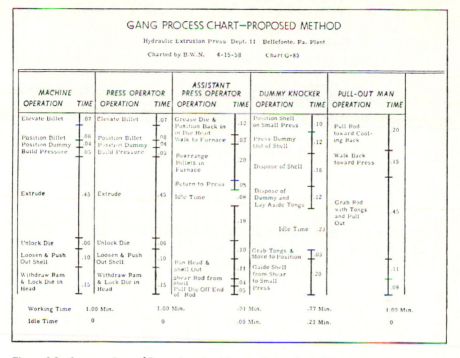

Figure 4-6 A gang chart. [*Reproduced with permission from* Time and Motion Study, *Fifth Edition, by B. W. Niebel (Homewood, Ill.: Richard D. Irwin, Inc., 1972 c.) p. 151.*]

to make use of the operator while the machine is involved in a "machine-controlled element." In this example, the operator is free to do other things while the grinder is making a grinding pass, and can therefore be effectively used to load a second machine during this free time. Obviously, there is a considerable reduction in labor time per unit produced if an operator can keep two machines going instead of one. Because of the competitiveness in the garment industry, multiple activity analysis is common and accepted in that industry. There are test operations in other industries, for example, in which a test employee sits idle for long periods of time waiting for temperature or some other factor to stabilize, yet multiple test stations are not commonly designed or employed according to the multiple activity approach in far too many instances.

The "gang chart" in Fig. 4-6 illustrates how the elemental activities of workers acting as a gang or crew can be analyzed to determine a minimum cycle time for a unit of production.

A wide variety of charts have been and are still in general use for analyzing the sequence of events of parts and product movement both within and between workstations.

Work Measurement

Time has always been one of the most important variables in engineering and science, as well as manufacturing. Galileo's experiments with falling bodies, for example, depended greatly on measurements of distance and time. Although time has been an important variable throughout history, it was Taylor who offered the concept of measuring the time of human activity as a means of controlling labor performance in industry. A watch is a device that, by means of gears powered by a spring through an escapement mechanism, rotates hands that display elapsed time. Because a watch measures time and accomplishes nothing else, it is understandable that stopwatch timestudy was the first work measurement technique developed.

In stopwatch timestudy, the analyst breaks down an operation into elements, as suggested by Taylor. Then with an operator performing the operation a number of times, the analyst observes elapsed time at the end of each element for the number of cycles of the study. The analyst also observes the rate of activity of the operator and records a "rating factor," which is the observed pace of the operator as compared to the analyst's concept of normal pace for the operation under study. Figure 4-7 is a representative timestudy sheet on which the observed times and rating factor or factors are recorded. Additional descriptive information about the operation is added either at the workplace or elsewhere. The timestudy analyst then calculates the standard time for the operation. The calculation, with reference to Fig. 4-7, is as follows:

$$ST = NT + A$$

$$NT = \sum_{i=1}^{n} (RF_i \times \overline{T}_i) \quad i = 1,2,3,4$$

where ST = standard time
NT = normal time
A = allowances (14.3% of NT in this example)
RF_i = rating factor for the ith element
$\overline{T}_i$ = average observed time for the ith element
Therefore,

$$NT = (0.90 \times 0.07) + (1.05 \times 0.15) + (1.00 \times 0.24) + (0.90 \times 0.10)$$
$$= 0.06 + 0.16 + 0.24 + 0.09$$
$$= 0.55 \text{ min.}$$
$$ST = 0.55 + (0.143 \times 0.55)$$
$$= 0.63 \text{ min.}$$

The above calculation is typical of the way in which timestudies are performed today. Even this simple example offers some insight into the need for research and standardization of methodology in this traditional area. The use of the rating factors 0.90, 1.05, 1.00, and 0.90 would suggest that, on the average, the employee's rate of activity during a number of cycles varies as indicated by line A in Fig. 4-8. Cyclical variation as indicated by line A is rather unlikely over a short time duration, however.

Time Study Observation Sheet															

Identification of operation *Assemble 24" x 36" chart blanks* **Date** *10/9*

Began timing: *9:26*   **Ended timing:** *9:32*   **Operator** *109*   **Approval** *PML*   **Observer** *MWS*

Element Description and Breakpoint			Cycles										Summary			
			1 $_{0.00}$	2	3	4	5	6	7	8	9	10	ΣT	$\overline{T}$	RF	NT
1	*Fold over end (grasp stapler)*	T	.07	.07	.05	.07	.09	.06	.05	.08	.08	.06	.68	.07	.90	.06
		R	.07	.61	.14	.67	.24	.78	.33	.88	.47	.09				
2	*Staple five times (drop stapler)*	T	.16	.14	.14	.15	.16	.16	.14	.17	.14	.15	1.51	.15	1.05	.16
		R	.23	.75	.28	.82	40	.94	47	.05 [4]	.61	.24				
3	*Bend and insert wire (drop pliers)*	T	.22	.25	22	.25	.23	.23	.21	.26	.25	.24	2.36	.24	1.00	.24
		R	.45	.00 [1]	.50	.07 [2]	.63	.17 [3]	.68	.31	.86	.48				
4	*Dispose of finished chart (touch next sheet)*	T	.09	.09	.10	.08	.09	.11	.12	.08	.17	.08	1.01	.10	.90	.09
		R	.54	.09	.60	.15	.72	.28	.80	.39	.03 [1]	.56 [5]				
5		T												0.55 normal minute for cycle		
		R														
6		T														
		R														
7		T														
		R														
8		T														
		R														
9		T														
		R														
10		T														
		R														

(0.55 X 0.143)

Normal cycle time ___0.55___ + Allowance ___or 0.08___ = Std. time ___0.63___ min./pc.

Figure 4-7 A representative timestudy chart. [*From Krick (8, p. 246).*]

A gradual drift in the rate of activity as indicated by line *B* is a great deal more plausible. The calculations are not consistent with the assumption of drift as compared to cyclical variation.

In the period 1910 to 1940, there was considerable misuse of timestudy by individuals neither trained in its use nor concerned with the employee's best interests. Far too many managements allowed and even encouraged misapplication, which resulted in a practice known as "rate cutting." Referring to Fig. 4-7, for example, if an employee assigned to this task performed at a rate consistent with this standard,

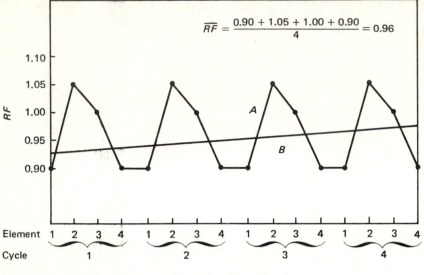

$$\overline{RF} = \frac{0.90 + 1.05 + 1.00 + 0.90}{4} = 0.96$$

Figure 4-8 Rating factors for a few cycles.

management might have assumed that it was "too loose," in which case they would have the operation restudied, resulting in a lower standard (e.g., 0.58 minute). Each reduction in the standard required the employee to work faster for the same take-home pay. Rate cutting understandably led to much labor and union dissatisfaction with timestudy. Today, a permanent standard is normally "guaranteed" by management, which means that it cannot be reduced as long as the job and method remain in use in the company.

In seeking to identify the basic motions employed in human work, Gilbreth invented the technique known as micromotion study. Today in this method a constant speed (i.e., 1,000 frames per minute) 16-millimeter industrial camera is typically employed and the time between each successive frame is therefore 0.001 minute. By examining and counting successive frames, usually with the aid of a frame counter attached to the projector, one can break down detailed human activity, such as high-volume benchwork, into rather detailed elements. Figure 4-9 is a "simo chart," which is used for displaying a micromotion study analysis. By filming various alternative ways of performing an operation, and analyzing the respective charts, it is possible to devise combinations of motion sequences that minimize unit cycle times. The employee is then trained to perform the operation by using the devised method.

Micromotion study may involve a considerable amount of film, and a technique known as memomotion study, similar to time-lapse photography, employs a longer time interval between successive frames. It is often more appropriate than micromotion study for medium- and long-cycle operations.

Another type of work measurement technique, which involves a level of detail similar to that of micromotion study, is commonly known as "predetermined time systems." Nine of these systems are reviewed by Barnes (4, pp. 472–473). By far, the predetermined time system in most common use today is MTM (methods time measurement). Figure 4-10, which is typical of tables in the MTM system, gives times in "TMUs" for reaching defined distances under varying conditions. A TMU is a "time-measurement unit" and is equal to 0.0006 minute. A predetermined time system, therefore, is a classification system that provides times for detailed elements of work activity under varying conditions. An analyst trained in MTM breaks down an operation into a sequence of elemental activities as classified by the MTM system. The

Simo chart						
Operation:	*Assemble Tab Shaft*					
Part: NA 37124			Operator: R. Rees			
Department: *assembly*			Date: 7/10			
Analysis: *Smalley*			Film No.: 16 - 48			
Left-Hand Description	Time in Minutes	Symbol	Time Scale	Symbol	Time in Minutes	Right-Hand Description
To shaft	0.007	R		D	0.007	
Shaft	0.016	G	0.010 — 0.020	R	0.016	To key
	0.006	D	0.030	G	0.006	Key
To assembly point	0.014	M	0.040	M	0.014	To shaft
Support assembly		H	0.050	P	0.009	To shaft
				RL	0.002	
			0.060	R	0.009	To collar
			0.070	G	0.007	Collar
				M	0.010	To assembly
			0.080	P	0.009	To assembly
			0.090	RL	0.003	
			0.100	R	0.008	To screwdriver

Figure 4-9 A simo chart. [*From Krick (8, p. 100).*]

Table 1   Reach—R

Distance moved, in.	Time, TMU				Hand in motion		Case and description	
	A	B	C or D	E	A	B		
¾ or less	2.0	2.0	2.0	2.0	1.6	1.6	A	Reach to object in fixed
1	2.5	2.5	3.6	2.4	2.3	2.3		location, or to object in
2	4.0	4.0	6.9	3.8	3.5	2.7		other hand or on which
3	5.3	5.3	7.3	5.3	4.5	3.6		other hand rests.
4	6.1	6.4	8.4	6.8	4.9	4.3		
5	6.5	7.8	9.4	7.4	5.3	5.0	B	Reach to single object in
6	7.0	8.6	10.1	8.0	5.7	5.7		location which may vary
7	7.4	9.3	10.8	8.7	6.1	6.5		slightly from cycle to cycle.
8	7.9	10.1	11.5	9.3	6.5	7.2	C	Reach to object jumbled
9	8.3	10.8	12.2	9.9	6.9	7.9		with other objects in a group
10	8.7	11.5	12.9	10.5	7.3	8.6		so that search and select
12	9.6	12.9	14.2	11.8	8.1	10.1		occur.
14	10.5	14.4	15.6	13.0	8.9	11.5		
16	11.4	15.8	17.0	14.2	9.7	12.9	D	Reach to a very small object
18	12.3	17.2	18.4	15.5	10.5	14.4		or where accurate grasp is
20	13.1	18.6	19.8	16.7	11.3	15.8		required.
22	14.0	20.1	21.2	18.0	12.1	17.3	E	Reach to indefinite location
24	14.9	21.5	22.5	19.2	12.9	18.8		to get hand in position for
26	15.8	22.9	23.9	20.4	13.7	20.2		body balance or next motion
28	16.7	24.4	25.3	21.7	14.5	21.7		or out of way.
30	17.5	25.8	26.7	22.9	15.3	23.2		

Figure 4-10   An MTM reach table. [*From Maynard (10, p. 5-18).*]

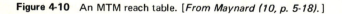

sum of the times for these detailed activities becomes the basis for establishing the standard time for the operation. Figure 4-11 is an example MTM analysis for a simple operation.

Methods time measurement and other predetermined time systems have their greatest utility in repetitive short-cycle operations. Yet a larger proportion of activity today has shifted from short-cycle repetitive work, such as an assembly operation on an electronics assembly line, to less repetitive medium- and long-cycle activity. A much greater proportion of workers today perform servicing or monitoring functions as equipment designs reach higher and higher levels of automation. One technique that has been developed for analyzing nonrepetitive work is called "work sampling."

Work sampling is nothing more than statistical sampling theory applied to industrial situations. In a basic course in probability theory it is often convenient to consider drawing a sample of colored balls from an urn containing balls of known characteristics (e.g., 400 white and 100 black balls). In work sampling, instead of drawing balls from an urn, random minutes are drawn from a population of possible minutes in a day. Based on the composition of activity of the sampled minutes, it is possible to make probabilistic statements about the composition of activity for all the minutes of the day. Instead of black and white balls, activity may have been classified, for example, as either working or idle.

The proportion of observed minutes of a particular class of activity to the total sample drawn is a best estimate of the percentage occurrence of that type of activity over the period of the study. If an employee is observed for 100 randomly selected minutes and for 90 of these observations is busy, our best estimate of the percentage of total time the employee is busy is $^{90}/_{100} = 90$ percent.

If the observer wanted to be more sure that the percent calculated was an accurate reflection of the true underlying percent occurrence he would likely increase the sample size. The following equation is commonly employed to relate the level of confidence with how well a sample represents the underlying percent of occurrence:

$$Sp = K_\alpha \sqrt{\frac{p(1-p)}{n}} \qquad (4\text{-}1)$$

where S = relative accuracy
      p = percent occurrence
      K_α = number of standard deviations from the mean for a given confidence level assuming a normal distribution
      n = sample size

Equation (4-1) is derived from what is called the "central limit theorem" in

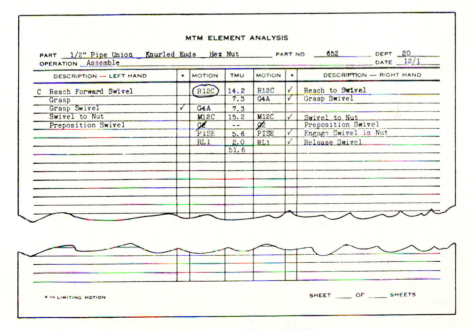

Figure 4-11 An MTM analysis. [*From Maynard (10, p. 5-33).*]

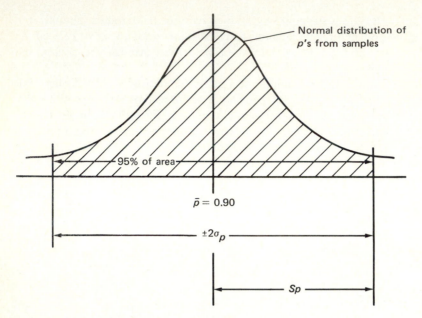

Figure 4-12   A distribution of average percent occurrence.

statistics. The theorem basically says that a distribution of means of samples, irrespective of the distribution of the population from which they are drawn, will approach a normal distribution as a limit as sample size increases. A distribution of proportions from samples behaves in a similar fashion.

Let us assume that a particular activity does involve 90 percent busy, 10 percent idle activity. The central limit theorem says then that if one were to take a number of samples from that activity and estimate "percent busy" from each, the estimates of the proportion would be distributed as shown in Fig. 4-12. It can be shown that

$$\sigma_p = \sqrt{\frac{p(1-p)}{n}}$$

and with the above constitutes the derivation of Eq. (4-1). An example will illustrate its use, but first Eq. (4-1) must be solved for n, which gives:

$$n = \frac{K_\alpha^2 (1-p)}{S^2 p} \tag{4-2}$$

Assume that a sample of size 50 was taken, and for 40 of the minutes the employee was observed to be working, how large should the sample size be to provide a relative accuracy of ±10 percent at a confidence level of approximately 95 percent? A K_α of ±2 contains 95.4 percent of the area under a normal curve between these limits and is close to the required 95 percent; therefore, it is often assumed that $K_\alpha = 2$. Our estimate of $p = {}^{40}\!/_{50} = 0.80$, and $S = 0.10$, therefore:

$$n = \frac{2^2(1 - 0.80)}{0.10^2(0.80)} = 100$$

This calculation indicates that the sample of 50 was not large enough for the desired level of accuracy and confidence level; therefore, sampling should continue. When a total of 100 samples have been taken, the calculation should be repeated until the estimated size of n required as calculated is less than the total size of the accumulated sample taken.

Work sampling is used for analysis of such nonrepetitive activities as engineering time, as indicated in Fig. 4-13. Service functions such as maintenance work, or long-cycle tasks such as manual washing of an airplane, can sometimes be analyzed

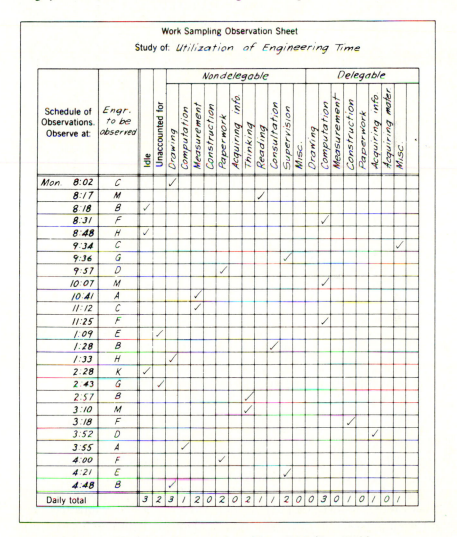

Work Sampling Observation Sheet
Study of: *Utilization of Engineering Time*

Schedule of Observations. Observe at:	Engr. to be observed	Idle	Unaccounted for	Nondelegable: Drawing	Computation	Measurement	Construction	Paperwork	Acquiring info	Thinking	Reading	Consultation	Supervision	Misc.	Delegable: Drawing	Computation	Measurement	Construction	Paperwork	Acquiring info	Acquiring mater.	Misc.
Mon. 8:02	C			✓																		
8:17	M											✓										
8:18	B	✓																				
8:31	F																✓						
8:48	H	✓																				
9:34	C																						✓
9:36	G												✓										
9:57	D							✓														
10:07	M																✓						
10:41	A					✓																
11:12	C					✓																
11:25	F																✓						
1:09	E		✓																			
1:28	B													✓								
1:33	H				✓																	
2:28	K	✓																				
2:43	G		✓																			
2:57	B										✓											
3:10	M										✓											
3:18	F																		✓			
3:52	D																				✓	
3:55	A			✓																		
4:00	F							✓														
4:21	E													✓								
4:48	B			✓																		
Daily total		3	2	3	1	2	0	2	0	2	1	1	2	0	0	3	0	1	0	1	0	1

Figure 4-13 A work sampling observation sheet. [*From Krick (8, p. 290).*]

best by employing work sampling. For work sampling to qualify as a work measurement technique, it is necessary not only to classify the nature of the activity during observations, but also to rate the pace of activity observed. The rated observations, along with a production count for a defined period of time, offer a means of determining a standard time for a unit of production. Such techniques have been successfully employed, for example, in determining how long it should take to replace a broken pane of glass, or replace a valve in a factory. Because of the increasing proportion of indirect labor activities such as maintenance, as compared to direct labor, the ability to measure work in these indirect areas is becoming more important, and in many companies it will be the key to effective labor control in the future.

A number of means have been developed to assist the work sampling analyst in the above calculations. Figures 4-14 and 4-15 are examples of tables and nomographs, respectively, used to simplify calculations. In employing these two figures, however, note that the "absolute error" in Fig. 4-14 and the "precision interval" in Fig. 4-15 are identical in meaning, and are defined as follows:

$$A = Sp$$

where A = absolute error (or precision interval)
S = relative error
p = percent occurrence

In the previous example, the absolute error was equal to $0.10(0.8) = 0.08$ or 8 percent (i.e., 10 percent of 80 percent = 8 percent, therefore the range of p, 95 percent of the time, was to lie between 72 and 88 percent).

Over the years, a work measurement technique known as "standard data" has evolved. Standard data employs summarized data from previous timestudies, thereby making it unnecessary to restudy work elements that have been timed sufficiently in the past. For example, a reach of 12 inches could have been an element in countless timestudies performed on a variety of operations in a plant. If a new operation requires a timestudy and involves a reach of 12 inches, it is not necessary to timestudy that element again. Figure 4-16 offers an example of standard data used for molding operations. Generally, timestudies should be taken in such a way that the elemental data they generate will be useful in meeting future standard data needs.

FACILITIES PLANNING AND DESIGN

For most industrial engineers, methods engineering connotes workstation design and the work measurement associated with such designs. Facilities planning and design, however, extends the analysis to include the design of the entire productive system. Plant layout (1, 12, 15, 17), a familiar term in industrial engineering, connotes design of a plant or other productive facility; however, corporate industrial engineering responsibilities today often extend this analysis to the determination of what facilities are needed, where, and in what sizes, to serve corporate goals. This is facilities planning and design, and it obviously includes plant sizing and location. Plant layout, concerned with determining the best arrangement of the appropriate number of various entities

Table for Determining the Number of Observations for a Given Absolute Error or Absolute Degree of Accuracy and Value of p, 95% Confidence Level

Per Cent of Total Time Occupied by Activity or Delay, p	Absolute Error					
	$\pm 1.0\%$	$\pm 1.5\%$	$\pm 2.0\%$	$\pm 2.5\%$	$\pm 3.0\%$	$\pm 3.5\%$
1	396	176	99	63	44	32
2	784	348	196	125	87	64
3	1,164	517	291	186	129	95
4	1,536	683	384	246	171	125
5	1,900	844	475	304	211	155
6	2,256	1003	564	361	251	184
7	2,604	1157	651	417	289	213
8	2,944	1308	736	471	327	240
9	3,276	1456	819	524	364	267
10	3,600	1690	900	576	400	294
11	3,916	1740	979	627	435	320
12	4,224	1877	1056	676	469	344
13	4,524	2011	1131	724	503	369
14	4,816	2140	1204	771	535	393
15	5,100	2267	1275	816	567	416
16	5,376	2389	1344	860	597	439
17	5,644	2508	1411	903	627	461
18	5,904	2624	1476	945	656	482
19	6,156	2736	1539	985	684	502
20	6,400	2844	1600	1024	711	522
21	6,636	2949	1659	1062	737	542
22	6,864	3050	1716	1098	763	560
23	7,084	3148	1771	1133	787	578
24	7,296	3243	1824	1167	811	596
25	7,500	3333	1875	1200	833	612
26	7,696	3420	1924	1231	855	628
27	7,884	3504	1971	1261	876	644
28	8,064	3584	2016	1290	896	658
29	8,236	3660	2059	1318	915	672
30	8,400	3733	2100	1344	933	686
31	8,556	3803	2139	1369	951	698
32	8,704	3868	2176	1393	967	710
33	8,844	3931	2211	1415	983	722
34	8,976	3989	2244	1436	997	733
35	9,100	4044	2275	1456	1011	743
36	9,216	4096	2304	1475	1024	753
37	9,324	4144	2331	1492	1036	761
38	9,424	4188	2356	1508	1047	769
39	9,516	4229	2379	1523	1057	777
40	9,600	4266	2400	1536	1067	784
41	9,676	4300	2419	1548	1075	790
42	9,744	4330	2436	1559	1083	795
43	9,804	4357	2451	1569	1089	800
44	9,856	4380	2464	1577	1095	804
45	9,900	4400	2475	1584	1099	808
46	9,936	4416	2484	1590	1104	811
47	9,964	4428	2491	1594	1107	813
48	9,984	4437	2496	1597	1109	815
49	9,996	4442	2499	1599	1110	816
50	10,000	4444	2500	1600	1111	816

Figure 4-14 A work sampling table. [*From Barnes (4, p. 528).*]

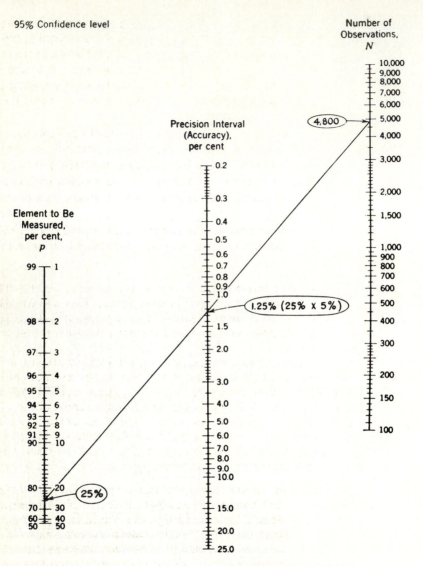

95% Confidence level

Number of
Observations,
N

Precision Interval
(Accuracy),
per cent

Element to Be
Measured,
per cent,
p

Figure 4-15 A work sampling nomograph. [*From Barnes (4, p. 522).*]

needed in the design of a productive facility, includes material handling. It must, because material handling cost is one of the key criteria in evaluating the success of a plant layout design.

A typical sequence of steps of analysis in the design of a plant is indicated in Fig. 4-17. The starting point for an industrial engineer assigned to such a study is typically a product design at a stated rate of production (i.e., volume). The design of the product is normally the responsibility of a product engineer, and the rate of production is normally determined at a level of management that bridges both the production and the marketing functions of the company. The responsibility of the

SQUEEZER MOLDING SPECIFICATION

DATE _May 13 -_   BY _P. Gr._   PART NO _47943_

PATTERN _Metal_   SIZE _12 x 14_

LOOSE ___ MOUNTED _✓_   PART NAME _Gear_

SPLIT ___ FOLLOW BOARD ___   WEIGHT _4½_   CASTING PER MOLD _1_ | STAND.

METAL	FLASK SIZE	COPE DRAG	DEPTH OF DRAW IN INCHES								RAISE DEPTH	STAND.
			0	½	1	1½	2	2½	3	3½	r	
IRON	12" X 12"	COPE	.30	.50	.65	.80	(.90)	1.00	1.10	1.20	.30	.90
	12" X 14"	DRAG	—	4.06	4.18	4.29	4.40	4.46	4.57	4.68		4.29
BRASS	12" X 12"	COPE	.35	.60	.70	.80	.90	1.00	1.10	1.20	.30	
	12" X 14"	DRAG	—	3.05	3.15	3.25	3.35	3.40	3.50	3.60		
	10" X 19"	COPE	.45	.70	.80	.90	1.00	1.10	1.20	1.30	.40	
		DRAG	—	4.05	4.15	4.25	4.35	4.40	4.50	4.60		

MISCELLANEOUS DATA	PER	1	2	3	4	5	6	
SET SOLDIERS	SOLDIER	.15	.25	.40	.50	.60	.68	
SET JOB NAILS	NAIL	.07	.14	.20	(.25)	.30	.35	.25
VENT (COPE, DRAG, OR SIDE)	VENT	.08	(.15)	.22	.30	.35	.40	15
BLACKEN MOLD	100 SQ IN	.06	.10	.15	.20	.23	.26	
WATER MOLD AFTER DRAW	10 LIN. IN.	.05	.12	.16	.20	.25	.29	
CUT GATE	GATE	.90	1.50	2.00	2.50	2.90	3.25	
SET, CUT AND REAM RISERS	RISER	.45	.90	1.30	1.70	2.00	2.30	
SPECIAL RAM (DRAG OR COPE)	100 CU. IN.	.30	.55	(.80)	1.00	1.15	1.30	.80
PUT PASTE ON CORES	SQ. INCH.	.05	.10	.15	.18	.20	.22	

CORE SETTING CLASSIFICATION	SKETCH OF TYPE	VOL. RANGE CU. IN.	CORES SET PER MOLD								
			1	2	3	4	5	6	7	8	
STOCK CORES		0-10	.10	.20	.25	.30	.35	.40	.45	.50	
		11-20	.15	.25	.30	.35	.40	.45	.50	.55	
		21-50	(.20)	.30	.35	.40	.45	.50	.55	60	.20
REGULAR BLOCK OR CYLINDER CORES		0-50	.20	.30	.40	.45	.50	.55	.60	65	
		51-100	.25	.40	.45	.50	.55	.60	65	.70	
		101-200	.35	.45	.50	.55	.60	.65	.70	.75	
IRREGULAR BLOCK OR CYLINDER CORES		0-50	.30	.35	.45	.50	.55	60	.65	.70	
		51-100	.40	.45	.50	.55	60	.65	.70	.75	
		101-200	.50	.55	.60	.65	.70	.75	.80	.85	

TOTAL STANDARD PER MOLD _6.59_

DIVIDED BY _1_   CASTINGS = STANDARD PER CASTING _6.6_

Figure 4-16   Standard data for squeezer molding. [*From Maynard (10, p. 3-157).*]

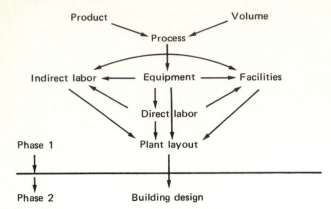

Figure 4-17   Major steps in the design of a plant.

industrial engineer, then, is to design a production facility that will produce the specified product at the stated production volume at a minimum cost.

In considering processes there are occasions when a change is discovered in the material or design of the product that improves the product while reducing its cost, or at least not increasing its cost intolerably, such that the change is adopted. That is, product designs change somewhat as a plant design is developed; but, for the most part, the overall goal of the industrial engineer is cost reduction—design of a process to produce a specified product at a given rate of production at a minimum cost.

For a given product at a stated volume of production, it is normally assumed that there is a most appropriate process for producing it. As volumes increase higher levels of mechanization become feasible, whereas at low levels of production general-purpose equipment is often employed. As one considers higher levels of mechanization and consequent specialization, process methods change as well as materials. For example, molded plastic parts might be justified if sufficient volume is demanded, but if it is not, metal parts may be more appropriate.

The process then establishes the requirement for equipment. The number of pieces of equipment of a particular type required in a specific area of the plant is a function of scrap rates, operation times, equipment utilization, machine operator performance, and other related factors, such as the type of plant layout employed.

The sum of operation times for equipment, along with an appropriate allowance for expected unproductive time, establishes a concurrent requirement for direct labor. By operation time is meant the time an employee takes to perform an operation, whereas unproductive time refers to all the remaining time the employee spends (e.g., discussing the world series game played the day before).

At this point, product and volume have led to a process, which dictates equipment which, in turn, requires operators. What remains are the people necessary to support these activities. Equipment and machine operators need maintenance support, janitors, material handlers, and many other indirect employees. Consider also that all the equipment and laborers thus far mentioned establish a need for staff personnel to plan and manage the total facility; this means, for example, accountants, quality control engineers, and production supervisors. And finally, equipment, personnel, and material need to be covered, or at least controlled. The determination of the correct number and arrangement of all these entities, indicated as phase 1 in Fig. 4-17, is called "plant layout" and is normally performed by industrial engineers. At this point, the industrial engineer often engages the services of an architect, or the construction division of the company, to design the cake cover (i.e., the building) to house the designed production system.

The previous discussion might suggest that plant design is a set of steps, for example 26 steps, so that having completed all the steps to step 13, one can assume that the job is halfway done. Unfortunately, this is not possible. Although the sequence of events described above is typical, it is at step 13, for example, that one gains the insight and perspective to improve on steps 2, 5 and 7, so back one goes to do it again. In general, plant design is first macro, then micro, and ultimately macro. That is, the first concern is the relative spatial arrangement of major plant areas, including aisle planning. Then detailed layout of entities within an area is undertaken, and then ultimately area layouts are combined, and adjusted, to arrive at a total layout contained within a predefined general overall outer shape.

One of the first concerns in designing a plant layout is the choice of plant layout type to employ. There are three main plant layout types: (1) process, (2) product, and (3) fixed location, as depicted in Fig. 4-18.

The process layout is typically employed where a large number of products are being processed simultaneously, particularly if the volume of each is relatively low. A production facility for books, for example, is typical of this job-shop type of layout. By grouping similar types of equipment together, it is possible to attain equipment type utilizations which are relatively high (e.g., if 18.6 cutters are needed, 19 cutters will result in little overall cutter nonutilization).

The product layout, however, offers the advantage that whatever types of equipment a product needs, they are arranged in sequence to suit the product. This results in much less product distance traveled, because the equipment is arranged in the desired sequence. Also, in the product layout the material stays in one area, which eliminates the need to "expedite" it from one area to another. Possibly more important, because such production lines are paced at some designed rate, control of production is limited to seeing that the expected amount of production is coming off the end of the line. If it is, the entire production line can be assumed to be producing as it should. Of course, such inflexibility offers the danger that if any part of the line has problems, the entire production line may be affected.

It would seem, therefore, that the main cost trade-off in choosing between the process and product types of layout involves comparing the cost effects of the utilization of equipment and of material handling, also including consideration of control effects. If the volume is sufficient to permit acceptable utilization of equipment, the product layout becomes superior to the process layout. In most plants there is a combination of both types. In plants involving parts fabrication and assembly, fabrication tends to employ the process layout, whereas assembly areas often employ the product type of layout. In the final analysis, the combination that produces the desired volume of product at least total cost is preferred. Marketing is

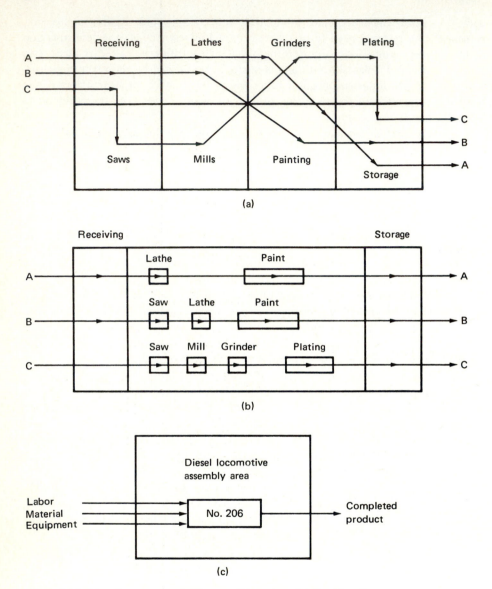

Figure 4-18 Plant layout types. (a) Process, (b) product, (c) fixed location.

concerned with maximizing income, industrial engineering is concerned with minimizing cost, and management is gambling that there is a sufficient difference in its favor.

A small example of some of the factors that need to be considered in determining the number of pieces of equipment needed in a product type of layout is indicative of calculations necessary in plant design. Figure 4-19 illustrates a process involving two sequential steps. Assume the required output is 200 units per hour, departmental labor efficiency is 85 percent, and standard times and scrap rates for steps 1 and 2 are as indicated in Fig. 4-19. How many step 1 and step 2 machines are needed? Make three assumptions in this example: (1) the process is such that when an employee is idle his machine is idle, (2) there is sufficient storage space between steps 1 and 2 that the steps are independent with respect to blocking effects, and (3) parts are tested for defectiveness after all processing is completed at each step. The calculation proceeds as follows:

Step 2

$$\text{Input} = \frac{200}{0.80} = 250$$

$$\text{Machines required} = \frac{250(2.2)}{60(0.85)} = 10.8 \simeq 11$$

Step 1

$$\text{Input} = \frac{250}{0.90} \simeq 278$$

$$\text{Machines required} = \frac{278(1.3)}{60(0.85)} = 7.1 \simeq 7$$

This is a simplified example. Maintenance time on the machines was not considered, normal absentee effects of an employee may or may not have been included in the departmental labor efficiency, and so on. The example is indicative, however, of calculations required in the design of a plant.

As was mentioned earlier, a plant design is first macro in nature. One of the problems encountered at this stage is the determination of the relative spatial location of major areas of the plant. A method known as the "activity relationship chart," originally developed by Muther (14), can be used to accomplish this. Figure 4-20 illustrates a completed activity relationship chart for eight areas in a small plant. The relationships are transferred to equal-sized blocks, as shown in Fig. 4-21, and then cut apart and rearranged spatially in light of the desired proximity relationships, as illustrated in Fig. 4-22. The next step is usually to change the sizes of the areas to more appropriately represent their scaled individual sizes. This author prefers to use squares that conform in size to the estimated size of each department. These templates are

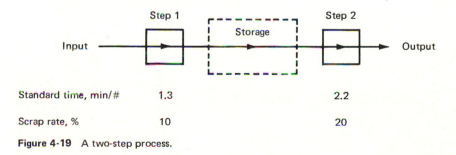

| Standard time, min/# | 1.3 | 2.2 |
| Scrap rate, % | 10 | 20 |

Figure 4-19 A two-step process.

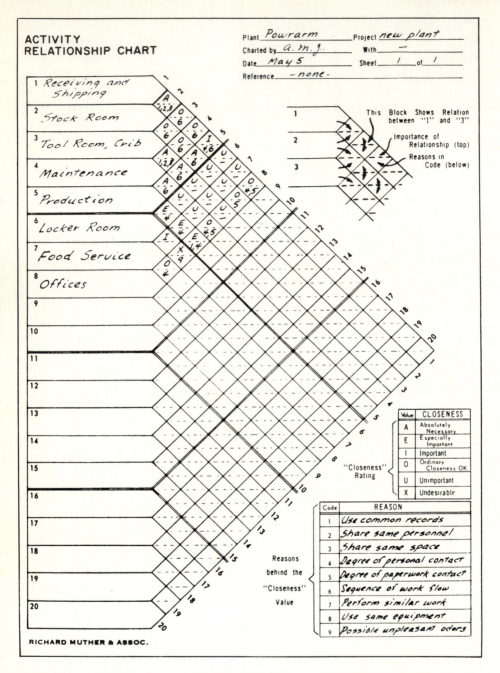

Figure 4-20 A typical activity relationship chart. (*From James M. Apple:* Plant Layout and Materials Handling, *2d ed., p. 175. Copyright © 1963, The Ronald Press Company, New York.*)

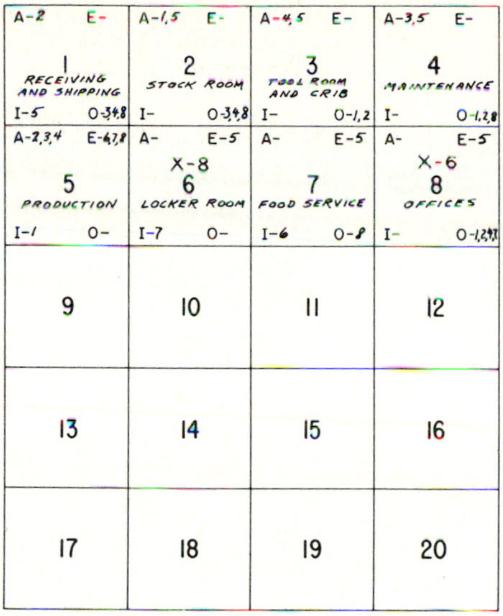

Figure 4-21 grid:

A-2 E-	A-1,5 E-	A-4,5 E-	A-3,5 E-
1 RECEIVING AND SHIPPING I-5 O-3,4,8	2 STOCK ROOM I- O-3,4,8	3 TOOL ROOM AND CRIB I- O-1,2	4 MAINTENANCE I- O-1,2,8
A-2,3,4 E-6,7,8	A- E-5 X-8	A- E-5	A- E-5 X-6
5 PRODUCTION I-1 O-	6 LOCKER ROOM I-7 O-	7 FOOD SERVICE I-6 O-8	8 OFFICES I- O-1,2,7
9	10	11	12
13	14	15	16
17	18	19	20

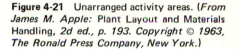

Figure 4-21 Unarranged activity areas. (*From James M. Apple:* Plant Layout and Materials Handling, *2d ed., p. 193. Copyright © 1963, The Ronald Press Company, New York.*)

Figure 4-22 grid:

A-2 E-	A-1,5 E-	A-3,5 E-	
1 RECEIVING AND SHIPPING I-5 O-3,4,8	2 STOCK ROOM I- O-3,4,8	4 MAINTENANCE I- O-1,2,8	
A- E-5 X-8	A-2,3,4 E-6,7,8	A-4,5 E-	
6 LOCKER ROOM, ETC. I-7 O-	5 PRODUCTION I-1 O-	3 TOOL ROOM I- O-1,2	
A- E-5	A- E-5 X-6		
7 FOOD SERVICE I-6 O-8	8 OFFICES I- O-1,2,4,7		

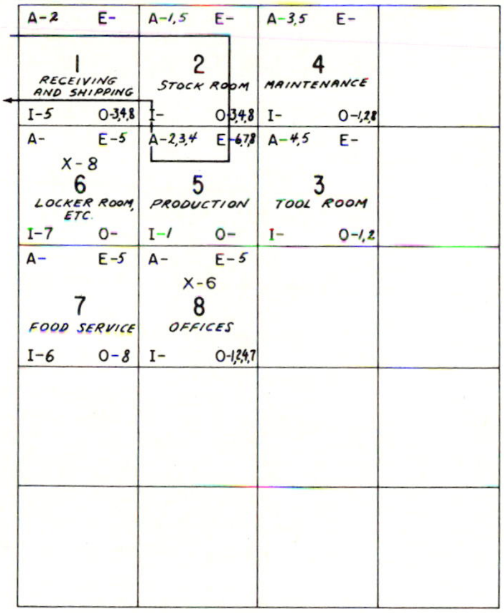

Figure 4-22 Arranged activity areas. (*From James M. Apple:* Plant Layout and Materials Handling, *2d ed., p. 194. Copyright © 1963, The Ronald Press Company, New York.*)

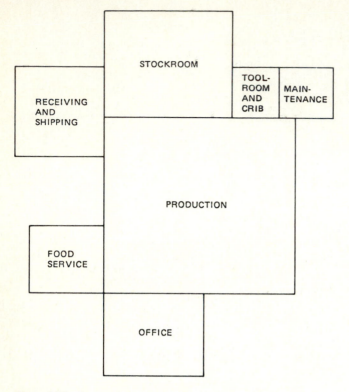

Figure 4-23   Arranged and scaled activity areas.

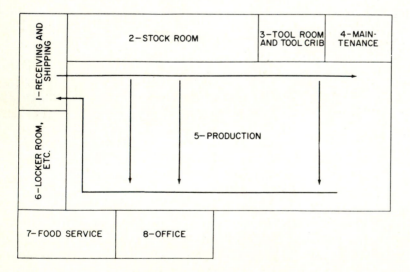

Figure 4-24   The final area layout. (*From James M. Apple:* Plant Layout and Materials Handling, *2d ed., p. 202. Copyright © 1963, The Ronald Press Company, New York.*)

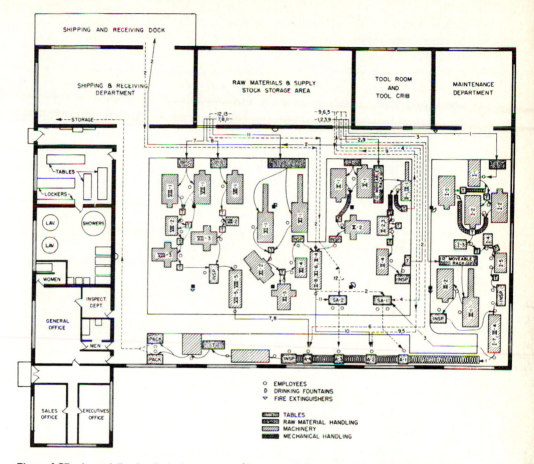

Figure 4-25 A partially detailed plant layout. (*From James M. Apple:* Plant Layout and Materials Handling, *2d ed., p. 353. Copyright © 1963, The Ronald Press Company, New York.*)

arranged spatially, maintaining consistency with the previous step, as indicated in Fig. 4-23. Finally, the general shapes of the plant areas are devised by adjusting this arrangement to conform to a simple exterior shape, as indicated in Fig. 4-24. The final detailed design of the plant should conform generally to this design, as indicated in Fig. 4-25.

This approach differs from the approach typically taken by individuals unfamiliar with these methods. An intuitive solution to a plant layout problem often evolves by detailing elements at the beginning of a process and proceeding until the sum of the details produces a layout. Such an approach can result in a large number of arrangements, depending on the sequence of decisions that are made. One wrong decision in this sequence can produce an overall layout with considerable deficiencies. Methods such as the activity relationship chart produce generally consistent results with respect to desired overall characteristics of the plant. The more macro the mistake, the greater the cost effect. This is the primary reason for analyzing from macro to micro in plant layout work.

Plant layout analysis has been and still is relatively unquantifiable as compared to many other areas within industrial engineering. A plant layout is a capstone design effort in that it requires consideration of all of the design and control aspects of a plant. It is fairly common in industrial engineering departments in universities for the plant layout course to be the last course in the industrial engineering sequence related to design and control, and it is often the senior design course concerned with combining all of these techniques in the form of a term project. The number of factors to be considered make quantification difficult. Figure 4-26 is a list of "marks of a good layout" from Apple (1, p. 11). Fourteen of the twenty-four "desirable characteristics" listed, directly related to material handling, are marked with an asterisk to draw attention to the importance material handling plays in plant layout. All twenty-four characteristics are important, however, and are interrelated, yet many other important factors could have been mentioned, such as flexibility and density.

Figure 4-27 is a layout for a plant to produce fishing reels in Las Cruces, New Mexico. The layout uses a Mylar $\frac{1}{4}$-inch gridded base and movable templates for equipment. A group of senior industrial engineering students labored over it for weeks, as the report that supports it attests. Many decisions had to be made, many criteria were considered, much calculation took place, yet the justification of their design in oral presentation was mostly qualitative, as is typically true. In many respects, good plant layouts tend to represent successful combinations of good principles. The

```
 * 1  Planned materials flow pattern
 * 2  Straight-line layout (or an adaptation thereof)
   3  Building constructed (or altered) around a preplanned layout design
 * 4  Straight, clear, marked aisles
 * 5  Backtracking kept to a minimum
 * 6  Related operations close together
   7  Production time predictable
   8  Minimum of scheduling difficulties
   9  Minimum of goods-in-process
  10  Easy adjustment to changing conditions
  11  Plans for expansion
  12  Maximum ratio of actual processing time to over-all production time
  13  Good quality with minimum inspection
 *14  Minimum materials handling distances
 *15  Minimum of manual handling
 *16  No unnecessary rehandling of materials
 *17  Materials handled in unit loads
 *18  Minimum handling between operations
 *19  Materials delivered to production employees
 *20  Materials efficiently removed from the work area
 *21  Materials handling being done by indirect labor
 *22  Orderly materials handling and storage
  23  Good housekeeping
  24  Busy employees, working at maximum efficiency
```

Figure 4-26 Marks of a good plant layout. (*From James M. Apple:* Plant Layout and Materials Handling, *2d ed., p. 11. Copyright © 1963, The Ronald Press Company, New York.*)

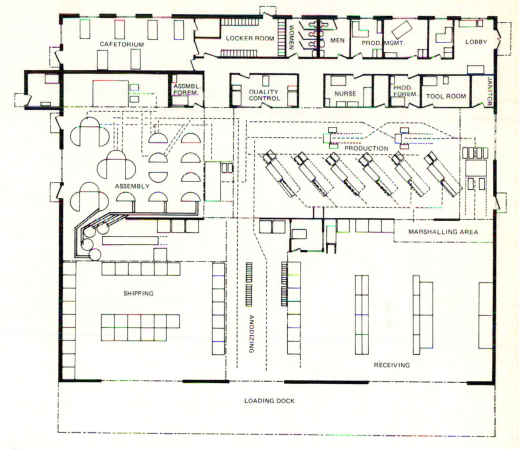

Figure 4-27 A detailed plant layout.

experienced plant layout engineer is constantly aware of a great many individual guiding principles of good plant layout practice. The engineer is also aware that many of these principles can be incompatible with one another, and why. In effect, plant layout and politics have "compromise" in common as an understood reality.

One method of evaluating a plant layout is to make a list of its advantages and disadvantages. To be successful in doing this, one must be capable of assuming the roles, one by one, of all the area managers in the plant affected by the layout and critically evaluating its effects on their activities. Only by this means is it possible to remove unnecessary and undesirable characteristics from a plant layout design. Ultimately, after much redesign, a point is typically reached where there is a minimum list of disadvantages. Removal of any of these, however, for the given design will produce an even greater disadvantage than the one being removed. At this point one probably has the best design for the general approach taken. Of course, this minimum list of disadvantages has to be valued along with the advantages the design offers in comparison to other totally different design approaches.

Computerized approaches have been developing in recent years to assist in limited specific areas of plant design. A method known as CRAFT (3) has been developed for determining desirable spatial arrangements of production departments in an attempt to minimize the total material handling cost incurred between all departments. An initial layout is given, and the cost per linear foot of material handled between each pair of departments is also given. The heuristic (i.e., the program) switches departments in the layout to effect a reduction in total material handling cost. It iteratively (i.e., successively) attempts additional desirable switches until no more can be found. The last layout produced offers the lowest total material handling cost and is considered best in this respect. Figures 4-28 and 4-29, respectively, are initial and final layouts for a sample problem, using the CRAFT method.

Another popular computerized plant layout method (9) is called CORELAP (computerized relationship layout planning). The input to CORELAP is a quantified activity relationship chart similar to that illustrated in Fig. 4-20. Instead of letters identifying the desired level of proximity, such as E or A, numbers between 0 and 10 are employed. The CORELAP heuristic starts with the area having the greatest sum of relationships, and then iteratively adds other areas, one at a time, to those already contained in the layout in such a way as to maximize the attainment of the desired relationships. Figure 4-30 is a final layout for a sample problem employing the CORELAP program. A number of other computerized programs have been developed in recent years (2, 7, 18). It should be noted, however, that the CORELAP heuristic is analogous to a relative location methodology that could have been performed by hand employing the basic activity relationship chart approach. Whereas the heuristic is limited solely to considering assigned relationship values and other limited data, such as maximum length-to-width ratios for areas, the same technique by hand can include consideration of a much broader range of factors at the time an additional area is added to the previously located departments. In fact, the addition of another area may dictate some readjustment of the previously arranged areas.

There has been a considerable increase in the use of quantitative methods in facilities planning and design. Operations research techniques in particular have made inroads in quantifying some aspects of this field. In later chapters some of the examples dealing with operations research will illustrate applications of these techniques in the facilities planning and design field.

Of all the quantitative techniques applied to facilities planning and design, simulation has been the most productive to date. Consider that a template layout, such as that illustrated in Fig. 4-27, offers only a static view of a producing system. If a photograph were made of a production process from a helicopter after the roof of the plant was removed, the photograph would be analogous to a layout, and would offer only a static view of where things were at a single point in time. Digital simulation makes it possible to evaluate the dynamic characteristics of a particular layout of a production process. Queueing theory analysis, to be discussed in the chapter on operations research, could offer an analytical approach to gaining this dynamic appreciation. The complexities of modeling a real-world production process, however, typically result in computer simulation of the system to gain these insights. The simulation program causes analogous flows of material and assemblies in time,

Location pattern

```
     1  2  3  4  5  6  7  8  9 10 11 12 13 14 15 16 17 18 19 20 21 22 23 24
 1   A  A  A  A  A  F  F  F  F  -  -  -  -  -  -  -  -  R  R  R  R  S  S  S
 2   A  A     A  A  F  F        F  -  -  -  -  -  -  L  -  R     R  R  S     S
 3   A        A  A  F        F  -  -  -  K  K  -  -  L  R  R  R  R  S  S  S
 4   A     A  A  A  F  F  F  F  -  J  J  K  K  -  -  L  R        R  S  S  S
 5   A  A  A  A  B  F  F  F  F  J  J  J  K  O  O  -  L  R        R  S  S  S
 6   B  B  B  B  B  F        F  J  M  M  M  K  O  L  L  P  T        T     T
 7   B  B     B  B  F  F  F  F  M  M  N  N  O  O  L  P  T        T  S     T
 8   B  B     B  B  F  F  F  G  M  M  N  N  O  O  O  P  T  T  T  T  T     T
 9   C  C  C  C  G  G  G  G  G  N  N  N  O  O  O  P  Q  -        -     S  T
10   C  C  C  G  G  G  G  G  G  M  N  N  N  O  O  O  Q  U              U     U
11   C  C  C  C  G  G  G  G  G  Q  Q  Q  Q  Q  Q  P  Q  U        U  U  S  U  U
12   C  C  C  D  G  G     G  G  Q  Q  Q  Q  Q  Q  Q  Q  U  U        W        U
13   D  D  D  D  H  G  G  G  G  H  H  H  H  H  H  H  H  U  U  U  V  W     U  U
14   D  D  D  H  H  H  H  H  H  H              Q  O  Q  Q  H  V        V  W  S  W
15   D  D  D  H  H  H  H  H  H  H              O  O  O  Q  H  V     V  V  W  S  W
16   E  E  E  E  H  H  H  H  H  H  H           Q  H  P  Q  H  V  V  V  W  W     W
17   E  E  E  E  H  H  H  H  H  H  H  H  H  H  H  P  H  T  V  V  V  W        S  W
18   E  E  E  E  H  H  H  H  H  H  H  H  H  H  H  H  H  T  T  T  T  S  T  T
19   E  E  E  E  H  H  H  H  H  H  H  H  H  H  H  H  H  T  V  V  V  W  W  W
```

Total cost   47.90   Est. cost reduction   0.0   Move A   Move B   Move C   Iteration 0

Figure 4-28 An initial CRAFT area layout.

93

Location pattern

	1	2	3	4	5	6	7	8	9	10	11	12	13	14	15	16	17	18	19	20	21	22	23	24
1	A	A	A	A	A	F	F	F	F	Q	Q	J	J	J	H	H	H	H	H	H	H	C	C	C
2	A	A	A	A	A	F	F	F	F	Q	Q	Q	J	J	H	H	H	H			H	C	C	C
3	A	A	A	A	A	F	F	F	F	Q	Q	Q	J	J	H	H	L	H	H		H	C		C
4	A	A	A	B	B	F	F	F	F	Q	Q	Q	K	K	J	L	L	H	H	H	H	H	H	H
5	B	B	A	B	B	F	F	F	O	Q		Q	K	K	J		L	S	S	S	S	H		H
6	B	B	B	B	B	F	F	O	O	O	O	Q	K	K	J	V	V	S		S	S	H		H
7	B	B	B	B	P	P	P	O	G	G	O	Q	Q	Z	V	V	V	S	S	S	H	H		H
8	B	B	B	T	P	P	P	G	G	G	O	N	Q	Q	V		V	H	H	H	H	H		H
9	T	T	T	T	P	P	P	G	G	G	W	N	Z	Z	H	H	H	H	H	H	H	H	M	H
10	T	T	T	T	T	T	T	G	G	W	W	W	N	N	H	H	H				H	H		M
11	T	T	T	T	T	T	T	G	G	–	–	–	–	–	H	H	H	H	–	–	H	M	M	M
12	T	D	D	D	T	T	T	G	G	G	G	G	G	G	–	–	–	–	–	–	H	M	M	M
13	D	D	D	D	T	T	T	G			U	G	G	G	–		U	U	U	–	H	R	R	R
14	D	D	D	D	T	T	T	G		G	U	U	G	U	–	U	U	U	U	E	R	R		R
15	D	D	D	D	T	T	T	G		G	U	U	U	U	U		U	E	E	E	E	R		R
16	D	D	D	T	T	T	T	G		G	U	U	U	U	U		U	E	E	E	E	R		R
17	T	T	T	T	T	T	T	T		G	U	U	U	U	U		U	E	E	E	E	R		R
18	T	T	T	T	T	T	T	T		G	U	U	U	U	U						E	R	R	R
19	T	T	T	T	T	T	T	G		G	U	U	U	U	U						R	R	R	R

Total cost 32.24   Est. cost reduction   0.01   Move A P   Move B   Move C   Iteration 15

Figure 4-29 A final CRAFT area layout.

94

```
0 0 0 0 0 0   0  0  0  0  0  0  0  0  0  0  0  0  0  0 0 0 0 0 0
0 0 0 0 0 0   0  0  0  0  0  0  0  0  0  0  0  0  0  0 0 0 0 0 0
0 0 0 .0 0 0  0  0  0  0  0  0 25 25  0  0  0  0  0  0 0 0 0 0 0
0 0 0 0 0 0   0  0  0  0  0 24 25 25  0  0  0  0  0  0 0 0 0 0 0
0 0 0 0 0 0   0  0  0  0 21 21 20 19  0  0  0  0  0 27 0 0 0 0 0
0 0 0 0 0 0   0  0  0  0  0 17 19 19 13 14 18 27 27 27 0 0 0 0 0
0 0 0 0 0 0  31 31 31 31 31 11 16 13 13 18 18 27 27 27 0 0 0 0 0
0 0 0 0 0 0  31 31 31 31 28 28 22 15 26 12 27 27 27 27 0 0 0 0 0
0 0 0 0 0 0  31 31 31 31 23 23 23 26 27 27 27 27 27 27 0 0 0 0 0
0 0 0 0 0 0  31 31 31 31 23 23 23 27 27 27 27 27 27 27 0 0 0 0 0
0 0 0 0 0 0  31 31 31 31 31 31 31 27 27 27 27 27 27 27 0 0 0 0 0
0 0 0 0 0 0  31 31 31 31 31 31 31 31 31 31 30  0  0  0 0 0 0 0 0
0 0 0 0 0 0  31 31 31 31 31 31 31 31 31 31 30  0  0  0 0 0 0 0 0
0 0 0 0 0 0   0 31 31 31 31 31 31 31 31 31 29  0  0  0 0 0 0 0 0
0 0 0 0 0 0   0 31 31 31 31 31 31 31 31 31  0  0  0  0 0 0 0 0 0
0 0 0 0 0 0   0  0  0  0  0  0  0  0  0  0  0  0  0  0 0 0 0 0 0
0 0 0 0 0 0   0  0  0  0  0  0  0  0  0  0  0  0  0  0 0 0 0 0 0
```

Figure 4-30 A final CORELAP area layout.

accumulations that develop in relation to processing times, handling and storage restrictions, and many other effects. The completed simulation provides a preview of how the production process will operate over time. The typical result of simulating a production system, and adjusting it accordingly, is a considerable reduction of material in process and a system that generally performs as it was hoped it would. Without simulation of a process, storage needs are difficult to estimate and consequently are often overestimated in an effort to play it safe.

Figure 4-31 illustrates a "before and after" design for part of a production process. As a result of simulating the dynamics of this process, the design was greatly simplified, resulting in less initial and operating costs and a simpler, more reliable process.

Materials handling analysis is a subset of plant layout. Methods study, plant layout, and materials handling are all part of the design of a production facility and can hardly be treated separately—in the final analysis it is all one design. Materials handling is a field in which, because there is a considerable body of information about equipment and a general inability to quantify problems, the use of experience seems to offer the greatest hope in solving problems. There are more than 430 different types of material handling equipment, each type being represented by six to ten major manufacturers, each with a line of equipment for that type offering a range of capacities and options. To be effective, a materials handling consultant must be generally cognizant of the primary capabilities and limitations of this broad range of equipment. At the same time, the consultant must consider a broad range of guiding principles of materials handling. Figure 4-32 lists principles in one of eight categories of principles of materials handling contained in Apple's book (1, p. 210). Each principle has its value in a particular environment, yet some of the principles are mutually inconsistent. The materials handling engineer must select the best combination of equipment in light of a best combination of materials handling

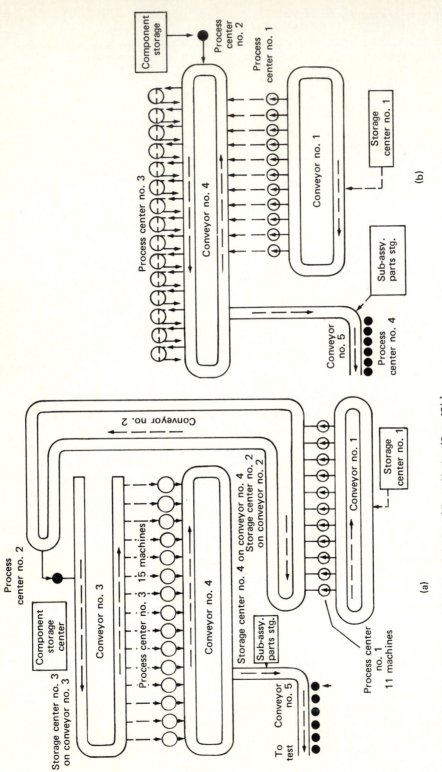

Figure 4-31 Conveyor designs. (a) Before; (b) after. [*From Hurley (6, p. 49).*]

principles for a particular solution. Typically, the best solution is a least-cost method of handling the required material. It is indeed rare when a more costly alternative is selected if all cost effects have been considered.

HUMAN ENGINEERING

Most industrial engineering curriculums contain at least one course, taught either in the industrial engineering department or the psychology department, with a title such as human engineering, human factors, human-machine systems, industrial psychology, or engineering psychology. These courses are concerned with understanding humans as part of a human-machine system. Industrial engineering systems are "people systems," and an understanding of the physical and mental capabilities and limitations of humans is necessary in designing these systems. Equally important is the fact that people systems only work if people want them to; therefore, the body of knowledge concerned with human motivations, drives, and attitudes requires some measure of technical appreciation, if not thorough understanding, if an industrial engineer is to be effective.

Relating to Methods.
a Plan for straight line moves.
b Combine inspection, storage, and processing within the transportation.
c Keep handling to a minimum.
d Reduce lengths of all moves.
e All handling should be analyzed for possible improvement by elimination, combination, or simplification.
f Avoid transfers from floor to container, or vice versa, or from container to container.
g Assemble product on crate bottom or pallet to facilitate handling.
h Use equipment to supply or remove materials at the proper rate.
i Provide for automatic processing of parts.
j Provide for mechanical handling to and from the workplace.
k Plan for mechanical removal of scrap.
l Plan for storage of a minimum amount of materials at the workplace.
m In planning individual operations, remember the necessary relationships with preceding and succeeding operations or work areas.
n Allow area for storage of the maximum amount of work to be on hand at an individual work area—ahead of the operator, or awaiting removal.
o Handle as many pieces as practical in one unit.
p Never pile anything directly on the floor without a container, skid, or pallet underneath.
q Deliver materials to the right place on the first move.

Figure 4-32 Some principles of materials handling. (*From James M. Apple:* Plant Layout and Materials Handling, *2d ed., p. 210. Copyright © 1963, The Ronald Press Company, New York.*)

In designing a production system, the choice between human and machine for performing various functions has to be resolved. Figure 4-33 attempts to identify the functions which are performed best either by man or machine. There is little question that, through mechanization and computerization humans are less involved in physical and mental drudgery. This is as it should be, if only from the point of view that the general physical and computational abilities of humans are poorer than those of

Man vs. Machine	
Man excels in	**Machines excel in**
Detection of certain forms of very low energy levels	Monitoring (both men and machines)
Sensitivity to an extremely wide variety of stimuli	Performing routine, repetitive, or very precise operations
Perceiving patterns and making generalizations about them	Responding very quickly to control signals
Detecting signals in high noise levels	Exerting great force, smoothly and with precision
Ability to store large amounts of information for long periods—and recalling relevant facts of appropriate moments	Storing and recalling large amounts of information in short time-periods
Ability to exercise judgment where events cannot be completely defined	Performing complex and rapid computation with high accuracy
Improvising and adopting flexible procedures	Sensitivity to stimuli beyond the range of human sensitivity (infrared, radio waves, etc.)
Ability to react to unexpected low-probability events	Doing many different things at one time
Applying originality in solving problems: i.e., alternate solutions	Deductive processes
Ability to profit from experience and alter course of action	Insensitivity to extraneous factors
Ability to perform fine manipulation, especially where misalignment appears unexpectedly	Ability to repeat operations very rapidly, continuously, and precisely the same way over a long period
Ability to continue to perform even when overloaded	Operating in environments which are hostile to man or beyond human tolerance
Ability to reason inductively	

Figure 4-33 Man versus machine. [*From Wesley E. Woodson and Donald W. Conover:* Human Engineering Guide for Equipment Designers, *2d ed., pp. 1–23. Copyright © 1954 & 1964 by The Regents of the University of California; reprinted by permission of the University of California Press, Berkeley.*]

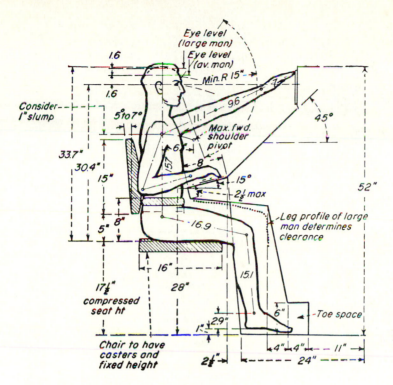

Figure 4-34 Recommended console dimensions.

machines, particularly if consistency and reliability are important factors. The role of humans today is shifting toward employing their analytical decision-making ability and judiciously using their sensory capabilities. Humans still do the coffee tasting, the bad-orange sorting, run the bridge crane, and decide when to shoot a burglar. Certain combinations of analytical and sensory ability will remain human functions for a long time to come.

In designing a workstation, the size and variability in size of those who will perform the job must be taken into consideration. Figure 4-34 illustrates recommended design dimensions based on anthropometric data. There is a wealth of information in human engineering, and these few examples are intended to offer some insight into the field. Quite often in many practical situations the desired anthropometric or other data are assumed to exist. The problem becomes one of finding them and evaluating their applicability.

The physical abilities and limitations of humans represent a significant body of useful data in workstation design. Figure 4-35, for example, provides data on maximum pedal thrust as a function of the relative position of an operator and a pedal. It is typical of very useful, detailed, specific information in this field.

With respect to elements of design related to mental functions, Fig. 4-36 details recommended choices of types of controls for different categories of control

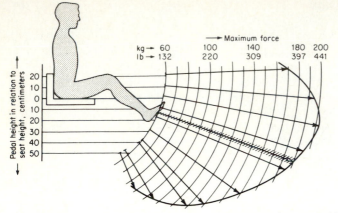

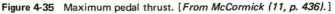

Figure 4-35 Maximum pedal thrust. [*From McCormick (11, p. 436).*]

Function	Application	Type of control
Selection between two alternatives	To start or stop an equipment ON or OFF sequence To insert momentary signal	Toggle switch Bat-handle switch Push button Foot switch Thumb switch Push-pull control Trigger Pointer knob (with detents) Round knob (reostat with on-off switch combined)
Selection among three or, more alternatives	To choose alternate modes of operation To select channels To select ranges	Pointer knob (with detents) Toggle switch Bat-handle switch
Precise adjustment	Continuous adjustment over a wide range For "fine" tuning or calibration	Round knob
Gross adjustment	Continuous adjustment (i.e., throttle or accelerator) Metering valve Faucet	Round knob Lever Pedal Wheel
Rapid adjustment	Slewing (such as an electronic cursor)	Hand crank Toggle or bat-handle (operating on electric drive)
Large force application	Braking Steering	Lever Wheel Pedal Rudder bar
Multiple (continuous positioning	Vehicle position and attitude Electronic coordinate-data pick-off	Joystick Combination wheel-joystick Pantograph Rolling ball Pressure stick

Figure 4-36 Choosing the right control for the job. [*From Wesley E. Woodson and Donald W. Conover:* Human Engineering Guide for Equipment Designers, *2d ed., pp. 2–92. Copyright © 1954 & 1964 by The Regents of the University of California; reprinted by permission of the University of California Press, Berkeley.*]

functions. Such a chart in many cases summarizes exhaustive tests to identify the superior choice of alternatives under varying conditions. A wealth of human engineering data has evolved from military tests of human performance—for example, at Wright Patterson Air Force Base, Ohio, since World War II.

It is intuitively known that a person feels warmer as humidity increases. Figure 4-37 is a chart developed for determining "effective temperature" as a function of both dry- and wet-bulb temperatures. The temperature that an employee "feels" is the temperature that must be considered in accounting for temperature effects on human performance in situations where perceived rather than actual temperature is the controlling factor.

When the attributes of a work environment are understood, it is possible to predict various effects concurrent with work performance as a function of the environment. Figure 4-38 illustrates the number of accidents that occurred as a

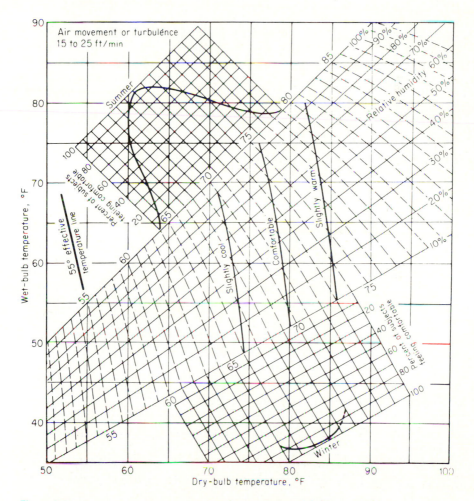

Figure 4-37 Scale of effective temperature. [*From McCormick (11, p. 485).*]

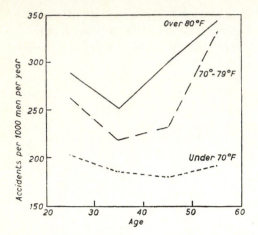

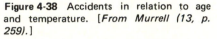

Figure 4-38 Accidents in relation to age and temperature. [*From Murrell (13, p. 259).*]

function of temperature and age, based on a study of coal miners in England. The extent to which data of this type can be useful in a different occupation is questionable; the trends it suggests are likely to exist, however, in similar work environments.

SUMMARY

Methods engineering, plant layout, and human engineering are treated under separate subheadings in this chapter, which might suggest that they are three separate areas of study. In reality, what is designed in industry is a production system, and it draws from all of these areas simultaneously. All three are highly interrelated. It is also incorrect to speak of materials handling and plant layout as if they were separable; generally, they are not. Even the words "production system" are meant by the author to possess a broad meaning, and include systems for handling laundry in a hospital or for evaluating tax forms in an Internal Revenue Service regional center.

This chapter represents the heart of traditional industrial engineering. The techniques employed are, for the most part, conceptual or graphical. Control techniques, such as inventory control or quality control, to be discussed in the next chapter, are a great deal more analytical in character. Of course, in the final analysis, techniques should be judged by what they produce. Production system design has been a major factor in the development of the standard of living enjoyed in the United States today.

There is little question that some quantitative techniques, particularly simulation and statistics, have had an impact on production system design. These disciplines have had a considerable upgrading effect on the area. The developing analytical frontier in quantification in facilities planning is discussed in *Facility Layout and Location* by Francis and White (5).

As will be discussed in more detail in later chapters, it is likely that the considerable volume of work being performed today by industrial engineers using traditional methods will be performed in the future by industrial engineering technicians skilled in modernized techniques. This will come about when the system of higher education in the United States senses and responds to the ever-increasing need for industrial engineering technicians.

REFERENCES

1 Apple, James M.: *Plant Layout and Materials Handling,* 2d ed., The Ronald Press Company, New York, 1963.
2 Apple, James M., and M. P. Deisenroth: "A Computerized Plant Layout Analysis and Evaluation Technique (PLANET)," in *Proceedings of the 23rd Annual Conference and Convention,* American Institute of Industrial Engineers, Anaheim, Calif., 1972, pp. 121–127.
3 Armour, G. C., and E. S. Buffa: "A Heuristic Algorithm and Simulation Approach to Relative Location of Facilities," *Manage. Sci.,* vol. 9, no. 1, pp. 294–309, 1963.
4 Barnes, Ralph M.: *Motion and Time Study,* 6th ed., John Wiley and Sons, Inc., New York, 1968.
5 Francis, Richard, and John White: *Facility Layout and Location,* Prentice-Hall, Inc., Englewood Cliffs, N.J., 1974.
6 Hurley, O. R.: "Simulation Finds the One Best Conveyor Layout," *Mod. Mater. Handling,* vol. 18, no. 10, pp. 47–49, October, 1963.
7 Krejeirik, M.: "RUGR Algorithm," Technical Paper, Computer Aided Plant Layout and Design Seminar, Helsinki, March, 1969 (reprinted in *Comput. Aided Des.,* 1969).
8 Krick, Edward V.: *Methods Engineering,* John Wiley and Sons, Inc., New York, 1962.
9 Lee, R. C., and J. M. Moore: "CORELAP–Computerized Relationship Layout Planning," *J. Ind. Eng.,* vol. 18, no. 3, pp. 195–200, 1967.
10 Maynard, H. B. (ed.): *Industrial Engineering Handbook,* 2d ed., McGraw-Hill Book Company, New York, 1963.
11 McCormick, Ernest J.: *Human Factors Engineering,* 3d ed., McGraw-Hill Book Company, New York, 1970.
12 Moore, James M.: *Plant Layout and Design,* The Macmillan Company, New York, 1969.
13 Murrell, K. F. H.: *Ergonomics,* Chapman and Hall, London, 1969.
14 Muther, Richard: *Systematic Layout Planning,* Industrial Education Institute, Boston, 1961.
15 Muther, Richard: *Practical Plant Layout,* McGraw-Hill Book Company, New York, 1955.
16 Niebel, Benjamin W.: *Motion and Time Study,* 5th ed., Richard D. Irwin, Inc., Homewood, Illinois, 1972.
17 Reed, Ruddell, Jr.: *Plant Layout,* Richard D. Irwin, Inc., Homewood, Ill., 1961.
18 Seehof, J. M., and W. O. Evans: "Automated Layout Design Program," *J. Ind. Eng.,* vol. 18, no. 12, pp. 690–695, 1967.
19 Woodson, Wesley E., and Donald W. Conover: *Human Engineering Guide for Equipment Designers,* 2d ed., University of California Press, Berkeley, 1964.

REVIEW QUESTIONS AND PROBLEMS

1 What is a time standard?
2 Differentiate between an operation process chart and a flow process chart.
3 What is the relationship between principles of motion economy and intraoperation handling?
4 What is a rating factor?
5 If the rating factors in Fig. 4-7 had been 1.00, 0.90, 0.85, and 0.90, respectively, for elements 1 through 4, what standard time would be calculated?
6 If, instead of the average rating factors for elements shown in Fig. 4-7, the analyst observed that for the first five cycles the rate was 1.00 and for the last five cycles 1.10, how might the standard time be calculated best using this information? What would the standard time be by this method?
7 How many minutes does it take to reach a very small object 20 inches away, according to the MTM system?
8 Using Eq. (4-2), calculate the number of observations that would be necessary if the percent occurrence of an activity of interest was 40 percent, and a relative accuracy of ±5 percent was desired at a confidence level of 95 percent (i.e., $K_\alpha \simeq 2$)?
9 Determine n for problem 8 by using the table in Fig. 4-14.
10 Determine n for problem 8 by using the nomograph in Fig. 4-15.
11 What are standard data?
12 What is meant by the statement, "Plant design is first macro, then micro, and ultimately macro"?
13 Assume that a three-step process possesses the following characteristics:

	Step 1	Step 2	Step 3
Standard time, min/#	2.0	3.0	4.0
Scrap rate, %	10	15	10

If 100 good units are needed per hour from the process, departmental efficiency is 80 percent, and the same assumptions apply as in Fig. 4-19, how many step 1, step 2, and step 3 machines would be required?

14 What is the purpose of employing the activity relationship chart approach in plant layout?
15 What particular quality of a plant layout design is typically better understood as a result of digital simulation of the design?

Chapter 5

Production Systems Control

Therefore, if any man objects to time study, the real objection is not that it makes him nervous. His real objection is that he does not want his employer to know how long it takes him to do his job.

F. W. Taylor

Methods engineering, materials handling, plant layout, and human factors are all necessary considerations in developing a best design for a productive system. Once the system is installed, industrial engineering attention usually shifts to devising the best methods for operating the system. Inventory, production, quality, and cost control all represent evolved qualitative and quantitative approaches for maximizing the economic utility of a productive system. In all but rare instances, these approaches attempt to minimize the unit cost associated with transforming incoming materials into a completed, well-specified product at a desired quality level. With rare exceptions, the goal of industrial engineering is cost reduction—how to produce something for less.

INVENTORY CONTROL

Most productive systems contain inventories. In a manufacturing plant inventories consist of basic materials, goods in process, and finished stock; in a hospital they may well be disposable hypodermic needles and aspirin tablets.

It costs money to maintain an inventory; therefore, inventories are inherently undesirable in the sense that they do not contribute to the direct transformation of materials and represent a cost of doing business. Inventories do provide, however, essential decoupling between unequal flow rates.

Assume for the moment that you are in the bologna sandwich business. You run the business alone; it takes you 5 minutes to make a bologna sandwich and you work 8 hours a day. Each day you make 96 bologna sandwiches, most of which are sold around lunchtime, with a few being sold in the morning and close to closing in late afternoon. The refrigerator in which you store your sandwiches is a buffer between a constant production rate and a variable demand rate. If you could sell bologna sandwiches at a constant rate, you would not need the refrigerator to store the completed sandwiches. That is what inventories are all about.

Around 1915, F. Harris developed what came to be known as the Wilson formula. Wilson publicized his work more than Harris did. It represents the typical starting point for the development of inventory models.

Figure 5-1 illustrates this simple and highly idealized inventory model. Assume that at time zero exactly Q units of material are on hand, and as time progresses there is a constant demand or issuance of the material from stock. This decreases the quantity in stock in a linear fashion, as indicated in Fig. 5-1. Assume also that when the remaining stock reaches a level R at time t_1, an order is placed for Q units of material to be delivered to the plant in L days. Time L is the number of days which,

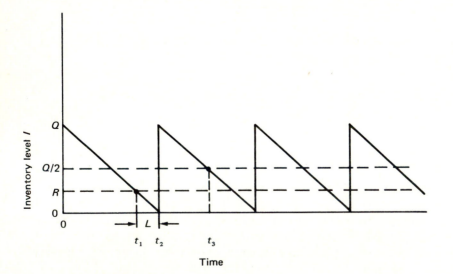

Figure 5-1   The Harris inventory model.

based on the daily reduction rate, will result in exact depletion of the remaining R units of inventory at the time Q units of material are received. Assume also that there is an annual requirement A, a unit material cost M, an inventory holding cost per unit per year H, an order preparation cost P, and an inventory level I.

This simple inventory model deals with the three costs—material, holding, and order preparation. Holding costs refer to all costs associated with providing facilities for and maintaining an inventory. Order preparation cost refers to an assumed constant cost associated with placing and receiving an order of materials. If orders are placed very frequently and the annual requirement A is fixed, Q is small, therefore holding costs are small; but the cost of placing numerous orders is high. If orders are placed infrequently, order placement costs for the year are low; but a larger quantity of material Q must be ordered each time, resulting in high inventory holding costs. The Harris model can be employed to determine the optimum lost size Q_0 to purchase, to minimize the sum of all costs associated with order preparation and storage of the material for the year. Total annual costs are

$$C = \text{material cost} + \text{inventory holding cost} + \text{order preparation costs}$$

Annual inventory holding cost is the product of average inventory on hand $Q/2$ and the annual unit holding cost H, because the inventory on hand varies linearly between Q and 0, as indicated in Fig. 5-1 at time t_3. The annual order preparation cost is the product of the cost of placing an order P and the number of times orders are placed per year A/Q. Therefore, total annual cost is

$$C = MA + H\frac{Q}{2} + P\frac{A}{Q} \tag{5-1}$$

To determine a minimum for C as a function of Q employing differential calculus, one differentiates the function with respect to Q and sets it equal to zero. The value of Q for minimizing C is then determined from this equation. Differentiating C with respect to Q and setting it equal to zero yields

$$\frac{dC}{dQ} = \frac{H}{2} - \frac{PA}{Q^2} = 0 \tag{5-2}$$

$$Q_0 = \sqrt{\frac{2PA}{H}} \tag{5-3}$$

Note that the term MA dropped out in the differentiation step, Eq. (5-2). This indicates that the optimum order quantity is independent of material cost. An annual quantity A of material will be purchased during the year and will represent the same cost regardless of how frequently material is purchased. The solution then only considers the variable cost V during the year for holding cost and order preparation costs.

Consider the following example:

$A = 4{,}500$ units

$H = \$2$

$P = \$20$

$M = \$10$

From Eq. (5-3):

$$Q_0 = \sqrt{\frac{2(20)4{,}500}{2}}$$

$$= 300$$

Table 5-1 illustrates total annual variable cost V for different values of Q. It is apparent in reviewing the total annual variable cost column of Table 5-1 and Fig. 5-2 that total variable cost reaches a minimum when $Q = 300$. The quantity Q_0 is often referred to as the EOQ (economic order quantity).

The Harris model is a highly idealized model in the sense that a minimum number of cost factors are considered and all elements of the model are deterministic. In practice, at least three aspects of an inventory model would likely be probabilistic: (1) daily demand, (2) lead time, and (3) lost demand due to back orders. By probabilistic is meant that the number of units demanded each day, for example, would likely vary from day to day. If we were to collect data on the number of units demanded each day for 100 days, the histogram of Fig. 5-3a might represent our best estimate of typical demand.

One common technique for analyzing inventory systems containing stochastic elements is digital simulation. Note in Table 5-2 that the histogram data can be readily converted first into probabilities of daily demand P for varying levels of D and then into cumulative probabilities of demand P' for levels of demand D or less. A very powerful technique in digital simulation, known as the Monte Carlo technique, utilizes an empirically derived cumulative density function, such as P', to generate a sequence of representative values for a random variable (e.g., a stochastic variable such as daily demand D).

Table 5-1   Total Annual Variable Costs as a Function of Lot Size

Lot size Q	Inventory handling cost $HQ/2$	Order preparation costs PA/Q	Total annual variable cost V
0	$ 0	$ ∞	$ ∞
100	100	900	1,000
200	200	450	650
300	300	300	600
400	400	225	625
500	500	180	680
600	600	150	750

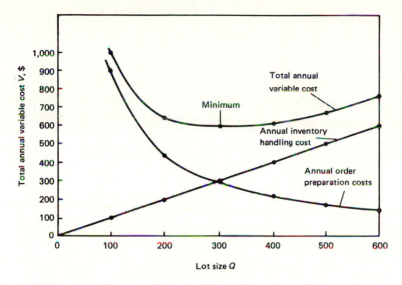

Figure 5-2 Annual variable costs as a function of lot size.

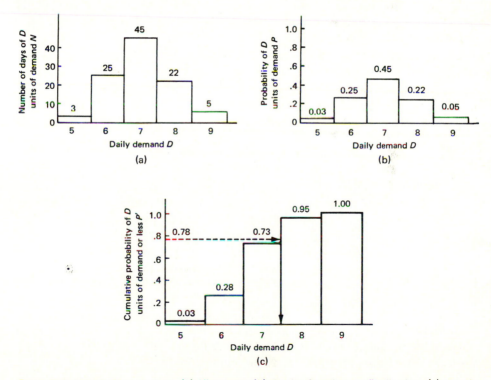

Figure 5-3 Probabilistic demand. (a) Histogram, (b) density function or distribution, (c) cumulative distribution.

Table 5-2   Probabilistic Demand

Daily demand D	Number of days of D units of demand N	Probability of a daily demand of D P	Cumulative probability of a demand of D or less P'
5	3	.03	.03
6	25	.25	.28
7	45	.45	.73
8	22	.22	.95
9	5	.05	1.00
	100	1.00	

The technique employs uniformly distributed (i.e., equally likely) numbers over some range. In employing Fig. 5-3c, entries are drawn from a table of random numbers, such as Table 5-3, in a consistent manner to correspond on a one-to-one basis with the 0 to 1 range of P' in Fig. 5-3c. For this problem, entries will be drawn from the first two columns of digits from top to bottom in Table 5-3 and converted to decimals to correspond with the 0 to 1 scale of Fig. 5-3c. The first thirty numbers drawn from Table 5-3 provide the sequence of demands indicated in Table 5-4, employing the cumulative density of Fig. 5-3c. These thirty generated values of demand are assembled in histogram form in Fig. 5-4. Note the similarity of this histogram to that of Fig. 5-3a. If histograms of larger and larger samples were plotted, it would be noted that as the sample increases in size, the distribution of values generated comes closer and closer to exactly matching Fig. 5-3a. What should be obvious at this point is that Monte Carlo sampling of an empirically derived distribution produces a sample whose distribution comes closer and closer to exactly matching the distribution from which it was drawn as the sample increases in size. The extent of the lack of correspondence for smaller samples is what is known as "sampling error" and is of a statistically calculable magnitude.

Similar sequences of values could be generated for the variables lead time and lost demand. A digital simulation containing these random variables along with other deterministic variables can simulate typical behavior of an inventory system over time. The accumulated results from such a simulation can then be evaluated to determine the desirability of the inventory policies employed. By the use of digital simulation one can discover the performance behavior of a candidate inventory model before the model and its attendant policies are ever applied in an actual industrial environment. In this way, it is possible to know how the inventory control technique will work before it is applied, minimizing the risks involved. The author, for example, simulated a multimillion-dollar computer-controlled stacker crane storage system (2) to determine the performance capability of the system before it was installed. Because of the nonintuitive nature of the dynamic characteristics of queueing (i.e., flow) systems, analyses of this nature often yield invaluable information.

Table 5-3 Table of Random Digits*

78466	83326	96589	88727	72655	49682	82338	28583	01522	11248
78722	47603	03477	29528	63956	01255	29840	32370	18032	82051
06401	87397	72898	32441	88861	71803	55626	77847	29925	76106
04754	14489	39420	94211	58042	43184	60977	74801	05931	73822
97118	06774	87743	60156	38037	16201	35137	54513	68023	34380
71923	49313	59713	95710	05975	64982	79253	93876	33707	84956
78870	77328	09637	67080	49168	75290	50175	34312	82593	76606
61208	17172	33187	92523	69895	28284	77956	45877	08044	58292
05033	24214	74232	33769	06304	54676	70026	41957	40112	66451
95983	13391	30369	51035	17042	11729	88647	70541	36026	23113
19946	55448	75049	24541	43007	11975	31797	05373	45893	25665
03580	67206	09635	84612	62611	86724	77411	99415	58901	86160
56823	49819	20283	22272	00114	92007	24369	00543	05417	92251
87633	31761	99865	31488	49947	06060	32083	47944	00449	06550
95152	10133	52693	22480	50336	49502	06296	76414	18358	05313
05639	24175	79438	92151	57602	03590	25465	54780	79098	73594
65927	55525	67270	22907	55097	63177	34119	94216	84861	10457
59005	29000	38395	80367	34112	41866	30170	84658	84441	03926
06626	42682	91522	45955	23263	09764	26824	82936	16813	13878
11306	02732	34189	04228	58541	72573	89071	58066	67159	29633
45143	56545	94617	42752	31209	14380	81477	36952	44934	97435
97612	87175	22613	84175	96413	83336	12408	89318	41713	90669
97035	62442	06940	45719	39918	60274	54353	54497	29789	82928
62498	00257	19179	06313	07900	46733	21413	63627	48734	92174
80306	19257	18690	54653	07263	19894	89909	76415	57246	02621
84114	84884	50129	68942	93264	72344	98794	16791	83861	32007
58437	88807	92141	88677	02864	02052	62843	21692	21373	29408
15702	53457	54258	47485	23399	71692	56806	70801	41548	94809
59966	41287	87001	26462	94000	28457	09469	80416	05897	87970
43641	05920	81346	02507	25349	93370	02064	62719	45740	62080
25501	50113	44600	87433	00683	79107	22315	42162	25516	98434
98294	08491	25251	26737	00071	45090	68628	64390	42684	94956
52582	89985	37863	60788	27412	47502	71577	13542	31077	13353
26510	83622	12546	00489	89304	15550	09482	07504	64588	92562
24755	71543	31667	83624	27085	65905	32386	30775	19689	41437
38399	88796	58856	18220	51016	04976	54062	49109	95563	48244
18889	87814	52232	58244	95206	05947	26622	01381	28744	38374
51774	89694	02654	63161	54622	31113	51160	29015	64730	07750
88375	37710	61619	69820	13131	90406	45206	06386	06398	68652
10416	70345	93307	87360	53452	61179	46845	91521	32430	74795
99258	03778	54674	51499	13659	36434	84760	76446	64026	97534
58923	18319	95092	11840	87646	85330	58143	42023	28972	30657
39407	41126	44469	78889	54462	38609	58555	69793	27258	11296
29372	70781	19554	95559	63088	35845	60162	21228	48296	05006
07287	76846	92658	21985	00872	11513	24443	44320	37737	97360
07089	02948	03699	71255	13944	86597	89052	88899	03553	42145
35757	37447	29860	04546	28742	27773	10215	09774	43426	22961
58797	70878	78167	91942	15108	37441	99254	27121	92358	94254
32281	97860	23029	61409	81887	02050	63060	45246	46312	30378
93531	08514	30244	34641	29820	72126	62419	93233	26537	21179

*From The Rand Corporation, *A Million Random Digits with 100,000 Normal Deviates,* p. 180, Santa Monica, Calif.

Table 5-4 A Generated Sequence of Daily Demands

Day	Number from table	Scaled number	Corresponding demand D	Day	Number from table	Scaled number	Corresponding demand D
1	78	0.78	8	16	05	0.05	6
2	78	0.78	8	17	65	0.65	7
3	06	0.06	6	18	59	0.59	7
4	04	0.04	6	19	06	0.06	6
5	97	0.97	9	20	11	0.11	6
6	71	0.71	7	21	45	0.45	7
7	78	0.78	8	22	97	0.97	9
8	61	0.61	7	23	97	0.97	9
9	05	0.05	6	24	62	0.62	7
10	95	0.95	8	25	80	0.80	8
11	19	0.19	6	26	84	0.84	8
12	03	0.03	5	27	58	0.58	7
13	56	0.56	7	28	15	0.15	6
14	87	0.87	8	29	59	0.59	7
15	95	0.95	8	30	43	0.43	7

It should be understood that the Harris model and the few additional elements considered to introduce the concept of a stochastic inventory model represent a very limited introduction to inventory theory and systems. Inventory control is really a subset of a larger area of control commonly referred to as production control.

PRODUCTION CONTROL

Production control and production planning are terms that are used interchangeably in some industries, while in other industries or companies they may have distinctly different connotations. Generally speaking, production planning at least suggests a larger scope than production control. Most production control departments carry out

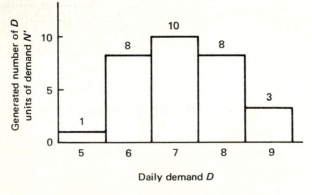

Figure 5-4 Histogram of sampled values for demand.

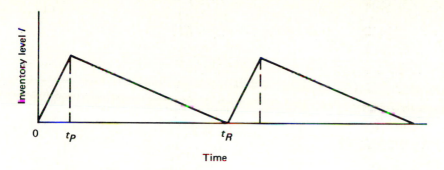

Figure 5-5 Production lot size model.

a staff function involving blue-collar workers concerned primarily with the day-to-day execution of production plans. Production planning may be done by white-collar production staff executives or engineers concerned with defining the overall production plan of a plant or a combination of plants, consistent with the overall short- and long-term production goals of a company or corporation. In summary, the purpose of production planning is to determine what needs to be produced over a period of time to be consistent with established corporate goals. Production control is concerned with detailed planning and execution for today, tomorrow, and next week to see that the maximum capability of the productive system is utilized.

An example of how production control might do this, which is, analogous to the Harris model, is by determining an optimum production lot size for a machine. Assume that a machine is used to make five different products, and that every time the machine is adjusted to change from one product to another a setup cost is involved. Assume also that the longer the machine produces one product before shifting to another, the larger an inventory of that product must be stored because of the longer time before the machine will return to producing that product again. The cost trade-off then, is between setup cost, which does not produce units of product, and inventory holding costs.

Let t_P be the production period for a product, and t_R the time from starting production on the product until the machine is returned to begin production on that product again, as indicated in Fig. 5-5. Let D be the daily demand rate, P the daily production rate, A the annual requirement, S the setup cost, H the unit annual holding cost, V the total annual variable cost, and Q the production lot size.

Inventory reaches a peak at time t_P, and is equal to $t_P(P-D)$. Average inventory is, therefore, equal to $t_P(P-D)/2$. The quantity of material produced is $Q = t_P P$; therefore, $t_P = Q/P$. Substituting for t_P above, average inventory then is

$$\frac{Q(P-D)}{2P} = \frac{Q}{2} - \frac{QD}{2P} = \left(1 - \frac{D}{P}\right)\frac{Q}{2} \qquad (5\text{-}4)$$

and consequently,

$$V = S\frac{A}{Q} + H\left(1 - \frac{D}{P}\right)\frac{Q}{2} \tag{5-5}$$

Differentiating V with respect to Q and setting the result equal to zero provides the basis for determining the value of Q that results in a minimum annual variable cost for production setups and the holding of inventory.

$$\frac{dV}{dQ} = -S\frac{A}{Q^2} + \frac{H}{2}\left(1 - \frac{D}{P}\right) = 0 \tag{5-6}$$

$$Q_0 = \sqrt{\frac{2SA}{H(1 - D/P)}} \tag{5-7}$$

Optimum lot size is Q_0, and the optimum number of production runs per year for the product is $N_0 = A/Q_0$.

If a number of products are being run sequentially on the same machine, and the utilization is fairly high, the optimum lot size as calculated above may result in an infeasible solution. Assume that the eight products of Table 5-5 are to be produced sequentially on a particular machine. It is common in industry to assume 22 working days per month. If 14 days are allowed for holidays and vacation shutdown, planning can be done assuming 250 working days per year. The number of production days required per year for the eight products, as indicated in Table 5-5, is 235 days. However, if the eight products were to be made in sequence in optimum lot size quantities with respect to the cost parameters associated with each product, one cycle of all eight products would require 48.3 production days, as indicated in Table 5-6. In column (7) of Table 5-6 it can be observed that material storage for four of the products (i.e., products B, C, F, and H) would become depleted during the 48.3-day cycle following production of each of these specific products. Therefore, it is obvious that individually determined optimum lot sizes can produce an infeasible solution.

Table 5-5 Product Data for a Particular Machine

Product	Annual requirement A_i	Daily production rate P_i	Production days needed n_i	Annual unit holding cost H_i	Setup cost S_i
A	4,000	400	10	$0.20	$110
B	14,000	400	35	0.50	80
C	5,200	100	52	0.65	35
D	6,000	300	20	0.35	55
E	1,200	150	8	0.25	25
F	13,000	200	65	0.60	90
G	8,000	400	20	0.30	15
H	12,500	500	25	0.60	20
			235		$430

Table 5-6 Lot Sizes Determined Separately

	(1)	(2)	(3)	(4)	(5)	(6)	(7)
					Production		Demand
			$N_{0_i}{}^*$		days required	$D_i{}^\dagger$	days for Q_{0_i}
Product	A_i	$Q_{0_i}{}^*$	(1)/(2)	P_i	(2)/(4)	(1)/250	(2)/(6)
A	4,000	2,140	1.87$\ddagger$	400	5.4	16.0	134
B	14,000	2,280	6.14	400	5.7	56.0	41$\S$
C	5,200	840	6.19	100	8.4	20.8	40$\S$
D	6,000	1,430	4.19	300	4.8	24.0	60
E	1,200	498	2.41$\ddagger$	150	3.3	4.8	104
F	13,000	2,290	5.68	200	11.4	52.0	44$\S$
G	8,000	2,950	2.71$\ddagger$	400	7.4	32.0	92
H	12,500	962	13.00	500	1.9	50.0	19$\S$
					48.3		

*See Eq. (5-7).
†Assume 250 production days per year.
‡Consider running every other cycle.
§Optimum lot size not adequate for 48.3-day cycle.

One approach to arriving at a desirable and feasible solution to this problem is to determine an optimum common cycle time and associated optimum number of runs per year with respect to cost factors for all eight products jointly.

Note that the second term on the right side of Eq. (5-5) is related to annual inventory handling cost for a product. By substituting A_i/N for Q in Eq. (5-5), the annual holding cost for any single product is

$$\frac{H_i\,(1 - D_i/P_i)\,A_i}{2N}$$

and the total annual inventory holding cost for all n products is therefore

$$\frac{1}{2N} \sum_{i=1}^{n} H_i A_i \left(1 - \frac{D_i}{P_i}\right)$$

Similarly, the sum of annual setup costs for all n products is

$$N \sum_{i=1}^{n} S_i$$

The total annual variable costs then are

$$V = N \sum_{i=1}^{n} S_i + \frac{1}{2N} \sum_{i=1}^{n} H_i A_i \left(1 - \frac{D_i}{P_i}\right) \tag{5-8}$$

For the purpose of differentiating with respect to N, to determine an optimum joint number of runs per year, it may be helpful to think of

$$\sum_{i=1}^{n} S_i = \phi \quad \text{and} \quad \sum_{i=1}^{n} H_i A_i \left(1 - \frac{D_i}{P_i}\right) = \lambda$$

which reduces Eq. (5-8) to

$$V = \phi N + \frac{\lambda}{2N}$$

Differentiating with respect to N yields

$$\frac{dV}{dN} = \phi - \frac{\lambda}{2N^2} = 0$$

Therefore,

$$N_0 = \sqrt{\frac{\lambda}{2\phi}} = \sqrt{\frac{\sum_{i=1}^{n} H_i A_i (1 - D_i/P_i)}{2 \sum_{i=1}^{n} S_i}} \tag{5-9}$$

and

$$Q_i = \frac{A_i}{N_0}$$

Solution of Eq. (5-9) for the product data of Table 5-5 produces $N_0 = 5.55$. Table 5-7 illustrates lot sizes for products employing the optimum common number of runs per year N_0. Note that by employing this approach, the number of demand days for each product is the same (i.e., 45), and in this instance is in excess of the cycle time of 42.4 days, ensuring feasibility for all eight products. Magee and Boodman (3) offer a rule of thumb that if the optimum N_0 for a product determined separately is less than half of the joint N_0, as indicated for products A, E, and G in Table 5-6, these products might preferably be run every other cycle or even less frequently, to further reduce annual variable costs. It must be kept in mind, however, that this increases the cycle time when the product is run and must, therefore, be adjusted judiciously. This problem should provide some insight into a typical production planning or control problem.

Another problem often dealt with in production control is that of assembly-line balancing. The most common method in practice is to first attempt to distribute work

Table 5-7 Lot Sizes Determined Collectively

Product	(1) Lot size determined collectively $Q_j = A_j/N_o$	(2) P_j	(3) Production days required (1)/(2)	(4) D_j	Demand days for Q_j (1)/(4)
A	721	400	1.8	16.0	45
B	2,520	400	6.3	56.0	45
C	937	100	9.4	20.8	45
D	1,080	300	3.6	24.0	45
E	216	150	1.4	4.8	45
F	2,340	200	11.7	52.0	45
G	1,440	400	3.6	32.0	45
H	2,250	500	4.5	50.0	45
			42.4		

elements to stations along an assembly line in such a manner that each station has approximately the same sum of elemental times. After the line is run for a short period of time, it usually becomes apparent that one employee on the line is having to work extra hard to keep up with the line or that work is beginning to collect at one workstation. The industrial engineer or timestudy analyst noting this would then remove some elements of work from this station, if that were justified, and add them to another station along the line that appears to be low on total elemental time. This juggling of time elements on an assembly line might result in one employee installing the master brake cylinder on an automobile and setting a muffler clamp on an automobile frame, because the muffler installer farther down the line needs time relief and the master brake cylinder installer has excess time available.

In some cases, the number of elemental times is rather large and their respective precedence relationships are sufficiently complex that some organized procedure is needed for determining an optimum assignment of elements to workstations. In the following example a heuristic first proposed by Moodie and Young (5) is used. "Heuristic" means that the approach produces good results but does not necessarily guarantee the best or optimum result. An "algorithm," on the other hand, is a procedure that, by definition, guarantees an optimum solution, as will be demonstrated in the next chapter with respect to solving linear programming problems. This particular heuristic, however, does provide a test for determining whether a better solution is possible, and it is rare when the heuristic produces a solution that is less than optimum.

Figure 5-6 illustrates the precedence relationships and elemental times for thirteen elements in a process. Assume that there are 8000 minutes of production time available for producing 1000 units of a product. A desired cycle time, therefore, would be:

$$\text{Cycle time} = \frac{8000 \text{ minutes}}{1000 \text{ units}} = 8 \text{ minutes/unit}$$

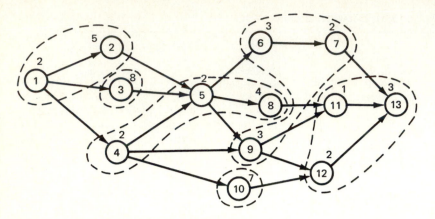

Figure 5-6 Precedence chart for elemental times.

The following terminology will be employed in this example:

Cycle time $= C$
Station number $= K$
Element number $= i$
Elemental time $= E_i$
Station time $= S_K$
Balance delay $= C - S_k$

In Fig. 5-6, the number within a circle is the element number i, and the number above the circle is the elemental time E_i. The arrows in the figure show which elements must be completed before an element can be initiated. For example, element 5 cannot begin until elements 2, 3, and 4 are completed; and elements 6, 8, and 9 cannot begin until element 5 has been completed. The sum of all thirteen elemental times is 44 minutes. With a desired cycle time of 8 minutes an optimal allocation of elements would result in six workstations with 4 minutes of balance delay left in the system (i.e., 6 × 8 = 48; 48 − 44 = 4). Any feasible grouping of elements into six workstations for the purpose of this problem is assumed to represent an optimal solution.

The heuristic (5) is as follows:

1 Note the rows of the predecessor matrix P that contain all zeros, and assign the largest element possible indicated by these rows if more than one exists. Check the element number to indicate that it has been assigned.

2 Note the element number in the row of the follower matrix F that corresponds to the assigned element, and go to the rows of matrix P indicated by these numbers and replace the assigned element's identification number with a zero.

3 Continue assigning elements to each station following the restriction that

$$\text{Maximum } E_i(K) \leqslant S_i \leqslant C$$

Proceed until the P matrix contains all zeros.

In this example, the heuristic is applied to Table 5-8 to produce the station assignments indicated in Table 5-9. The number of columns of the P matrix is determined by the largest number of elements preceding any element in the precedence diagram. In this example elements 2, 3, and 4 precede element 5; therefore, three columns are required for the P matrix. In a like manner, the maximum number following any element in Fig. 5-6 is three; therefore, three columns are needed in the F matrix. The crossouts in Table 5-8 appear as they would at the time station number one element assignments have been made.

When all assignments have been made, all numbers other than zeros will have been crossed out and replaced by zeros. The station assignments of Table 5-9 are indicated by the dotted line zones of Fig. 5-6. Note that element 2 was selected to be added to element 1 in station one because 6 minutes of assignable time remained and element 2 had the largest possible assignable time available. If the sum of balance delay in this example had been 8 or greater, it would not be apparent that the solution obtained was an optimum solution.

It should also be noted that the heuristic as stated results in a well-defined sequence of assignments, which can be easily programmed. A computer program could be readily developed to make station assignments for a large number of elements and precedence relationships. This example is designed to demonstrate how a simple heuristic can be devised to assist in making element assignments to workstations, including a means of checking the feasibility of using a smaller number of stations with a different assignment. Simple heuristics such as this one represent the "tricks of the trade" in applying industrial engineering on a day-to-day basis.

Another class of techniques in the general area of production planning and

Table 5-8   Line Balancing Data

i	E_i	P^*			$F^\dagger$		
1	2	0	0	0	2	3	4
2✓	5	1̶	0	0	5	0	0
3✓	8	1̶	0	0	5	0	0
4	2	1̶	0	0	5	9	10
5	2	2̶	0	3	6	8	9
6	3	5	0	0	7	0	0
7	2	6	0	0	13	0	0
8	4	5	0	0	11	0	0
9	3	4	5	0	11	12	0
10	7	4	0	0	12	0	0
11	1	8	9	0	13	0	0
12	2	9	10	0	13	0	0
13	3	7	11	12	0	0	0
	44						

*The numbers in the rows of matrix P are the element numbers preceding element i for that row.

†The numbers in the rows of matrix F are the element numbers following element i for that row.

Table 5-9 Line Balancing Station Summary

K	i	E_i	S_K	$C - S_K$
1	1	2		
	2	5	7	1
2	3	8	8	0
3	4	2		
	5	2		
	8	4	8	0
4	10	7	7	1
5	6	3		
	9	3		
	7	2	8	0
6	12	2		
	11	1		
	13	3	6	2
				4

control are commonly referred to as project management techniques. The two most common of these are CPM (critical path method) and PERT (project evaluation and review technique). CPM was developed in industry at the same time that PERT was being developed in the U.S. Navy, specifically the Polaris submarine program. Papers were first published describing both techniques in 1959 (4, 7).

Assume that a tank is to be designed and assembled in the field, and that Table 5-10 represents the project engineer's best estimate of the time required for each task and the key precedence relationships for each task. As indicated in Table 5-10, field assembly of the tank takes 15 days, but cannot begin until the tank is transported to the site (activity H) and tank site preparation (activity I) has been completed. Figure 5-7 is a basic CPM diagram for the tank project, indicating the precedence relationships by use of connected arrows.

One of the primary reasons for performing a CPM analysis of a project is to determine the "critical path," which is that sequence of project tasks through the network which determines the minimum completion time for the project. The minimum time for the tank project is 115 days, which is the sum of task times for tasks A, B, D, F, J, L, M, O, and P. No other sequence of tasks through the network requires a longer time. If the project engineer is concerned with completing the project as soon as possible, which is usually the case, particular attention must be paid to tasks on the critical path. Essentially, a delay in any of these tasks will represent a like delay in the total completion time of the project.

For a large-scale project, such as the research and development of the first Polaris submarine or the building of a sports stadium, only a small fraction of the tasks to be

Table 5-10   Tank Installation Project Data

Activity symbol	Activity	Time (days)	Precedence
A	Design tank	30	
B	Design controls	10	A
C	Prepare parts list for tank	5	A
D	Prepare parts list for controls	3	B
E	Procure tank parts	40	C
F	Procure control parts	60	D
G	Partially fabricate tank	5	E
H	Transport tank to site	7	G
I	Prepare site	10	B
J	Transport controls to site	3	F
K	Field assembly of tank	15	H,I
L	Field assembly of controls	2	J
M	Install controls to tank	4	K,L
N	Inspect and test tank	3	K
O	Inspect and test controls	2	M
P	Document and cleanup	1	N,O

performed are on the critical path. Identification of and attention to critical path tasks often result in a dramatic improvement in total project completion time; the Polaris submarine project was completed 18 months ahead of its initial schedule, which was previously unheard of for this type of project.

In analyzing the network for Fig. 5-7 it is common practice to record earliest start (ES), earliest completion (EC), latest start (LS), and latest completion (LC) times on the network, as indicated in Fig. 5-8. Both i and j are assigned arbitrary, yet unique, node numbers to numerically identify the beginning and ending points for each task. The letters X and t represent a task designation symbol and a corresponding task time, respectively. The CPM network for the tank project including these data is shown as Fig. 5-9.

To determine earliest start and earliest completion times, one starts at the beginning of the network, adding the task time for an activity to the earliest start time to provide the earliest completion time. If two tasks precede a third task, the later completion time determines the earliest start time for the third task (e.g., task K).

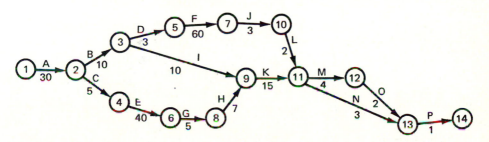

Figure 5-7   A basic CPM diagram for the tank project.

Figure 5-8 Standard CPM network symbol.

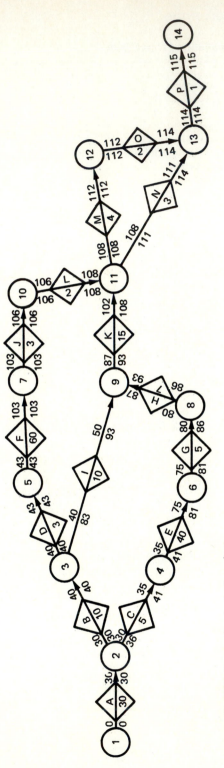

Figure 5-9 CPM diagram for the tank project.

In a similar manner, latest completion and latest start times are determined by starting at the end of the network and working backward. If more than one task follows a third task, for example task K, the smaller latest start time of the two following tasks determines the latest completion time of the preceding task (i.e., LC for task K is 108).

The earliest and latest start and completion times recorded on Fig. 5-9 provide useful information to the project engineer. Slack time is calculated as follows:

$$\text{Slack time} = LS - ES \quad \text{or} \quad LC - EC$$

For task J, for example, slack time is zero. This says that task J is on the critical path and, if delayed, will delay the total project. Slack for task N, however, is 3 days. This indicates that if task N is delayed 3 days, the total project would not be delayed, as noted by analyzing the later tasks (i.e., N, O, and P) in Fig. 5-9.

The assignment of node numbers i and j facilitates tabular analysis of a CPM network on a computer, as indicated in Table 5-11. The preceding graphical networks of Figs. 5-7 and 5-9 are essentially superfluous if a computer solution is employed. It has been reported by practitioners that for relatively small networks (i.e., less than 100 activities), a time-scaled network, as indicated in Fig. 5-10, is often quite adequate for graphically displaying and subsequently controlling a project, without need of a computerized solution. This is an example of a situation in which the use of a computer may not provide a sufficient return to justify the inconvenience that inputting data may involve. It is not uncommon today to observe someone trying to provide a computer solution to a problem for which a noncomputerized technique may well be superior, everything considered.

Table 5-11 Tabular Analysis of Tank Project

Activity	From i	To j	t	Forward pass ES	Forward pass EC	Backward pass LS	Backward pass LC	Slack
A	1	2	30	0	30	0	30	0
B	2	3	10	30	40	30	40	0
C	2	4	5	30	35	36	41	6
D	3	5	3	40	43	40	43	0
E	4	6	40	35	75	41	81	6
F	5	7	60	43	103	43	103	0
G	6	8	5	75	80	81	86	6
H	8	9	7	80	87	86	93	6
I	3	9	10	40	50	83	93	43
J	7	10	3	103	106	103	106	0
K	9	11	15	87	102	93	108	6
L	10	11	2	106	108	106	108	0
M	11	12	4	108	112	108	112	0
N	11	13	3	108	111	111	114	3
O	12	13	2	112	114	112	114	0
P	13	14	1	114	115	114	115	0

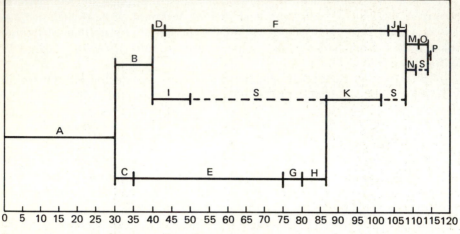

Figure 5-10 Time-scaled CPM network for the tank project.

In CPM, all tasks or activities are assumed to be deterministic, with respect to both the sequence of tasks and their duration. PERT was originally developed with the hope of specifying tasks in a simple yet reasonably accurate stochastic manner. Three times were estimated for each task: the optimistic time A, the most likely time M, and the pessimistic time B. These three times were then used to specify a beta distribution with a mean $\overline{X}$ and a variance S^2 estimated as follows:

$$\overline{X} = \frac{1}{6}(A + 4M + B)$$

$$S^2 = \left[\frac{1}{6}(B - A)\right]^2$$

Figure 5-11 illustrates a typical beta distribution employing the A, B, and M estimates. Times A and B were to be estimated on the assumption that more extreme values than A (or B) would be expected to occur one out of a hundred times.

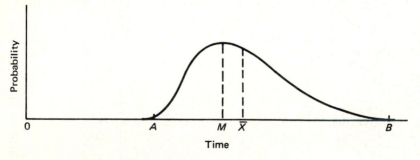

Figure 5-11 A beta distribution in PERT.

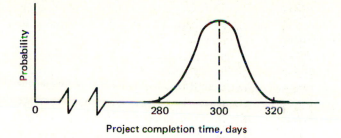

Figure 5-12 Density function for project completion time.

By having all tasks specified as density functions, it was possible to produce a density function for overall project completion time, as indicated in Fig. 5-12. This density might indicate, for example, that it would be 99 percent likely that the project would be completed in more than 280 days, or 99 percent likely that the project would not take more than 320 days, or 98 percent likely that it would take between 280 and 320 days, with 300 days being most likely. Employing CPM, the estimate of completion time would simply be 300 days, assuming the constant task times selected in CPM summed to the sum of $\overline{X}_i$ for the same activities in PERT.

The three-estimates concept of PERT seems to have essentially disappeared in practice. Most practitioners seem to have adopted the single-estimate approach of CPM. Two possible reasons for this trend are: (1) primary practical interest has been and still is in providing a "best estimate" of actual completion time, and (2) the beta distribution assumption may have been too great a simplifying assumption, introducing considerable error in some applications and causing the general applicability of the beta distribution assumption to be suspect.

There is little question today of the need for employing some project management technique, similar to CPM or PERT, to control the complex scheduling of limited resources for key activities. These techniques have provided a measure of control that makes timely completion of complex projects possible, predictable, and even expected, barring unusual unforeseen factors beyond predictability or control. Bad weather, for example, is predictable, and it can be included as lost days due to bad weather based on accumulated data in project planning; a hurricane or earthquake at a specific location and time, however, is far less predictable.

Production planning and control are primarily concerned with specifying who will do what when in a production activity. Another area of control—quality control—is primarily concerned with establishing procedures to assure that a prespecified quality level of finished product is produced.

QUALITY CONTROL

In the days of Taylor, every plant, with rare exceptions, had an inspection department. The function of the department was to see that unacceptable material was removed throughout the process of making a product, from incoming material to the finished product. In those early days, statistical sampling was essentially unknown in manufacturing. Consequently, key points were often selected throughout the process at which 100 percent inspection was performed. Because employees were often paid on the basis of the number of good units produced, and production managers were evaluated on the number of units shipped, inspection department employees and managers have traditionally endured often unfriendly if not hostile work environments.

Quality control today is essentially applied statistical sampling and includes the administrative procedures for assuring overall quality objectives at a minimum cost. As was mentioned earlier in this text, no single course in a typical industrial engineering curriculum is more important than a course in probability and statistics. This is particularly true in the area of quality control. Quality control, in terms of its technical content, is simply the application of sampling theory in an industrial or productive environment. As important as probability and statistics are to the education of an industrial engineer, no attempt will be made to describe the content of such a course in this text. It can be noted in this section on quality control that statistics is a prerequisite topic to quality control.

Probably the most commonly known technique in quality control is the $\bar{X}$ and R chart. This chart is employed when it is desired to maintain statistical control of some single variable of importance for a product. Assume that round bar stock in 20-foot lengths is being fed into a screw machine and one of the tools in the screw machine removes material from the outer diameter of the rod to produce a smaller final outer diameter on the completed piece, as indicated in Fig. 5-13. Assume also that twelve samples of size 4 (i.e., 48 parts) have been produced and measured, producing the data input to Table 5-12.

The $\bar{X}$ and R chart of Fig. 5-14 displays the way the mean and range of the samples varies in time. Assume for the moment that considerable tool wear results from producing dimension A, and consequently successive pieces tend to have larger diameters as the tool wears. Assume also that after the parts for sample 9 were completed the tool was adjusted to account for wear, causing the diameter to be reduced significantly. $\bar{X}$ and R charts are employed to detect when a process variable, in this case dimension A, has gone beyond its normal operating range. The $\bar{X}$ chart is concerned with the mean value for a sample, whereas the R chart is concerned with displaying the dispersion of individual values within a sample as estimated by the sample range.

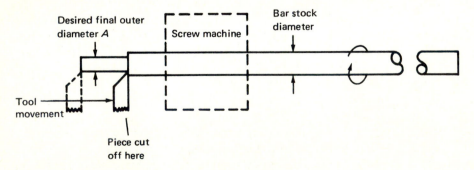

Figure 5-13   Final outer diameter A in a screw machine.

Table 5-12 $\overline{X}$ and R Data and Calculations for Dimension A

	Sample number											
	1	2	3	4	5	6	7	8	9	10	11	12
Diameters, in.	3.5	3.6	3.6	3.6	3.5	3.7	3.7	4.0	4.5	3.5	3.7	3.7
	3.5	3.5	3.7	3.7	3.5	3.8	4.0	4.4	4.6	3.6	3.7	3.6
	3.6	3.6	3.5	3.6	3.6	3.9	4.2	4.5	4.2	3.6	3.6	3.8
	3.5	3.6	3.6	3.6	3.6	3.9	4.1	4.2	4.1	3.5	3.6	3.8
Total	14.1	1.43	14.4	14.5	14.2	15.3	16.0	17.1	17.4	14.2	14.6	14.9
$\overline{X}_i$	3.5	3.6	3.6	3.6	3.6	3.8	4.0	4.3	4.4	3.6	3.7	3.7
High value	3.6	3.6	3.7	3.7	3.6	3.9	4.2	4.5	4.6	3.6	3.7	3.8
Low value	3.5	3.5	3.5	3.6	3.5	3.7	3.7	4.0	4.1	3.5	3.6	3.6
R_i	0.1	0.1	0.2	0.1	0.1	0.2	0.5	0.5	0.5	0.1	0.1	0.2

$$\overline{\overline{X}} = \Sigma \overline{X}_i/n = 45.4/12 = 3.78$$
$$\overline{R} = \Sigma R_i/n = 2.7/10 = 0.27$$
$$\text{UCL}\overline{X} = \overline{\overline{X}} + A_2\overline{R} = 3.78 + (0.729 \times 0.27) = 3.98^*$$
$$\text{LCL}\overline{X} = \overline{\overline{X}} - A_2\overline{R} = 3.78 - (0.729 \times 0.27) = 3.58^*$$
$$\text{UCL}R = D_4\overline{R} = 2.282 \times 0.27 = 0.62^*$$
$$\text{LCL}R = D_3\overline{R} = 0 \times 0.27 = 0^*$$

*Values for A_2, D_3, and D_4 are for samples of size 4 from an ASTM table (1).

The upper control limits (UCL) and lower control limits (LCL) are calculated from historical data for the process, in this example the twelve samples. They establish boundaries on the range of the sample mean and range that would normally (i.e., 99 percent of the time) not be exceeded. When the boundaries are exceeded, as indicated by the circles for the data points of samples 1, 7, 8, and 9, it suggests that a shift of some kind due to an "assignable cause" has occurred, and adjustment is likely to be necessary.

The reason for plotting means of samples rather than individual values goes back to desirable properties of the central limit theorem discussed briefly in Chap. 4. It suffices to say that the reliable determination of control limits depends on the assumption that the means of samples are distributed normally irrespective of the distribution of the individual values from which the samples were drawn.

$\overline{X}$ and R charts have been employed in quality control since Shewhart (6) brought forth these concepts in the early 1930s. Quality control has grown to include a considerable array of statistical techniques for evaluating the quality of manufactured products. A very common, necessary, and time-consuming process in industry is checking the quality of incoming purchased materials.

A 100 percent inspection is much less popular today than in previous times. One reason is the cost; another is the realization today that 100 percent inspection is really a misnomer. To prove it to yourself, count the number of times the letter "e" appears on this page, and then repeat the count a few more times. You will likely find that you do not get the same number each time; 100 percent inspection is difficult to perform reliably.

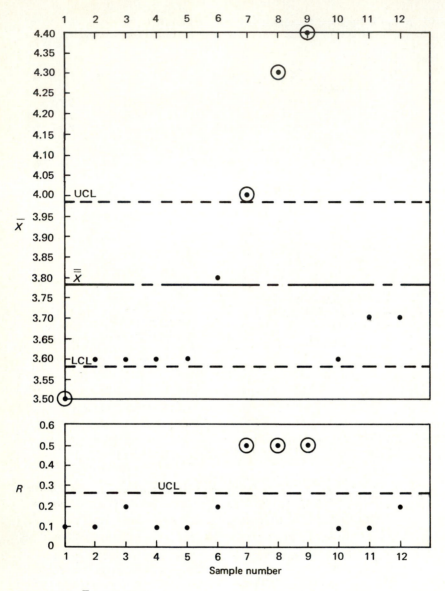

Figure 5-14 $\bar{X}$ and R chart.

For both cost and reliability reasons, then, statistical sampling is commonly employed for evaluating the quality of incoming material. There has been considerable development over the years, particularly in the military, with respect to sampling plans. Table 5-13 demonstrates the use of a typical sampling plan. It must first be assumed that of all sampling plans, this particular one is most appropriate for the sampling task to be performed.

Assume that a batch of 1,000 units is to be sampled to determine whether they

Table 5-13 Master Table for Normal Inspection (Double Sampling)—MIL-STD-105D (ABC Standard)

Acceptable quality levels (normal inspection)

Sample size code letter	Sample	Sample size	Cumulative sample size
A			
B	First / Second	2 / 2	2 / 4
C	First / Second	3 / 3	3 / 6
D	First / Second	5 / 5	5 / 10
E	First / Second	8 / 8	8 / 16
F	First / Second	13 / 13	13 / 26
G	First / Second	20 / 20	20 / 40
H	First / Second	32 / 32	32 / 64
J	First / Second	50 / 50	50 / 100
K	First / Second	80 / 80	80 / 160
L	First / Second	125 / 125	125 / 250
M	First / Second	200 / 200	200 / 400
N	First / Second	315 / 315	315 / 630
P	First / Second	500 / 500	500 / 1,000
Q	First / Second	800 / 800	800 / 1,600
R	First / Second	1,250 / 1,250	1,250 / 2,500

The AQL columns (each with Ac and Re) are: 0.010, 0.015, 0.025, 0.040, 0.065, 0.10, 0.15, 0.25, 0.40, 0.65, 1.0, 1.5, 2.5, 4.0, 6.5, 10, 15, 25, 40, 65, 100, 150, 250, 400, 650, 1,000.

Representative double-sampling acceptance (Ac) and rejection (Re) number pairs (First row / Second row), arranged diagonally across the AQL columns:

	Ac Re (First)	Ac Re (Second)
	0 2	1 2
	0 3	3 4
	1 4	4 5
	2 5	6 7
	3 7	8 9
	5 9	12 13
	7 11	18 19
	11 16	26 27
	17 22	37 38
	25 31	56 57

↓ = use first sampling plan below arrow. If sample size equals or exceeds lot or batch size, do 100% inspection.

↑ = use first sampling plan above arrow.

Ac = acceptance number.

Re = rejection number.

† Use corresponding single sampling plan (or alternatively, use double sampling plan below, where available).

129

meet the acceptable quality level (AQL) of 4.0; this means that if not more than 4 percent of the material is defective, the batch is to be considered acceptable. The AQL then represents the acceptable percent defective for values up to 10. Above 10, the AQL represents the desired upper limit for the acceptable number of defects per hundred units. The most commonly employed general inspection level (i.e., level II) will be assumed to apply in this case, for which sample size code J is recommended. Examination of Table 5-13 indicates that for sample size code J and an AQL of 4, the accept-reject values for a first sample of size 50 are 3 and 7, respectively. If inspection of the random sample of 50 parts from the batch produces 3 or less defects the batch is accepted, if 7 or more defects are found the batch should be rejected, and if 4 to 6 defects are found the sample size should be increased by inspecting 50 more parts. Based on the total of 100 parts inspected, if 8 or less defects are found the batch should be accepted, and if 9 or more defects are found the batch should be rejected. This demonstrates how a typical sampling plan is employed in industry.

The emphasis today in quality control or quality assurance is not only to evaluate the quality of the manufactured product, but also to effectively employ these data to search out the root cause of unacceptable material and correct such undesirable conditions where feasible. Whereas inspection merely sorts good material from bad, quality control is concerned with identifying and correcting the sources of unacceptable products.

Inventory, production, and quality control are all concerned with developing procedures that ultimately affect overall cost control. Some control techniques can only be classified as cost control techniques, however.

COST CONTROL

Accounting is the language of cost control in manufacturing, and to be effective in appreciating cost effects as recorded in accounting one must know the language. Possibly as important as the language are the conventions embedded within the historical tradition of accounting. The recording and reporting of cost has long been a function of accountants, and only those familiar with their ways can properly decipher their results. Industrial engineers have traditionally been required to understand accountants and their methods, both in acquiring an education and in practice. Accounting is one of the dimensions in which industrial engineers have served in the role of translator to technical management, and vice versa.

As one example of accounting convention, it is common in cost accounting to prorate costs to products in some manner if costs were not originally recorded as required by each product in their consumption. Power cost in a plant, for example, may be paid as a single bill to the supplying electric utility by the accounts payable clerk within the accounting organization. In an effort to distribute this cost to reflect product costs for some period, a cost accountant would typically prorate this cost on some convenient basis. The square footage of plant space can readily be calculated for each product in the plant. Consequently, it is not uncommon for a product to bear its ratio of power cost based on the ratio of plant space it occupies.

If products throughout the plant consume an essentially like amount of power per unit of floor space, proration on the basis of floor space may well be appropriate. If, however, as is often the case, some products require a much greater amount of power per unit of space than do other products, proration on the basis of space may distort the relative power cost allocation to products. In such a case measurement of typical power demand in a product area by use of clamp-on portable ammeters could provide a much more equitable basis for power cost allocation. In too many instances, accountants are too far removed from the production process to appreciate inequities they may have inadvertently introduced, and technical management may be too far removed from the allocation of product expense to appreciate the basis for product cost reports they receive. Industrial engineers often play an important role in bringing understanding between such disparate groups.

The following two examples are offered to demonstrate how standard accounting practice sometimes has to be overlooked or reevaluated in the best interests of the plant. The first example involves a company that had purchased a very expensive numerically controlled milling machine. The parts demand for the machine and operating costs were so great that it was decided that the cost of making any part on the machine had first to be compared with outside vendor quotations for producing the part. Only parts that could be produced more cheaply on the in-plant machine would be made on the machine, and all expenses for operating the machine would be borne by the parts being produced on the machine as an indirect charge.

A few years later, one part that constituted approximately one half of the load of the machine was no longer needed. The remaining parts, which had constituted the other half of the load, had to absorb the expense cost of the part no longer needed; consequently, the indirect portion of part cost doubled for the other parts. This so increased their apparent product cost that vendor-produced parts were now sometimes cheaper, resulting in lower and lower utilization of the machine and a higher and higher indirect cost burden for the remaining parts. Ultimately, apparent costs became so high for the remaining parts that management decided to dispose of the machine. Fortunately, this occurred at the same time that another technical group had submitted a proposal to management to purchase a new numerically controlled milling machine. When management reviewed the situation, the loss of one part of the load of the machine was seen to have excessively reflected infeasible in-plant costs for the remaining parts. All that was needed to remedy the situation was to shift some marginal in-plant versus vendor parts back to in-plant production to provide an adequate basis for distributing the indirect expense. The example demonstrates how a make or buy decision may have to vary to suit the specific requirements at a particular time.

The second example shows how a method known as direct costing can aid the profitability of a plant in the short run. Assume for the purpose of this example that product pricing for a plant is calculated in the following manner. To the direct costs of material and labor to produce a product, one half of the cost of labor is normally added to cover all overhead costs, and an additional 10 percent of the total of the above costs is added for profit. A typical product would be priced as follows:

Product A Costs

Unit material cost	$10
Unit labor cost	20
Overhead (50% of unit labor cost)	10
Total manufacturing cost	40
Profit (10% of total manufacturing cost)	4
Unit price	$44

Based on typical accounting methods, if a potential purchaser offers to purchase 1,000 units of product A, which could be produced next month for a unit price of $41 each, the order would not be accepted. If the plant is scheduled at full capacity for next month, this action would seem to be justified.

Assume, however, that the plant is presently scheduled for 70 percent of normal capacity for next month, the order for 1,000 units of product A would bring the plant to 90 percent capacity, and the purchase price offered is less than $44. Should the offer be accepted?

Assume first that the offer is $40 per unit. Why would we accept an offer that does not add to profit? The answer is that in accounting a term called "contribution to profit and overhead" is used. Even though such an order will not produce profits in the traditional sense, it will help pay some of the overhead expenses of the plant; in fact, to the amount of 1,000 × $10 = $10,000.

To stretch the point a bit further, assume that the offer is $15 per unit; could there be any logical reason for accepting the offer? The answer is yes. If the plant has been operating at 90 percent capacity, dropping to 70 percent capacity would suggest that we either have to lay off employees next month for lack of work or pay them for doing nothing. Many employees today are so highly skilled that an employer wants to maintain full employment; otherwise, the employer has to run the risk of losing trained employees and has to suffer the considerable cost of retraining new ones if they can be found. If the employer intends to keep all the employees next month, the $15 unit purchase cost can be viewed as offering a $5 per unit "contribution to labor, overhead, and profit" to the extent of 1,000 × $5 = $5,000.

Of course, an order for 1,000 units at a $15 purchase price would only be accepted to keep the plant at near full capacity over the short run. The sales manager would likely make it clear to the lucky purchaser that this would be a one-time deal that might never happen again, and why. For the purchaser, it would be a little like Christmas in July. In the final analysis, they both could conceivably come out ahead on the deal.

The concepts mentioned here are certainly not unknown to accountants. The purpose of this discussion is to demonstrate that normal or standard accounting approaches need to be adjusted to fit the occasion, and that the industrial engineers who bridge the gap that sometimes exists between accounting or sales and technical management can sometimes see the need for adjustments because of their unique vantage point. This may partially explain why industrial engineers in IBM manufacturing plants have been referred to as the "business managers of the plant."

A more detailed coverage of cost control could include such topics as break-even charts and engineering economy (sometimes referred to as discounted cash flow methods), or a number of other related topics. For the purpose of this text, the topic of cost control is concluded here.

SUMMARY

Inventory, production, quality, and cost control are all part of the control system design responsibility. First a plant is designed, as discussed in Chap. 4, then procedures are devised for operating or controlling the design in an optimal manner. Some mathematical techniques have been developed in recent years which concentrate, in particular, in identifying optimal procedures. Operations research, the topic of the next chapter, can be thought of as a field primarily concerned with the application of optimizing procedures and approaches to productive systems.

REFERENCES

1 *ASTM Manual on Quality Control of Materials*, p. 115, American Society for Testing and Materials, Philadelphia, Pa.
2 "Hallmark's Brainy Warehouse," *Fortune,* pp. 91–93, August, 1973.
3 Magee, John F., and David M. Boodman: *Production Planning and Inventory Control,* McGraw-Hill Book Company, New York, 1967.
4 Malcolm, D. G., J. H. Rosenbloom, C. E. Clark, and W. Fazar: "Applications of a Technique for R & D Program Evaluation (PERT)," *Oper. Res.,* vol. 7, no. 5, pp. 646–669, 1959.
5 Moodie, C. L., and H. H. Young: "A Heuristic Method of Assembly Line Balancing for Assumptions of Constant or Variable Work Elemental Times," *J. Ind. Eng.,* vol. 16, no. 1, January-February, 1965.
6 Shewhart, Walter A.: *Economic Control of Quality of Manufactured Product,* D. Van Nostrand Co., New York, 1931.
7 Walker, M. R., and J. Sayer: "Project Planning and Scheduling," Report no. 6959, E. I. DuPont de Nemours, Inc., Wilmington, Del., March, 1959.

REVIEW QUESTIONS AND PROBLEMS

1 What is the fundamental purpose of maintaining an inventory?
2 a Using the Harris model, determine the EOQ for the following data:

$A = 10,000$ units
$H = \$5$
$P = \$10$
$M = \$4$

   b What is the optimum number of purchases per year N_0?
   c Plot a figure similar to Fig. 5-2 for the data above.
3 Assuming that the following data represent the number of days of lead time L that occurred on the last 50 orders that were placed, and using the fifth and sixth columns of Table 5-3 from the top down, develop a sequence of ten typical lead times using the Monte Carlo technique.

Days of lead time L	Frequency
3	20
4	17
5	6
6	4
7	3
	50

4 Determine the optimum lot size to run on a machine for the following data:

$A = 1,000$ units
$S = \$200$
$H = \$4$
$D = 10$ units/day
$P = 50$ units/day

5 **a** Determine an optimum joint number of runs per year N_0, assuming 250 production days per year and given the following product data for a machine:

Product	Annual requirement A_i	Daily production rate P_i	Annual unit holding cost H_i	Setup cost S_i
A	10,000	200	$1.50	$200
B	8,000	100	0.80	350
C	4,000	450	0.70	400
D	8,000	300	0.30	250
E	6,000	200	0.60	325
F	1,000	250	0.50	170

b Should any of the products be run every other cycle?

6 Assume that 10,000 minutes of production time are available to produce 1,000 units on a production line. The precedence relationships and times for the sixteen production elements are as indicated below.

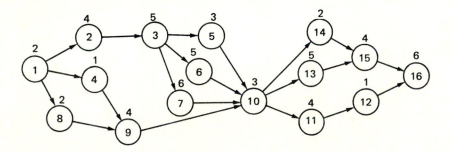

Determine station assignments of elements that will result in an optimum number of stations on the production line.

7 a The activities and their times and precedence relationships are indicated below for building and testing a boat.

Activity symbol	Activity	Time (days)	Precedence
A	Construct hull	10	
B	Construct cross beams	4	A
C	Attach transom	4	A
D	Lay the deck	5	B, C
E	Install the mast	2	B
F	Paint the boat	4	D
G	Install hardware	5	F
H	Install the engine	2	F
I	Test for leaks	2	H
J	Trial run	1	E, G, I

Which activities are on the critical path?

b If construction of cross beams is delayed a week, what effect will it have on the overall completion time?

8 The following data for ten samples of size 4 are available for initiating an $\bar{X}$ and R chart:

				Sample number					
1	2	3	4	5	6	7	8	9	10
6.5	5.3	6.2	6.1	7.2	7.3	6.8	6.3	7.2	8.2
6.8	6.2	6.3	5.9	6.9	6.9	6.9	9.1	8.4	7.2
7.3	7.1	6.8	5.8	5.7	6.2	5.9	7.2	9.2	6.8
6.4	7.4	5.8	6.4	5.9	6.4	8.1	6.8	8.1	6.3

Prepare an $\bar{X}$ and R chart for these data identifying the upper and lower control limits, and circle any sample means or ranges that are outside the control limits.

9 What is an AQL?

10 What is meant by direct costing?

Chapter 6

Operations Research

Evolution is progress from an indefinite, incoherent homogeneity toward a definite, coherent heterogeneity.

Herbert Spencer

There is a wide range of opinion as to just what operations research is or will be. It is not uncommon for industrial engineering traditionalists, particularly if they have been successful in the use of traditional industrial engineering techniques, to view operations research as possibly interesting but unnecessary most of the time. They may honestly believe that if operations research were employed, it would likely be only partially successful, if at all. This attitude is certainly not unique to industrial engineering; it cuts across all fields. It is a human trait commonly referred to as "resistance to change." Unfortunately, they would probably be right at least some of the time on the basis of many past applications of operations research. But more important, they would be as correct in their criticism of operations research as a bystander viewing the first flight of Orville and Wilbur Wright's Kitty Hawk would have been in evaluating the airplane as a competitive form of transportation. There is little question that operations research has been poorly applied in some instances in the past, but misapplication does not lessen the ultimate utility of operations research techniques.

136

The industrial engineering manager who refuses to allow operations research to be attempted under favorable circumstances may be correct in assuming that it will not be successful in that particular area. The manager who gambles all on operations research, but is too busy to take the time to understand its use, may also prove that operations research can be unsuccessful. However, the industrial manager who appreciates what operations research can do, and applies it well in situations for which it represents the appropriate viable technique, will make impressive inroads into improved design and operation of productive systems. Operations research is one of the forefronts of industrial engineering.

Those most unfamiliar with operations research often view it as a collection of techniques (i.e., linear programming, dynamic programming, queueing theory, and so forth). Operations research has failed occasionally in the past because someone familiar with an operations research technique tried to use it in a situation for which it was inadequate. Successful application often depends on either modifying an existing technique to overcome its inherent weakness, or developing a new technique. As a result, far too many operations research applications to date can be viewed as somewhat analogous to "cutting toothpicks with an axe." Unfortunately, it is far easier in practice to apply an existing technique than to improve on it. It takes considerably more ingenuity—sometimes more than exists at a particular place and time—to devise an appropriate operations research solution to a complex problem.

What follows is an attempt to demonstrate representative operations research techniques that are in common use today. Those selected are meant to be representative rather than exhaustive, and the selection has been limited to those within the likely mathematical capability of the intended reader of this text.

LINEAR PROGRAMMING

The one technique most commonly identified with operations research is linear programming. Dantzig (3) would be many operations researchers' choice for "father of operations research." His development of an algorithm (i.e., the simplex method) for solving linear programming problems is likely the most significant development to date in operations research. The generalized linear programming approach, for which the simplex method provides a solution technique, is of such generality that it is useful in solving resource allocation problems, which fit its assumptions, across a broad spectrum of application areas.

Before illustrating the generalized approach, two special cases of linear programming, the assignment algorithm and the transportation algorithm, will be illustrated. These two algorithms can be used more efficiently than the generalized approach for problems that possess the specific characteristics of these special cases.

The Assignment Algorithm

The assignment algorithm provides a means of assigning n resources to n tasks so as to maximize or minimize the sum of effectiveness of all task assignments. Consider the following example.

A department head would like to assign each of four subordinates to four assignments. An estimate of the number of days each worker would spend on each assignment with equal ultimate task effectiveness is as follows:

		Workers			
		I	II	III	IV
Tasks	A	6	7	10	9
	B	2	8	7	8
	C	8	9	5	12
	D	7	11	12	3

Table 6-1 is a three-phase assignment algorithm. The first step in attempting a solution to the problem is the application of phase 1 of the algorithm. Row reductions for the above problem would produce the following:

		Workers			
		I	II	III	IV
Tasks	A	0	1	4	3
	B	0	6	5	6
	C	3	4	0	7
	D	4	8	9	0

Column reductions would then produce

		Workers			
		I	II	III	IV
Tasks	A	0	[0]	4	3
	B	[0]	5	5	6
	C	3	3	[0]	7
	D	4	7	9	[0]

A complete assignment, one worker to each of the four tasks, can be made as indicated above, using only zero elements. The optimal assignment of workers to tasks referring to the initial matrix would be

Task	Worker	Task time (days)
A	II	7
B	I	2
C	III	5
D	IV	3
		17

No other assignment of the four workers, one each to a task, can be made with a sum of task times of less than 17 days.

In matrices of a larger magnitude, a complete assignment may well exist using only zero elements as above, but the larger number of zeros requires a procedure for identifying the set of zeros, one in each row and column, that represents the

Table 6-1 An Assignment Algorithm[*]

Phase 1

1 Subtract the smallest value in each row (or column) from all other elements in the row (or column) until there is at least one zero element in each column and row.

Phase 2

1 Examine rows successively until a row with exactly one unmarked zero is found. Mark (□) this zero, as an assignment will be made there. Mark (X) all other zeros in the column to show that they cannot be used to make other assignments. Proceed in this fashion until all rows have been examined.
2 Next, examine columns for single unmarked zeros, marking them (□) and also marking with an (X) any other unmarked zeros in their rows.
3 Repeat steps 1 and 2 successively until one of three things occurs:
   a There are no zeros left unmarked and we have a complete assignment.
   b The remaining unmarked zeros lie at least two in each row and column and we must complete by trial and error.
   c We do not have a complete assignment and must proceed to phase 3.

Phase 3

1 Mark all rows for which assignments have not been made.
2 Mark columns not already marked which have zeros in marked rows.
3 Mark rows not already marked which have assignments in marked columns.
4 Repeat steps 2 and 3 until the chain of marking ends.
5 Draw lines through all unmarked rows and through all marked columns.
6 Select the smallest element not covered by a line and:
   a Subtract its value from all uncovered elements.
   b Add its value to all assignments covered by two lines.
   c Repeat the previous value for all elements covered by one line.
   d Return to phase 2 with the resulting matrix.

[*]A modification of algorithms in Maurice Sasieni et al.: *Operations Research: Methods and Problems,* pp. 186–192, John Wiley & Sons, Inc., New York, 1959.

complete assignment. In these cases, phase 2 of the algorithm can be used to attempt to identify a complete assignment from among existing zero elements.

The following reduced matrix, considering zero elements only, can be used to illustrate phase 2 of the algorithm:

	I	II	III	IV	V
A		0			
B	0	0			
C	0		0		0
D	0		0		0
E			0	0	0

Application of steps 1 and 2 of phase 2 would result in the following:

	I	II	III	IV	V
A		[0]			
B	[0]	✗			
C	✗		0		0
D	✗		0		0
E			✗	[0]	✗

As indicated in step 3b of phase 2, arbitrary assignment of C to III and D to V results in the following optimal solution:

	I	II	III	IV	V
A		[0]			
B	[0]	✗			
C	✗		[0]		0
D	✗		0		[0]
E			✗	[0]	0

To illustrate phase 3 of the algorithm, assume a solution is needed for the following reduced matrix:

	I	II	III	IV	V
A	5	0	8	10	11
B	0	6	15	0	3
C	8	5	0	0	0
D	0	6	4	2	7
E	3	5	6	0	8

Application of phase 2 would produce

	I	II	III	IV	V
A	5	[0]	8	10	11
B	⊠	6	15	⊠	3
C	8	5	[0]	⊠	⊠
D	[0]	6	4	2	7
E	3	5	6	[0]	8

Note that this does not represent a complete assignment. Application of steps 1 through 5 of phase 3 results in the following:

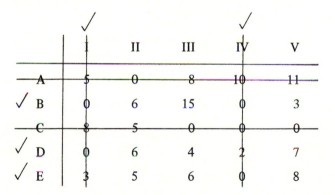

	I	II	III	IV	V
A	5	0	8	10	11
✓ B	0	6	15	0	3
C	8	5	0	0	0
✓ D	0	6	4	2	7
✓ E	3	5	6	0	8

Note that element BV is the smallest element not covered by a line. The application of step 6 of phase 3 produces the following:

	I	II	III	IV	V
A	8	0	8	13	11
B	0	3	12	0	0
C	11	5	0	3	0
D	0	3	1	2	4
E	3	2	3	0	5

Application of phase 2 to this revised matrix produces

	I	II	III	IV	V
A	8	[0]	8	13	11
B	⦻	3	12	⦻	[0]
C	11	5	[0]	3	⦻
D	[0]	3	1	2	4
E	3	2	3	[0]	5

The complete optimal assignment then is

A–II
B–V
C–III
D–I
E–IV

The assignment algorithm shown above is a minimization algorithm. To produce an optimum as a maximum it is only necessary to replace element values with complement values determined by subtracting each element from the largest element value in the original matrix. The following example will illustrate the procedure.

Assume that the expected profit from assigning ice cream sellers to four districts is as follows:

		Sellers			
		I	II	III	IV
	A	16	14	12	16
Districts	B	12	18	10	15
	C	13	16	14	20
	D	10	12	15	14

To produce a maximum, each element is subtracted from 20, which is the largest element value in the matrix, producing complement values as follows:

	I	II	III	IV
A	4	6	8	4
B	8	2	10	5
C	7	4	6	0
D	10	8	5	6

Phase 1 row and column reductions would then produce

	I	II	III	IV
A	[0]	2	4	0
B	6	[0]	8	3
C	7	4	6	[0]
D	5	3	[0]	1

Therefore, the maximal assignment would be

A–I
B–II
C–IV
D–III

This assignment produces a total profit of 69. No other assignment of the four ice cream sellers to the four districts will produce a greater profit. The generality of the technique should be apparent at this point. The assignments could represent the sum of effectiveness of any one of the following types of assignments:

1   Four nurses to four stations in a hospital
2   Four traffic controllers to four different air traffic tasks
3   Four police cars to four crime districts
4   Four heavy equipment operators to four pieces of equipment
5   Four pieces of equipment to produce four different products
6   Four bank auditors to four different types of banks in a district

It should be apparent that this list could be essentially infinite because of the level of generalization of the technique, in assigning n "resources" to n "tasks."

As an example of the use of the assignment method, consider an enterprising restaurateur who intends to attempt the following new concept in the restaurant business. The restaurateur has four locations in a city (i.e., W, X, Y, and Z). Three kitchens are to be located at three of the four locations, with one location remaining empty. Four kitchens will be labelled M, C, P, or N to represent Mexican food, Chinese food, pizzas, or no food, respectively. The restaurateur intends to produce the three different types of food in the three separate kitchens and deliver the prepared foods from all three kitchens to all five of the restaurant locations (i.e., 1, 2, 3, 4, and 5) throughout the city. The four available kitchen locations and five restaurant locations are shown in Fig. 6-1. Locations are shown with respect to the nearest street crossing to simplify distance calculations by employing unit block increments (i.e., there are three block distances between locations X and 1, and five between Z and 1). The relative number of trips that would be needed between kitchens and restaurants over some defined period of time is given in Table 6-2. The restaurateur would like to

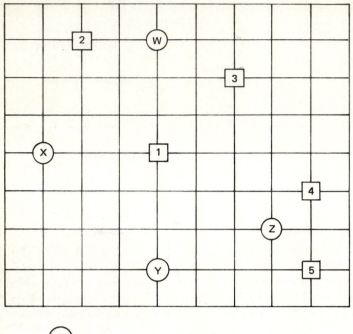

○ Candidate kitchen locations

▢ Present restaurant locations

Figure 6-1 Kitchen and restaurant locations.

locate the kitchens, one at a location, so that the sum of delivery distances from all kitchens to all restaurants is a minimum.

This problem is analogous to one proposed by Moore (10) in locating new machines at candidate locations in a factory so as to minimize the material handling distance for all products flowing between existing and new machines. Figure 6-2*a* illustrates the matrix multiplication of trips and distances to produce a matrix that shows the effect on distance of locating each kitchen at each possible location. An

Table 6-2   Relative Number of Trips Required Between Kitchens and Restaurants Over Some Defined Period

		Restaurants				
		1	2	3	4	5
	M	10	12	4	8	6
Kitchens	C	3	5	10	12	4
	P	7	3	5	4	8
	N	0	0	0	0	0

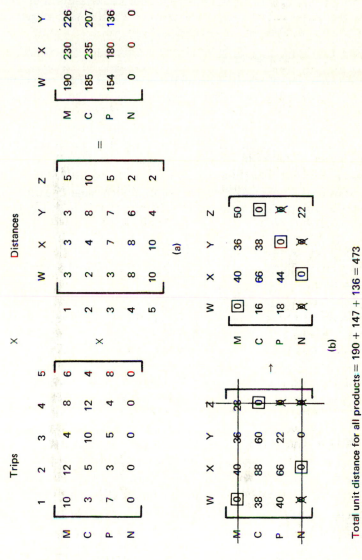

Total unit distance for all products = 190 + 147 + 136 = 473

Figure 6-2 Matrix multiplication of trips and distances and assignment method solution.

assignment method solution of this matrix, as shown in Fig. 6-2*b*, identifies the best location for each kitchen to minimize the total distance traveled for all required deliveries (i.e., 473 unit block distances). No other assignment of kitchens to possible locations will result in a smaller total distance. This example suggests the generality of the technique with respect to potential applications.

The Transportation Algorithm

Another very useful special case of the generalized linear programming problem is commonly referred to as the transportation problem. The transportation algorithm is employed when m sources are supplying n destinations, and individual cost coefficients per unit of flow between each source and destination are known and are linear as a function of the volume of flow. It is desired to allocate units from specific sources to specific destinations to result in a least-cost solution for all products delivered. The following example illustrates a typical simple transportation problem.

How many tons should be shipped from each mine to each processing plant so that mine capacities are not exceeded, each plant receives its required tonnage input,

Table 6-3 A Transportation Algorithm

Phase 1 (Vogel's approximation)

1 Develop a penalty rating value for each row and column, which is the difference between the least-cost and the next-to-least-cost elements in each row (or column). Make maximum possible assignments in decreasing order of the size of the penalty rating until all assignments are made.

Phase 2

1 Obtain an initial assignment by some prior method (e.g., Vogel's approximation).
2 Determine U_i and V_j values for rows and columns starting with $U_1 = 0$. Determine remaining U's and V's by using the equation

$$U_i + V_j = C_{ij}$$

employing only cells in which an assignment has been made.
3 Determine t_{ij} values for all remaining cells by using U's and V's developed in step 2 above as follows:

$$t_{ij} = U_i + V_j - C_{ij}$$

4 If all t_{ij}'s are zero or negative, an optimal solution has been reached. If there are positive t_{ij}'s, bring the largest positive t_{ij} into solution by employing step 5.
5 Identify a "transfer route" starting with the largest positive t_{ij} above, moving alternately up or down and left or right, on assigned cells only, returning to the largest positive t_{ij} cell above.
6 Alternately add and subtract a value θ from the assigned quantities in cells identified in step 5. The value θ is equal to the smallest quantity presently assigned to cells from which θ is to be subtracted. When θ is added or subtracted to cell quantities, one new assignment is created in the cell that possessed the largest positive t_{ij}, and one of the cells in which the quantity was reduced to zero should no longer possess an assignment.
7 Return to step 2.

and a least cost of delivered ore results? Shipping cost per ton between specific mines and plants, mine capacities, and plant requirements are assumed to be as follows:

Cost of shipping to plant
($/ton)

		A	B	C	D	
	1	$3	$1	$4	$5	50
Mine	2	7	3	8	6	50
	3	2	3	9	2	75
		40	55	60	20	175

Mine capacities (tons)

Plant requirements (tons)

Table 6-3 contains the transportation algorithm that will be employed in solving this problem. Phase 1 of the algorithm is generally known as Vogel's approximation. It permits determination of a good initial starting point as a first step in searching for an optimal solution. Phase 2 permits ultimate determination of an optimal solution to any transportation problem.

The setup for the above problem would be as follows:

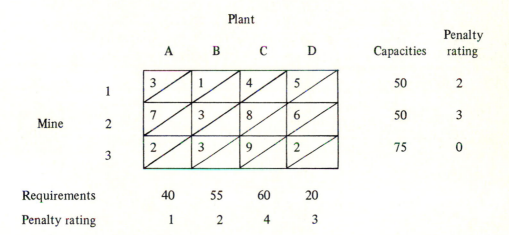

		A	B	C	D	Capacities	Penalty rating
	1	3	1	4	5	50	2
Mine	2	7	3	8	6	50	3
	3	2	3	9	2	75	0
Requirements		40	55	60	20		
Penalty rating		1	2	4	3		

The maximum penalty rating is 4; therefore, as much material will be allocated to the lowest-cost cell in this row or column as possible. This heuristic simply allocates material in order to preclude suffering high penalty costs associated with second-best allocations. In this instance the difference between $4 per ton and $8 per ton makes it desirable to move as much material as possible at $4 per ton with respect to plant C. The limit of how much can be allocated to cell 1C is determined by the lesser of (1) the capacity of mine 1, 50 tons, and (2) the requirement of plant C, 60 tons. Therefore, the next matrix below shows no additional capacity remaining at mine 1, and a remaining requirement at plant C of 10 tons, as follows:

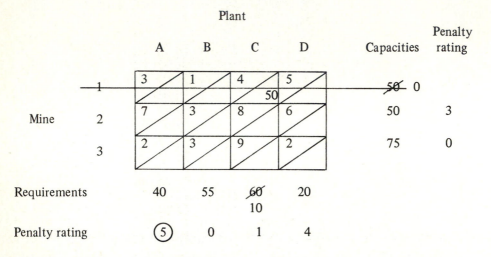

Recalculation of the penalty ratings, as shown on the above matrix, is necessary because row 1 has been removed from further consideration due to the allocation of all of the output of mine 1 to plant C. Because of the new penalty rating of 5 with respect to plant A, as indicated above, a maximum allocation is next made to this plant. Because 40 is less than 75, 40 tons are allocated to plant A from mine 3, reducing the remaining capacity from mine 3 to 35 tons and removing plant A from further consideration by reducing its remaining requirement to zero, as follows:

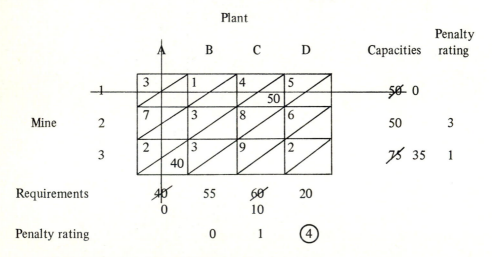

This process is continued until all mine capacity is allocated to meet the requirements of all plants, as shown below:

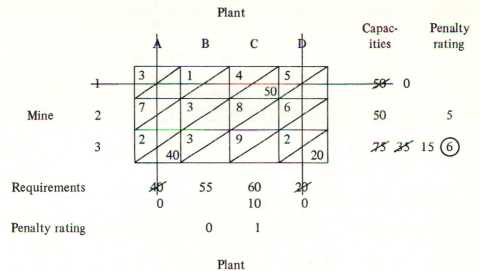

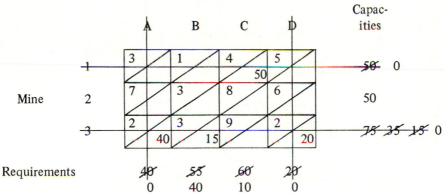

The only possible remaining assignments are 40 tons from mine 2 to plant B and 10 tons from mine 2 to plant C, as follows:

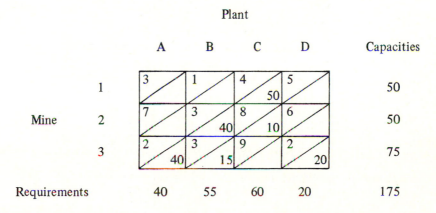

The total cost of shipping the 175 tons from the three (m) mines to the four (n) plants would be

Mine to plant	Cost per ton	Tons	Cost
1–C	$4	50	$200
2–B	3	40	120
2–C	8	10	80
3–A	2	40	80
3–B	3	15	45
3–D	2	20	40
			$565

The above assignment, obtained by employing Vogel's approximation, happens to be optimum. The test for optimality can be made by continuing to phase 2. The use of phase 2 will either prove that the application of phase 1 produced an optimal solution, as was the case above, or permit improvement in allocations until the optimum is reached.

Assume that phase 1 had produced the following allocation:

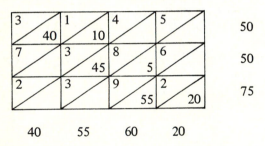

Then U's and V's would be added as follows, employing steps 1 and 2 of phase 2:

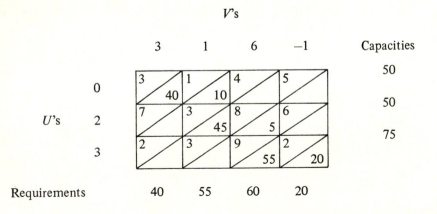

The determination of t_{ij} for the remaining cells, as indicated in step 3 of phase 2, would add to the above matrix as follows:

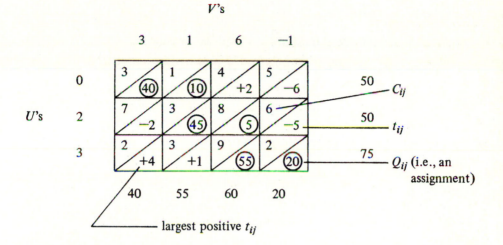

largest positive t_{ij}

Step 4 of phase 2 is the test for optimality. There is a positive nonzero t_{ij} in the matrix above; therefore, optimality has not been reached and reallocation is necessary. Only assigned cells and the cell with the largest t_{ij} are considered in the reallocation, as indicated below:

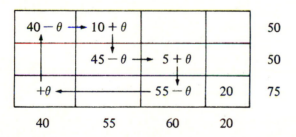

The quantity to be reallocated is θ. Starting by assigning $+\theta$ to the cell with the largest t_{ij}, one alternately adds $-\theta$ and $+\theta$ around the loop by successive vertical and horizontal transfers, returning ultimately to the t_{ij} cell. Thus, any quantity θ added to a row or column will be negated by a $-\theta$ quantity in the same row or column. Row 1 will still possess 50 allocated units, for example, regardless of the size of θ. Since reallocation of units will improve the solution as indicated by a positive t_{ij} in a candidate cell, a maximum reallocation is made to this cell. The maximum quantity is determined by the $-\theta$ cell in the loop with the smallest previous assignment. In this example, 40 is less than 45 or 55; therefore, 40 units will be reallocated. Obviously, if θ were larger than 40, the $40 - \theta$ cell would become a negative quantity, which would represent a nonfeasible solution. The reallocation, therefore, would result in assignments as follows:

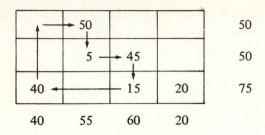

Applying steps 2 through 7 until an optimal solution is found produces the following sequence of matrix operations:

V's

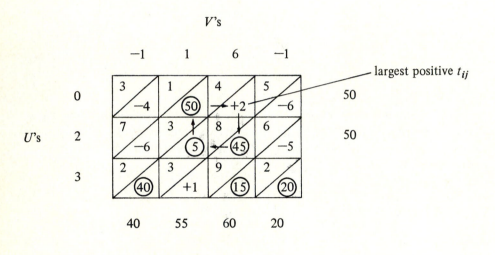

largest positive t_{ij}

	$50 - \theta$ → θ		
	$5 + \theta$ ← $45 - \theta$			$\theta = 45$
40		15	20

	5	45	
	50		
40		15	20

V's

	-3	1	4	-3

		3	1	4	5
U's	0	-6	⑤	㊺	-8
	2	7	3	8	6
		-8	㊾	-2	-7
	5	2	3	9	2
		㊵	$+3$	⑮	⑳

	$5 - \theta \longrightarrow 45 + \theta$		
	↑ ↓		$\theta = 5$
	$\theta \longleftarrow 15 - \theta$		

V's

	-3	-2	4	-3

		3	1	4	5
U's	0	$+0$	-1	㊿	-8
	5	7	3	8	6
		-5	㊿	$+1$	-8
	5	2	3	9	2
		㊵	⑤	⑩	⑳

	$50 - \theta \longrightarrow \theta$		$\theta = 10$
	↑ ↓		
	$5 + \theta \longleftarrow 10 - \theta$		

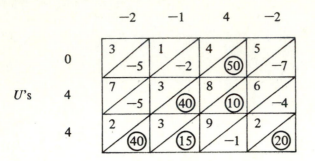

All t_{ij}'s in the above matrix are nonpositive; therefore, an optimal assignment has been reached. A number of iterations were required in this example because the assumed initial assignment was a relatively poor one.

As in the case of the assignment algorithm, a minimization solution of complement cell values can be employed to produce a maximum solution. Assume, for example, that the profit received from sending a unit of product from source i to destination j was as follows:

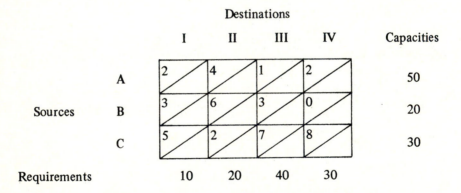

Minimization of the following complement values would determine the best allocation to produce a maximum profit:

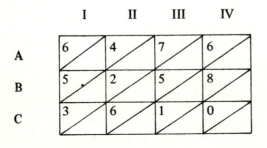

An alternate method is to bring cells into solution that have the largest negative t_{ij} until all t_{ij} values are nonnegative; this also produces a maximum assignment.

In some cases it is necessary to include a dummy (i.e., imaginary) source or destination in order to solve a transportation problem. Assume that in the prior example concerning three mines serving four plants that the requirement for plant C was only 40 tons. The matrix would be as follows:

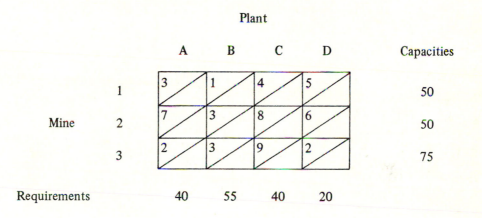

Note that the total capacity exceeds the total requirement by 20 tons. The transportation algorithm requires that the total requirement equal the total capacity; therefore, a nonexistent plant E is imagined to require 20 tons with shipping cost coefficients of zero for all mines. Actually, dummy plant E represents the 20 tons of capacity that will never leave the mines. The matrix would be as follows:

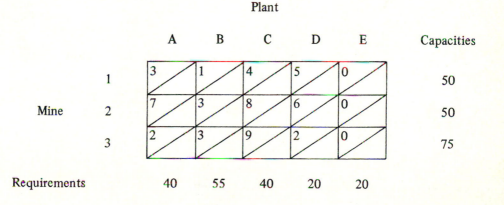

Use of phase 1 produces the following assignment:

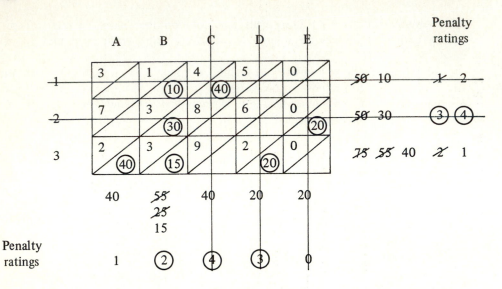

The test for optimality, step 4 of phase 2, indicates that an optimum has been reached, as follows:

The complete assignment would therefore be

Mine to plant	Cost per ton	Tons	Cost
1–B	$1	10	$ 10
1–C	4	40	160
2–B	3	30	90
2–E	0	20	0
3–A	2	40	80
3–B	3	15	45
3–D	2	20	40
			$425

Table 6-4 Dishwasher Problem Information

Items	Worker-seconds per unit			Daily requirement (units)
	New automatic	Old automatic	Hand dishwashing	
Pots	5	10	25	50
Plates	2	6	20	250
Silverware sets	5	12	25	200
Frying pans	10	13	30	20
Glasses	3	4	10	200
Cups	1	3	10	250
Capacity (units)	550	350	70	970

The transportation technique is also a very general one, with a broad range of potential applications. For example, assume that dishwashing in a small hospital can be performed by employing any one of the three techniques shown in Table 6-4. The numbers of units normally requiring washing each day are indicated as requirements; and the capacities of the new and old automatic dishwashers are given as 550 and 350 units, respectively. Because the total requirement is for 970 units, 70 units are indicated as a capacity for hand dishwashing to equate total capacity to total requirement.

Assume that it is desired to minimize total worker-seconds associated with the dishwashing function. Figure 6-3 illustrates the solution to the assignment of the new dishwasher, old dishwasher, and the hand dishwasher to specific items requiring washing. This allocation of the available resources results in a minimum of approximately 1.07 worker-hours associated with the dishwashing function, as indicated in Table 6-5.

The Generalized Linear Programming Problem

There is a general class of problems that require maximization or minimization of an objective function represented by a sum of products of linear nonnegative variables and associated coefficients. The variables are also constrained by a set of linear equalities or inequalities relating them in a specific way. The general formulation of a linear programming problem can be explicitly stated as follows:

Minimize (or maximize) $f(x) = C_1 X_1 + C_2 X_2 + \ldots + C_n X_n$

Subject to:

$$a_{11} X_1 + a_{12} X_2 + \ldots a_{1n} X_n \lesseqqgtr b_1$$

$$a_{21} X_1 + a_{22} X_2 + \ldots a_{2n} X_n \lesseqqgtr b_2$$

$$a_{m1} X_1 + a_{m2} X_2 + \ldots a_{mn} X_n \lesseqqgtr b_m$$

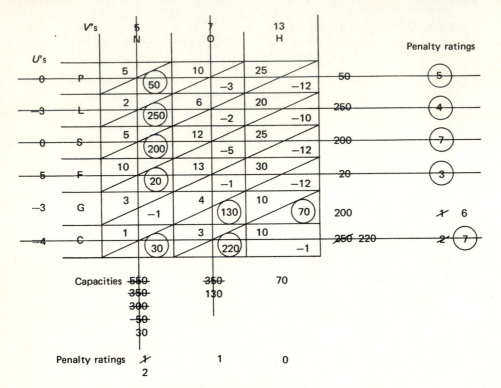

Figure 6-3 Assignment solution to dishwasher problem.

and

$$X_1, X_2, \ldots X_n \geqslant 0$$

The following example will indicate the type of problem involved. Assume that products A, B, and C can be sold for $5, $10, and $20 per unit, respectively. The unit costs of four possible resources, called inputs 1, 2, 3, and 4, for the manufacture of each product are $2, $1, $0.5, and $2, respectively, for the four inputs. Limitations on

Table 6-5 Dishwashing Worker-Hours Required

Total worker-seconds required	
50 × 5 = 250	30 × 1 = 30
250 × 2 = 500	130 × 4 = 520
200 × 5 = 1,000	220 × 3 = 660
20 × 10 = 200	70 × 10 = 700
	3,860 seconds

$$\frac{3{,}860}{3{,}600} \simeq 1.07 \text{ hours}$$

supply of inputs 1, 2, 3, and 4 are 100, 200, 400, and 100 units, respectively. Also, the following combinations of units of input are required to produce a unit of product A, B, and C:

Products

		A	B	C
	1	0	1	2
Inputs	2	1	2	1
	3	4	6	10
	4	0	0	2

The specific formulation of this problem would be as follows:

Maximize $f(x) = 2X_1 + 3X_2 + 6X_3$ (objective function)

Subject to the constraint equations:

$$X_2 + 2X_3 \leqslant 100 \tag{1}$$

$$X_1 + 2X_2 + X_3 \leqslant 200 \tag{2}$$

$$4X_1 + 6X_2 + 10X_3 \leqslant 400 \tag{3}$$

$$2X_3 \leqslant 100 \tag{4}$$

and

$$X_1, X_2, X_3 \geqslant 0 \tag{5, 6, 7}$$

The first coefficient in the objective function above (i.e., 2), denoted as C_1 in the general formulation, is the profit per unit of product A, where X_1 is the number of units of product A to be produced. It is determined as follows:

C_1 = product A unit profit = product A unit revenue − sum of product A unit input costs

$= 5 − [1(1) + 4(0.5)]$

$= \$2$ profit per unit of product A sold

Coefficients for C_2 and C_3, \$3 and \$6 respectively, were determined in a similar manner. The first constraint equation, Eq. (1) above, says that there is a limit to how many of products B and C can be made because each unit of product B requires one unit of input 1, and each unit of product C requires two units of input 1, and there are only 100 units of input 1 available. Equations (2) through (4) similarly constrain the use of inputs 2, 3, and 4; and Eqs. (5) through (7) are called nonnegativity constraints, requiring that the values of X_1, X_2, and X_3 be positive.

The simplex algorithm developed by Dantzig (3) is normally used to determine the values of $X_1, X_2, \ldots X_n$ that meet the constraints imposed, and maximize (or minimize) the objective function involved. In actual practice today, it is not necessary to manually employ the simplex method; there is considerable programming capability for solving linear programming problems. It is only necessary to be able to construct the objective function and constraint equation set and input them into any one of a variety of available linear programming programs, such as the IBM MPS (mathematical program systems) program (9).

A two-dimensional problem can be graphed to illustrate the general way in which the simplex algorithm successively approaches a maximum (or minimum) solution. The following example illustrates this general procedure:

Maximize $f(x) = X_1 + X_2$

Subject to:

$$2X_1 + X_2 \leqslant 12 \tag{1}$$
$$X_1 + 2X_2 \leqslant 16 \tag{2}$$
$$2X_1 \leqslant 8 \tag{3}$$

and

$$X_1, X_2 \geqslant 0$$

Note that by assuming the constraint equations to be equalities it is possible to plot them, and then define the region included by the inequality, as follows:

Assume

$$2X_1 + X_2 = 12 \tag{1}$$
$$X_1 + 2X_2 = 16 \tag{2}$$
$$2X_1 = 8 \tag{3}$$

For constraint (1) above

$$\text{if } X_1 = 0, \quad X_2 = 12$$
$$\text{and if } X_2 = 0, \quad X_1 = 6$$

as indicated in Fig. 6-4. Linear constraint lines to represent constraints (2) and (3) can be determined in a similar manner and are also shown in Fig. 6-4. The shaded area represents the common feasible solution zone because the three equations are in fact "less than or equal" constraints. That is, any point in the shaded area of Fig. 6-4 represents a feasible solution with respect to all the constraint equations.

To determine a solution to the problem it is necessary to identify the point or points in the feasible zone (i.e., the shaded area) that will produce a maximum for the objective function.

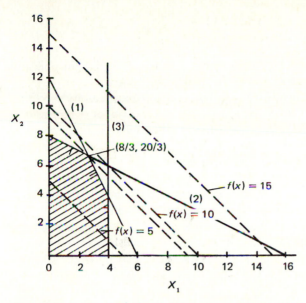

Figure 6-4 Graphical solution to linear programming problem.

To gain insight into how the simplex algorithm works, at least conceptually, lines can be added to Fig. 6-4 for values of the objective function $f(x) = 5$, 10, and 15 determined as follows:

$$
\begin{aligned}
&\text{if } f(x) = 5 && \text{and } X_1 = 0, X_2 = 5 \\
&&& \text{and if } X_2 = 0, X_1 = 5 \\
&\text{if } f(x) = 10 && \text{and } X_1 = 0, X_2 = 10 \\
&&& \text{and if } X_2 = 0, X_1 = 10 \\
&\text{if } f(x) = 15 && \text{and } X_1 = 0, X_2 = 15 \\
&&& \text{and if } X_2 = 0, X_1 = 15
\end{aligned}
$$

After plotting lines representing $f(x) = 5$, 10, and 15, it can be noted that they are parallel lines moving out from the origin for increasing values of $f(x)$. Note that a line parallel to the $f(x) = 10$ line would appear to possess a maximum objective function value of approximately 9 where constraints (1) and (2) intersect. Their intersection, then, determines the values of X_1 and X_2 that maximize the objective function. Solving these two equations simultaneously yields the following:

$$2X_1 + X_2 = 12 \tag{1}$$

$$2X_1 + 4X_2 = 32 \tag{2 $\times$ (2)}$$

$$\overline{\qquad 3X_2 = 20 \qquad}$$

$$X_2 = \frac{20}{3}$$

$$X_1 = \frac{12 - X_2}{2} = \frac{8}{3}$$

Therefore, the objective function is optimized if $X_1 = \frac{8}{3}$ and $X_2 = \frac{20}{3}$. The objective function reaches a peak feasible value of $9\frac{1}{3}$ [i.e., $f(x) = X_1 + X_2 = \frac{8}{3} + \frac{20}{3}$].

As can be noted in this example, the optimum occurs at constraint intersections. The simplex method is nothing more than an orderly search of n-dimensional intersections in an effort to identify the intersection farthest removed from the origin perpendicular to the plane of the objective function, yet still in the feasible zone, which produces a maximum value for the objective function.

In practice, recognizing the properties of a linear programming problem and knowing whether the linear assumptions are appropriate, are the most difficult aspects of applying an existing operations research technique such as linear programming. As in all fields, good judgment is an essential ingredient in the successful application of a technique. Also, as in most fields, techniques rarely include all the effects that come to mind as influencing the outcome. Judgment plays a crucial part in sensing when an effect that cannot be included in a particular technique can be ignored because it does not represent a significant effect with respect to the outcome.

In the two-dimensional problem above it was possible to determine the solution by graphical means. In the n-dimensional space of more typical problems, graphical methods are inadequate. The simplex algorithm mentioned earlier is a procedure that utilizes matrix algebra methods to successively search improved constraint intersections on the surface of an n-dimensional convex polygon (i.e., the feasible solution space of an n-dimensional constraint equation set) to determine the point or points on this polygon that maximize the objective function.

The following example from Hadley (5, pp. 464–465) represents a more typical linear programming problem and suggests the need for some organized procedure for dealing with complex problems of allocation:

A farmer has 100 acres which can be used for growing wheat or corn. The yield is 60 bushels per acre per year of wheat or 95 bushels of corn. Any fraction of the 100 acres can be devoted to growing wheat or corn. Labor requirements are 4 hours per acre per year, plus 0.15 hour per bushel of wheat and 0.70 hour per bushel of corn. Cost of seed, fertilizer, etc., is 20 cents per bushel of wheat and 12 cents per bushel of corn. Wheat can be sold for $1.75 per bushel and corn for $0.95 per bushel. Wheat can be bought for $2.50 per bushel, and corn for $1.50 per bushel.

In addition, the farmer may raise pigs and/or poultry. The farmer sells the pigs or poultry when they reach the age of one year. A pig sells for $40. He measures the poultry in terms of one "pig equivalent" (the number of chickens needed to bring $40 at the time of sale). One pig requires 25 bushels of wheat or 20 bushels of corn, plus 25 hours of labor and 25 square feet of floor space. An equivalent amount of poultry requires 25 bushels of corn or 10 bushels of wheat, 40 hours of labor and 15 square feet of floor space.

The farmer has 10,000 square feet of floor space. He has available per year 2,000 hours of his own time and another 3,000 hours from his family. He can hire labor at $1.50 per hour. However, for each hour of hired labor 0.15 hour of the farmer's time is required for supervision. How much land should be devoted to corn and how much to wheat, and in addition, how many pigs and/or poultry should be raised to maximize the farmer's profits?

This example was selected for two reasons: (1) it is indicative of the inherent complexity of many allocation problems, and (2) it should suggest the breadth of generality of the simplex algorithm. The farmer's problem above is basically one of inputs of production, constraints, and prices in relation to some objective function. There is some best number of bushels of corn and wheat and numbers of pigs and chickens to raise that will maximize the farmer's profits. For thousands of years farmers have grown what they have always grown, about the same way they did the previous year. Farmers who veered from tradition did it on their own. Sometimes what they tried worked and they were heros among their peers, but too often their curiosity cost them. The linear programming solution to the above problem, assuming linear constraints are appropriate and the data are correct, identifies the best combination of inputs to maximize profit.

Keep in mind that there is an infinite number of problems analogous to the one above. The following are a few that come to mind:

1   What products to make in a steel mill to maximize profits
2   What combinations of offensive weapons to maximize destruction of an opposing force
3   What combination of chemical pesticides to control pests
4   What combination of workforce inputs to staff a corporation
5   What assignment of products to machines to minimize costs of production

Linear programming is only one of a set of related techniques generally referred to today as mathematical programming techniques. They include zero-one programming, nonlinear programming, integer programming, and dynamic programming, to name a few. They represent a very powerful arsenal of techniques when used by individuals who understand how and why they work and what their limitations are. A further requirement, possibly the one most commonly violated, is the need for either extending or imaginatively employing a technique to meet the unique requirements of a particular application.

Unfortunately, there are far too many individuals employing operations research today who do not know how or why the techniques they employ work. It has been said that a little education can be a dangerous thing. Operations research, and particularly mathematical programming, is a complex discipline in which this adage applies far too often. Compounding the problem is a failure of those who specialize in operations research to control entry into the discipline. Those practicing operations research represent a variety of backgrounds. At present, neither certification nor registration is employed in limiting entry into the field, and it is not apparent that this would offer the screening necessary to assure competent practice.

The greatest problem in operations research today seems to be related to the breadth of expertise it requires. Those who perform this work should be well versed in operations research techniques and at the same time capable of collecting and analyzing data and working in and understanding the system environment they are attempting to improve. Far too few individuals applying operations research possess sufficient capability in both areas mentioned above. However, there are those being educated today (e.g., industrial engineers) whose education will permit development of the breadth of competence demanded once they have gained sufficient knowledge of the system in which they are employed. More will be said about this later.

One of the oldest techniques in operations research and still one of the most useful is queueing theory. "Queue" is not a common word in the United States; it is in England. English people know that when they stand in line at the butcher shop they are in a queue.

QUEUEING THEORY

Queues are so common that they are easily overlooked; in fact, they are almost impossible to avoid in the conduct of everyday life. For example, people form queues at the family bathroom (especially in the morning), a barber shop, a service station, a supermarket, while buying theater tickets, before bank tellers, in the hospital emergency room, and at highway toll booths. One learns very early in life to dislike waiting in line. A line represents a form of regimentation, in a sense a loss of freedom, and a waste of time. In industry, time is money; therefore, wasted time represents wasted money. To the $15 per hour tool and die maker, standing in line for a blueprint or a tool seems like work—in fact, rather pleasant work because it provides an opportunity to discuss the world series while getting paid. To the industrial engineer, queueing represents a need for cost control and often a possibility for cost savings. Consider also, particularly in the rapidly expanding service industries, the displeasure created in those who have to wait to be served.

Consider for a moment the far too common supermarket that fails to provide a sufficient number of cashiers, which results in long queues. Does this limitation of service capacity save the store money (i.e., is it economizing)? No, the customers in queues must be served sooner or later. What happens in this type of situation is that the customer is led to consider the following question: "Based on the expected waiting time in this line in this store, and on other factors, is it worth the loss of my time to shop here?" The modern neighborhood convenience stores came into existence because their higher prices were acceptable in light of (1) location, (2) longer hours, and in particular (3) shorter queues, in comparison to supermarkets. Some supermarkets seem to have realized this and are providing longer hours and shorter queues. The supermarket example was offered to stress the point that queueing is a fact of life, both on and off the job, in any industry.

Queueing theory was pioneered by the Danish engineer A. K. Erlang (2) in the telephone industry well over fifty years ago. Erlang drew attention to a family of probability density functions which he found useful in functionally expressing interarrival times of incoming telephone calls. This very useful family of mathematical functions bears his name today.

The study of queueing of calls in the telephone industry is at least partially responsible for today's efficient telephone systems. An analogous system is the flow of signals in a computer system. The study of these queues was an essential precursor to the development of the efficient computer systems of today. GPSS [general purpose simulation system (4)], which is probably the most common simulation language in use today, was originally developed by IBM engineers to simulate the flow of information (i.e., signals) in computer systems. In the subsequent development and use of the GPSS language it became apparent that the flow of cars in a bank drive-in system, for example, is quite analogous to the flow of signals in a computer. Consequently, only a small fraction of GPSS simulations today are of computer systems; the bulk are concerned with a potentially infinite variety of actual flow systems in everyday life.

In most productive systems, attention beyond process transformation steps is sooner or later directed toward "inventories" and "queues." As was mentioned earlier, inventories serve as a buffer between unequal rates of flow, as indicated in Fig. 6-5. The inventory assures that if the "rate out" is greater than the "rate in" for some defined period, an adequate supply exists at point B in the system. If some activity of variable time duration is performed on each item of flow at point B in the system, the queue that develops represents an inventory of goods in process. It is clear that queueing and inventory analysis are related in application in a typical flow system.

One of the guiding principles of good plant design is that a dense design of the production system should be developed. This results in minimum material distances, which translate to minimum material handling costs. Because material handling does not add "value in use" to a product, minimum material handling cost is probably the most important single criterion in plant design. Without belaboring the point further, it should be apparent that plant design, queueing theory, and inventory theory are all interrelated. This is precisely why modern industrial engineering involves a blend of traditional industrial engineering, operations research, management science, computer science, and statistics. Two characteristics of any queueing model that must be specified are the arrival and service distributions. Single-server queues with Poisson-distributed arrivals and exponentially distributed service times are the simplest to consider mathematically and normally represent the logical beginning point for queueing model development. The arrival (or service) distribution refers to the probabilistic description of the arrivals (or service times) in the form of a probability density function. The Poisson density function of Fig. 6-6 indicates that although 8 arrivals per hour is the most frequently occurring rate, the average arrival rate λ is 10 per hour, resulting in an arrival every $^{60}/_{10} = 6$ minutes on the average.

The exponential density function of Fig. 6-7 has been employed in numerous

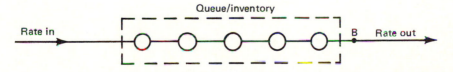

Figure 6-5   A flow system.

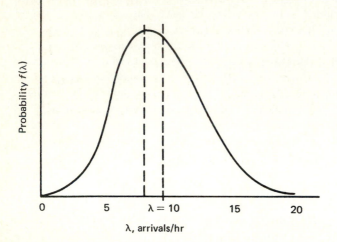

Figure 6-6 Poisson arrivals density function.

queueing models to describe service times. The mean service time $1/\mu$ from Fig. 6-7 is 8 minutes. If the arrival rate and service time distributions of Figs. 6-6 and 6-7, respectively, describe arrivals and service in a particular single-service system, it should be apparent that a queue should be expected to develop because arrivals on the average occur every 6 minutes whereas the average service time is 8 minutes. As this system functions over time, the queue would be expected to continue to grow in size. Such a system has no "steady-state" characteristics; it continues to grow indefinitely because of its lack of service capacity.

Assume in another case, however, that the mean arrival rate is 6 per hour, resulting in a mean time between arrivals of 10 minutes. The time between arrivals is

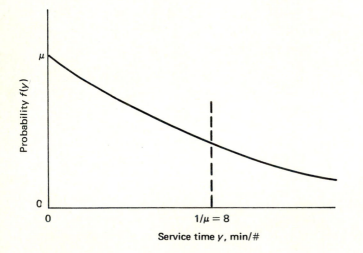

Figure 6-7 Exponential service times density function.

now greater than the mean service time of 8 minutes. In such a case (i.e., service time $<$ mean time between arrivals), steady-state queue statistics are expected. By steady state it is meant that the queue will not grow indefinitely but will vary in size, possessing some long-run mean size. Other characteristics of the system such as time in the queue and time in the system will also vary over time but have mean values in the long run. Obviously, if the mean time between arrivals is very large compared to the mean service time, the expected queue length will approach zero. However, if the mean time between arrivals is close to the mean service time, the expected queue length will be very large.

The following steady-state equations for Poisson-distributed arrivals and exponentially distributed service times will be offered without proof:

$$\lambda = \text{mean arrival rate}$$
$$\mu = \text{mean service rate}$$
$$P_0 = \text{probability that no units are in the system}$$
$$P_n = \text{probability that } n \text{ units are in the system}$$
$$m = \text{queue length}$$
$$n = \text{number of units in the system}$$
$$w = \text{waiting time of an arrival}$$
$$v = \text{total time an arrival spends in the system}$$
$$E(\ ) = \text{expected (mean) value of } (\ )$$

$$P_0 = 1 - \frac{\lambda}{\mu}$$

$$P_n = \left(\frac{\lambda}{\mu}\right)^n P_0$$

$$\sum_{n=0}^{\infty} P_n = 1$$

$$E(m) = \frac{\lambda^2}{\mu(\mu - \lambda)}$$

$$E(m|m > 0) = \frac{\mu}{\mu - \lambda}$$

$$E(n) = \frac{\lambda}{\mu - \lambda}$$

$$E(w) = \frac{\lambda}{\mu(\mu - \lambda)}$$

$$E(w|w > 0) = \frac{1}{\mu - \lambda}$$

$$E(v) = \frac{1}{\mu - \lambda}$$

Assume, for example, that cars arriving at a bank drive-in window are Poisson-distributed with a mean time between successive arrivals of 5 minutes. Assume also that the service time of the window teller is exponentially distributed with a mean time of 2 minutes.

1 What is the expected queue length?

$$\lambda = \frac{1}{5} = 0.2$$

$$\mu = \frac{1}{2} = 0.5$$

$$E(m) = \frac{\lambda^2}{\mu(\mu - \lambda)} = \frac{(0.2)^2}{0.5(0.5 - 0.2)} = \frac{0.04}{0.5(0.3)} = 0.26 \text{ cars}$$

2 What is the expected queue length during the times when there is at least one car waiting to be served?

$$E(m|m > 0) = \frac{\mu}{\mu - \lambda} = \frac{0.5}{0.5 - 0.2} = 1.667 \text{ cars}$$

The expected queue length of 0.26 cars for question 1 as compared to 1.667 cars for question 2 indicates that for much of the time there is no queue. The averaging in of these zero queue lengths with the expected queues when they do exist accounts for the difference in the two different queue lengths.

3 What is the probability that an arriving car will have to wait before being served by the teller?

$$P(\text{an arrival will wait}) = 1 - P_0$$

$$P_0 = 1 - \frac{\lambda}{\mu}$$

$$\therefore P(\text{an arrival will wait}) = \frac{\lambda}{\mu} = \frac{0.2}{0.5} = 0.4$$

4 What is the expected waiting time of an arrival?

$$E(w) = \frac{\lambda}{\mu(\mu - \lambda)} = \frac{0.2}{0.5(0.5 - 0.2)} = 1.333 \text{ minutes}$$

5 What is the expected total time a car will spend in the system?

$$E(v) = \frac{1}{\mu - \lambda} = \frac{1}{0.5 - 0.2} = 3.333 \text{ minutes}$$

The difference in time between the total time in the system of 3.333 minutes and the time waiting of 1.333 minutes is 2 minutes. This is the time in service and agrees with that given in the statement of the problem.

The example above demonstrates how queueing theory can be employed to estimate the expected steady-state parameters of queueing systems. The next example illustrates how queueing theory can be employed to assist in making economic comparisons in special cases.

Assume that two different tools, A and B, can be purchased for use in repairing machines, and that the rate of breakdowns necessitating repairs is Poisson-distributed with a mean of 5 per hour. Assume also that downtime on a machine costs the company $4 per hour for the period of time the machine is out of service. Tool A can be leased at a rate of $3.50 per hour and repairs machines at an exponentially distributed rate having a mean of 6 per hour. Tool B can be leased at the rate of $17.00 per hour and repairs machines at an exponentially distributed rate with a mean of 10 per hour. Which tool should be leased?

The total hourly cost of employing tool A is the sum of the hourly leasing cost and the hourly downtime cost. The average number of machines experiencing downtime (i.e., the number of arrivals in the queue) is

$$E_A(n) = \frac{\lambda}{\mu_A - \lambda} = \frac{5}{6 - 5} = 5$$

The total cost of employing tool A is $3.50 + 5($4) = $23.50 per hour. The average number of machines experiencing downtime when tool B is employed would be

$$E_B(n) = \frac{\lambda}{\mu_B - \lambda} = \frac{5}{10 - 5} = 1$$

The total hourly cost for tool B would be $17 + 1($4) = $21.00 per hour. Therefore, tool B should be leased. Authors have referred to queueing systems as "nonintuitive." This example demonstrates that property when one considers that $17 per hour for tool B seems costly compared to $3.50 per hour for tool A; yet, tool B is cost-effective because of the queue lengths machine A produces, which are not obvious from the data given in the problem. It is only after calculation of comparative total costs that the best choice becomes obvious.

Hillier (7) reported a study performed at a Boeing aircraft plant that demonstrates the nonintuitiveness of queueing problems. In this case, supervisors were expressing concern that their mechanics were spending too much time waiting in line at the tool crib to receive tools. Of course management, concerned about overhead costs, were interested in not having more tool crib attendants than necessary.

A study of the tool crib indicated that the mean time between arrivals was 35 seconds, while mean service time was 50 seconds. Based on these data it would seem obvious that one clerk would not be able to keep up, but that two could do so without difficulty. Consider, however, that an idle mechanic is a great deal more costly than an idle tool crib attendant. If one assumes, as Hillier did at the time, that the mechanic is paid $5 per hour and the clerk $2 per hour, estimates of the total daily idle time are $64.10, $31.00, and $40.00 for two, three, and four clerks, respectively. These results indicate that three clerks is the best choice and that even four clerks is a better choice than two clerks.

Obviously with four clerks, it is likely that management, observing the utilization of clerks, would question whether they had enough to do. Yet it is the ability of the fourth clerk to prevent the queue length from increasing significantly because of a combination of a temporarily high arrival rate and the service time involved that is of primary interest in this problem. The best solution in this type of situation is usually to provide excess service capacity when needed (e.g., three clerks) and secondary tasks, such as typing labels, assembling cardboard boxes, or other potentially intermittent duties, during slow periods in the tool crib.

Queueing systems are so common and diverse with respect to types and combinations of components that a system of classification of multiple-server queues was developed and is in common use today; the system was proposed in 1953 by Kendall and extended in 1966 by Lee. The Kendall-Lee system (11), detailed in Table 6-6, identifies the following six main characteristics essential to any multiserver queueing system, as illustrated in Fig. 6-8: (1) the arrival distribution, (2) the service distribution, (3) the number of parallel service channels, (4) the service discipline, (5) the number of units of flow permitted in the system, and (6) the source population.

Table 6-6   Kendall-Lee Multiple-Server Queue Classification System

General Notation   (a/b/c):(d/e/f)

Specific notation	Description
a	Arrival or interarrival distribution
b	Leaving or service time distribution
c	Number of parallel service channels
d	Service discipline
e	Maximum number allowed in the system
f	Calling source

Code for a and b	
M	Poisson arrival (or equivalent exponential interarrival or service times)
D	Deterministic interarrival or service times
E_K	Erlangian* or gamma interarrival or service time distributions
GI	General independent distribution of arrivals or interarrival times
G	General distribution of leaving or service times

Code for d	
FCFS	First come, first served
LCFS	Last come, first served
SIRO	Service in random order
GD	General service discipline

c, e, and f in general notation above are finite or infinite numbers

*The parameter K explicitly defines a unique member of the density family (e.g., if $K = 1$ the density function is exponential).

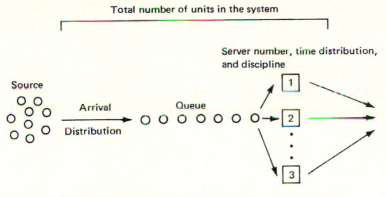

Figure 6-8 Multiserver queueing system components.

The following problem situation illustrates the use of the system. Assume that the processing of frankfurters in a plant is to be modeled. Assume also that after the frankfurters are made they enter any one of five rotary cooking bins, which cook them for 5 minutes and then dump them one at a time onto a conveyor belt that carries them to a packaging machine. Assume that the queue that develops in cooking the frankfurters is the part of the process to be coded. A possible notation for this system might be (M, D, 5):(SIRO, 100, ∞). This notation indicates that this is a multiple queue system possessing Poisson-distributed arrivals, discrete service time, five multiple servers, random entry into service, a maximum of 100 units in the system, and an essentially infinite source of arrivals. This example suggests the diversity of potentially different multiple-server queueing systems.

As would be expected, queueing formulations for multiple-server queues are more complex than those for single-server queues. In the following equations r represents the number of servers in the sysem. If the combined service capacity of the r servers exceeds the arrival rate into the system (i.e., $r\mu \geqslant \lambda$) the queue will not grow over time but instead will have predictable steady-state characteristics. The following multiple-server equations, assuming Poisson-distributed arrivals and exponentially distributed service times, are offered without proof:

$$P_0 = \frac{1}{\left[\sum_{n=0}^{r-1} 1/n! \, (\lambda/\mu)^n\right] + 1/r! \, (\lambda/\mu)^r \, [r\mu/(r\mu - \lambda)]}$$

for $n < r$

$$P_n = \frac{1}{n!} \left(\frac{\lambda}{\mu}\right)^n P_0$$

and for $n \geqslant r$

$$P_n = \frac{1}{r! r^{n-r}} \left(\frac{\lambda}{\mu}\right)^n P_0$$

$$E(m) = \frac{\lambda\mu (\lambda\mu)^r}{(r-1)! (r\mu - \lambda)^2} P_0$$

$$E(n) = \frac{\lambda\mu(\lambda/\mu)^r}{(r-1)!(r\mu - \lambda)^2} P_0 + \frac{\lambda}{\mu}$$

$$E(w) = \frac{\mu (\lambda/\mu)^r}{(r-1)!(r\mu - \lambda)^2} P_0$$

$$E(v) = \frac{\mu (\lambda/\mu)^r}{(r-1)!(r\mu - \lambda)^2} P_0 + \frac{1}{\mu}$$

Assume, as an example, that students instructed to see any one of three counselors in a high school arrive for counseling on a first-come, first-served basis at a Poisson-distributed arrival rate of 10 per hour, or on the average every 6 minutes. Assume also that all three counselors spend 15 minutes on the average distributed exponentially with each student. A check should first be made to determine whether service is adequate to expect steady-state characteristics.

$\lambda = 10$ per hour
$\mu = 4$ per hour per counselor
$r\mu = 3(4) = 12 \geqslant 10$

Therefore, there is sufficient service capacity and the steady-state equations are applicable. Assume that one wishes to know the percentage of time when there are no students either waiting to see or seeing any of the three counselors.

$$P_0 = \frac{1}{\left[\displaystyle\sum_{n=0}^{2} 1/n! \, (10/4)^n\right] + 1/3! \, (10/4)^3 \, [3(4)/(3(4) - 10)]}$$

$$\sum_{n=0}^{2} \frac{1}{n!} \left(\frac{10}{4}\right)^n = \frac{1}{0!} \left(\frac{10}{4}\right)^0 + \frac{1}{1!} \left(\frac{10}{4}\right)^1 + \frac{1}{2!} \left(\frac{10}{4}\right)^2 = 1 + \frac{10}{4} + \frac{100}{32} = \frac{212}{32}$$

Therefore

$$P_0 = \frac{1}{212/32 + 1,000(12)/3(2)64(2)} = 0.0449 \simeq 4.5 \text{ percent}$$

Queueing theory is very useful as an area of study for developing a theoretical understanding of how queueing systems function. Such a study will usually develop one's intuition at least to the extent of providing a qualitative understanding of the primary factors influencing the output characteristics of queueing systems.

Many practical systems that are of interest to industrial engineers represent flows of some resource. For this reason many systems of practical interest contain elements of queueing. Queueing systems are more often than not modeled by using simulation rather than the analytical queueing models discussed in this chapter. Many operations researchers would consider simulation an operations research technique. However, it is so often employed in the study of systems problems that it seems appropriate to introduce the technique in the next chapter, which deals specifically with systems.

The optimization techniques discussed so far have all been single-stage techniques; that is, given the data for the problem and the relationships for the variables involved, a single solution in time is obtained. There is a class of problems in which appropriate values for a sequence of decisions are needed. A technique known as dynamic programming, developed originally by Bellman (1), has proved useful in simplifying such problems.

DYNAMIC PROGRAMMING

All dynamic programming problems, as complex as they may become, have in common a multistage decision process, as indicated in Fig. 6-9. The more common solution approach, called a "backward pass," begins by giving consideration to the nth or last decision required in the sequences of decision. A functional representation of the variables involved is sought to determine the optimum solution for this last decision segment of the sequence of decisions. Next, consideration is given to the stage $n-1$ of the decision process. Based on the combined functional representation with respect to the last two stages of the decision sequence, a means of identifying the optimum decision for this stage, including the stage that follows it, is sought. At this point the functional representation permits identification of optimum decisions for the last two stages. This process is continued, one stage at a time, until the beginning of the network is reached. The representation for what constitutes an optimum solution for what remains in the decision sequence, which at this point is the entire network, provides the solution to the optimum set of sequential decisions required to optimize the total sequence of decisions.

The "stagecoach problem" provides a limited insight into this methodology common to all dynamic programming problems. Assume, as indicated in Fig. 6-10, that a stagecoach is at location a and is to take the shortest route to location m. The numbers on the path segments in the network represent relative distances. One way to

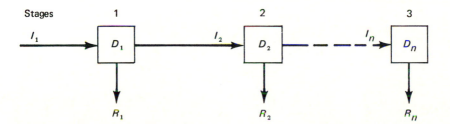

Figure 6-9 Sequential decision process.

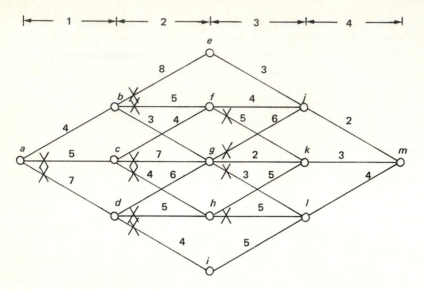

Figure 6-10 Stagecoach problem graph.

solve this problem, without employing dynamic programming, is by total enumeration and evaluation of all possible solutions. Table 6-7 lists all the possible paths through the network and the sum of distances for each path. It is apparent from this list that path *abgkm* is optimum, with a sum of distances of 12. Dynamic programming offers an approach to determining an optimal path without resorting to total enumeration.

Attention is given first to the last decision in the sequence of decisions. Note that in stage 4 of Fig. 6-10 there are three ways to get to location *m*. Next, in stage 3 there are three possible path segments entering each path that leads to location *m* in stage 4. The three path segments entering location *j* are from locations *e, f,* and *g*. The total distances to the end of the network from locations *e, f,* and *g* are 5, 6, and 8, respectively. When all paths are evaluated for stages 3 and 4, it is apparent that there were three possible paths connecting locations *g* and *m* as indicated in stage 3 of Fig. 6-10. They are *gjm, gkm,* and *glm,* with total distances respectively of 8, 5, and 7. It should be apparent that if location *g* is on the optimal path through the network, stages 3 and 4 of the optimum path will be *gkm*. Consequently, all path combinations

Table 6-7 Total Enumeration of Paths for the Stagecoach Problem

Path	Distance	Path	Distance	Path	Distance
abejm	17	*acfjm*	15	*adgjm*	21
abfjm	15	*acfkm*	17	*adgkm*	18
abfkm	17	*acgjm*	20	*adglm*	20
abgjm	15	*acgkm*	17	*adhkm*	20
abgkm	12	*acglm*	19	*adhlm*	21
abglm	14	*achkm*	17	*adilm*	20
		achlm	18		

including *gj* and *gl* need not be given further consideration in attempting to identify the optimal path and are lined out in Table 6-8. In a similar fashion, we can also eliminate path segments *fk* and *hl* in stage 3. Path segments eliminated from further consideration are crossed out in Fig. 6-10. Note also that for each of the five entry locations into stage 3 (i.e., *e, f, g, h,* and *i*) there is only one path to the end of the network remaining in the solution. Of course, if path segment *km* had had a distance of 1, path *fkm* would still be in solution along with path *fjm* because they would be equally optimal.

In stage 2 of Fig. 6-10 there are three possible path segments emanating from location *b*. The one passing through *g* is preferable to *be* or *bf*, which are therefore eliminated from further consideration. Path segments *cg, ch, dh,* and *di* can be eliminated as well.

At stage 1 in Fig. 6-10 it is apparent that the path beginning with segment *ab* is optimum. Tracing back through either Table 6-8 or Fig. 6-10, it is clear that the optimal path is *abgkm*, with a total distance of 12 units.

One might ask, "Why bother? Total enumeration is quicker." For problems of the size and degree of complexity of this example, dynamic programming has little to offer. For larger networks, however, the built-in process of elimination of dynamic

Table 6-8 Stage Calculations for the Stagecoach Problem

Stage	Path	Distance for this stage	+	Optimal distance for all remaining stages	=	Total distance
4	*jm*	2		0		2
	km	3		0		3
	lm	4		0		4
3	*ej*	3		2		5
	fj	4		2		6
	~~*gj*~~	~~6~~		~~2~~		~~8~~
	~~*fk*~~	~~5~~		~~3~~		~~8~~
	gk	2		3		5
	hk	5		3		8
	~~*gl*~~	~~3~~		~~4~~		~~7~~
	~~*hl*~~	~~5~~		~~4~~		~~9~~
	il	5		4		9
2	~~*be*~~	~~8~~		~~5~~		~~13~~
	~~*bf*~~	~~5~~		~~6~~		~~11~~
	cf	4		6		10
	bg	3		5		8
	~~*cg*~~	~~7~~		~~5~~		~~12~~
	dg	6		5		11
	~~*ch*~~	~~4~~		~~8~~		~~12~~
	~~*dh*~~	~~5~~		~~8~~		~~13~~
	~~*di*~~	~~4~~		~~9~~		~~13~~
1	*ab*	4		8		12
	~~*ac*~~	~~5~~		~~10~~		~~15~~
	~~*ad*~~	~~7~~		~~11~~		~~18~~

Table 6-9 Commodity Barge Data
(Maximum Barge Capacity = 11,000 ft³)

Commodity	Value per unit load (× $1,000)	Volume per unit load (× 1,000 ft³)
X	8	2
Y	11	3
Z	15	4

programming makes a solution possible, whereas total enumeration may well be beyond practical consideration.

In this example minimum distance was the "measure of effectiveness." The technique is equally effective with respect to other units of measure, such as cost or weight. Dynamic programming, like other mathematical programming techniques, has a broad range of potential applications because of its generality of application.

Assume, for example, that the management of a barge company has three basic commodities, X, Y, and Z, that it wishes to ship and it wishes to ship the maximum value of commodities that a particular barge can hold. Assume that each commodity is shipped in multiples of unit loads, and the barge under consideration has 11,000 ft³ of storage space. If the value and volume per unit load for each commodity are as given in Table 6-9, how many unit loads of each commodity should be shipped in a barge of this capacity to maximize total cargo value? Figure 6-11 illustrates a dynamic programming conceptualization of this problem involving three sequential decisions. Decisions D_1, D_2, and D_3 required in this problem must answer the question "How many unit loads of commodities X, Y, and Z should be shipped to maximize the total value of the cargo?" The number of unit loads of each cargo determines the return R_i, in this example contribution to cargo value, at each stage, and I_i represents the state or input at each stage. In this problem, the remaining space in multiples of 1,000 ft³ is considered the input; therefore, I_1 is 11.

In a backward pass the object is to determine, one stage at a time, the optimum decisions for the remainder of the network. In Table 6-10 R_i' identifies the maximum value for a given value of input I_i at each stage. For example, in stage 3 it can be noted in Table 6-10 that if 5,000 ft³ remain as input to this stage and one unit of commodity Z is loaded using 3,000 ft³ of this space, the value of the cargo is increased

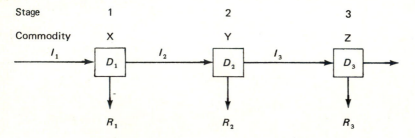

Figure 6-11 Commodity barge dynamic programming flowchart.

Table 6-10 Commodity Barge Calculations
(NP = Not Possible)

			Stage 3			
I_3	$D_3 = 0$		$D_3 = 1$	$D_3 = 2$		R_3'
0	0		NP	NP		0
1	0		NP	NP		0
2	0		NP	NP		0
3	0		NP	NP		0
4	0		15	NP		15
5	0		15	NP		15
6	0		15	NP		15
7	0		15	NP		15
8	0		15	2(15) = 30		30
9	0		15	2(15) = 30		30
10	0		15	2(15) = 30		30
11	0		15	2(15) = 30		30

		Stage 2			
I_2	$D_2 = 0$	$D_2 = \textcircled{1}$	$D_2 = 2$	$D_2 = 3$	R_2'
0	0	NP	NP	NP	0
1	0	NP	NP	NP	0
2	0	NP	NP	NP	0
3	0	⑪	NP	NP	⑪
4	15	11	NP	NP	15
5	15	11	NP	NP	15
6	15	11	2(11) = 22	NP	22
7	15	11 + 15 = 26	2(11) = 22	NP	26
8	30	11 + 15 = 26	2(11) = 22	NP	30
9	30	11 + 15 = 26	2(11) = 22	33	33
10	30	11 + 15 = 26	2(11) + 15 = 37	33	3
11	30	11 + 30 = 41	2(11) + 15 = 37	33	41

			Stage 1				
I_1	$D_1 = 0$	$D_1 = 1$	$D_1 = 2$	$D_1 = 3$	$D_1 = \textcircled{4}$	$D_1 = 5$	R_1'
11	0 + 41 = 41	8 + 33 = 41	2(8) + 26 = 42	3(8) + 15 = 39	4(8) + 11 = ㊸	5(8) = 40	43

by \$15,000. This is the maximum feasible increase in the value of the cargo that can be made by additions of cargo at this and remaining stages, given the quantity of space remaining at this and the remaining stages. Note that at stage 2, if $I_2 = 10$ (i.e., if 1,000 ft^3 was used at stage 1), the optimum cargo value contributable by stages 2 and 3 is \$37,000, obtained by loading two unit loads of commodity Y at \$11,000 per 3,000 ft^3 and one unit load of commodity Z at \$15,000 per 4,000 ft^3, exactly utilizing the 10,000 ft^3 available.

Stage 1 results in Table 6-10 indicate that the maximum cargo value possible is

$43,000, obtained by loading four unit loads of commodity X, one unit load of commodity Y, and none of commodity Z.

It should be apparent at this point that dynamic programming permits one to eliminate obviously undesirable path segment combinations at the earliest possible opportunity, thereby limiting the amount of overall calculation necessary in searching for a solution. At least in the past, practical problems in industry, even with the computational efficiencies offered by dynamic programming, have often strained the memory storage capacity of computer systems. Dynamic programming does, however, make it feasible today to solve many problems previously considered to be well beyond practical computational limits. Improvements in the design of computers, particularly in the last ten years, have produced similar gains for an even broader spectrum of problems.

SUMMARY

The techniques discussed in this chapter were linear programming, queueing theory, and dynamic programming. Zero-one programming, integer programming, nonlinear programming, simulation, game theory, and many others were not considered. The techniques that were discussed were offered as a fairly representative sample of operations research techniques. Of course, it should be understood that the examples offered were trivial and were designed only to provide insight into the nature of typical operations research techniques. Real-world problems are a great deal more complex than those posed here.

This chapter has also been wholly technique-oriented. There are those who feel that operations research is nothing more than a collection of techniques from which the operations research specialist draws the appropriate one to suit the occasion. Machol, in an OR/MS publication, illustrates this philosophy humorously (8, p. 26)[1]:

> I first heard the screwdriver story many years ago from Russell Ackoff. It seems that an OR worker was home for the weekend with nothing to do, and decided to help around the house. So he went down to the cellar and got a screwdriver, and then he went around the house tightening all the screws that were loose. Eventually there weren't any loose screws anymore, but he still had the screwdriver, so he went back down to the cellar and got a file; he came up, found a nail, filed a slot in the head of the nail, inserted the screwdriver, and started twisting.
>
> We all know people whose personal screwdriver (a solution looking for a problem) is LP or queue theory or what have you. Of course those (including you and me) who know lots of tools don't have to worry about the screwdriver syndrome, because we always pick from our armamentarium the tool that is most applicable.
>
> In some cases the screwdriver syndrome may be benign; i.e., it is still a disease, but not necessarily malignant. Thus, when a physicist constructs a traffic

---

[1] Reprinted from "The Screwdriver Syndrome," by Robert E. Machol, *Interfaces,* vol. 4, no. 3, May, 1974, p. 26, published by The Institute of Management Sciences.

model, he tends to consider the automobiles as molecules in a Maxwellian gas, and to compute mean free paths and the like. The electrical engineer considers it a servomechanism, and looks for instabilities as drivers react to the cars in front of them. A fluid-mechanics expert considers the stream of traffic to be a fluid flowing along the highway. The statistician views traffic as a stochastic process with the gaps between cars as samples from a probability density function.

Drawing on what one has learned is not uncommon. Those doing the teaching were trying to pass on what they had learned, but they likely had high hopes that what they were teaching would be extended and adapted in innovative ways. Much operations research work is more analytical than suggested by most of the examples discussed thus far. However, the restricted mathematical level of this text limits the use of analytic examples. This is unfortunate in that much of the best in operations research derives its value by forging new solution approaches to previously unsolved problems. The truly powerful operations researcher must be highly versed in mathematics to be able to conceive and solve mathematical models appropriate to situations as they arise. Unfortunately, as some operations researchers highly trained in mathematics have discovered, mathematical ability may be necessary but is certainly not sufficient. This is especially true in the real world, where it is necessary to find and understand problems well buried in an environment replete with incomplete data, attitudes greatly influenced by motives of personal gain, semantic hurdles, managerial fears of risk, and support by average people (i.e., some of whom are imperfect) making average salaries and exhibiting average scientific enthusiasm.

Pure operations researchers may have an advantage over industrial engineers or management scientists because in their university educations they were able to devote more of their energies to the acquisition of mathematical skills. Having done so, however, they are at a disadvantage compared to industrial engineers because of the weakness of their education in dealing with physical systems, which is also true for management scientists. Of course, operations researchers are also at a disadvantage compared to management scientists because of their limited understanding and appreciation of the management environment, and this is also the case for industrial engineers.

Who would know better what operations researchers' strengths and weaknesses are than a professional employment recruiter, particularly one specializing in operations research clientele. Halbrecht Associates, Inc., is one such organization. The following article suggests that operations research practitioners are of limited economic value if the only tool they can bring to bear in solving industrial problems is mathematics (6)[2] :

[2] Reprinted from "So Your Mother Wanted You to Be a Doctor, but a Ph.D. in OR?" *Interfaces,* vol. 2, no. 2, February, 1972, pp. 44–49, published by The Institute of Management Sciences.

So Your Mother Wanted You to Be a Doctor but a Ph.D. in OR?

Herbert Halbrecht

Mr. Herbert Halbrecht
Halbrecht Associates, Inc.
2 Greenwich Plaza
Greenwich, Connecticut 06830

Dear Sir:

I have just completed my junior year at Harvard College where I am majoring in Applied Mathematics to Economics. I am currently in the process of deciding whether to go to graduate school for a Ph.D. in Operations Research. Before making any decisions, I would like to find out about the job opportunities in Operations Research and related fields, i.e. Economics, Econometrics, Systems Engineering, etc. Could you please provide me with your list of current job opportunities for the Northeast area in these fields?

Also, I would greatly appreciate if you could give me some idea as to the starting salary of a Ph.D. in Operations Research and related fields, and how rapidly salaries advance. What is the average salary after fifteen years experience? What is the current job market for men in Operations Research and related fields? What do you think the job market for these fields will be like in the future?

Any information that you can provide will be greatly appreciated.

Yours truly,
Samuel Student

Dear Sam,

I have read your letter several times and have mixed emotions concerning how best to answer you. But I hope that, to the extent that you ask my advice, my reply is of assistance in your career planning.

Permit me, however, the freedom to answer your questions in essay form, in addition to providing some specifics, rather than merely quantitatively.

The first, and by far the most important, question is one only you can answer: what are your own long term interests in and motivations for considering Operations Research as a career?

Are you interested in OR as an end in itself? Are you truly research oriented, interested in the development of new methodologies, advancement of the state of the art, and the like? If so, a Ph.D. in operations research is vital, as it is if you contemplate an academic career teaching OR methodology. The same would be true in other fields as well, of course.

However, if, as I suspect from your questions, you may be interested in a career in industry, I have strong feelings in other directions. If you plan to enter the OR/MS field in industry, the first thing I would do would be to transfer out of the Department of Mathematics, particularly before the graduate level, and get an MBA with a quantitative science option or major.

Although there is now, and I expect there will continue for quite some time to be, a great demand for operations researchers, the nature of the demand is clearly changing. Let me oversimplify a bit and say that at least two career paths emerge as possibilities. One path is that of the technician immersed in methodology, concerned with the *how* of a technical answer to a problem. The other is that of the businessman, the entrepreneur, who, while also interested in *how*, is more preoccupied with the *why* of a problem. The latter goes beyond an algorithm to find out what is the true marketing, production, advertising, finance, or corporate planning problem. He has an interest in the substantive, functional activity, and OR is one of the tools of his trade of businessman/manager. The former, on the other hand, is much more preoccupied with methodology and analysis and it is of only casual interest that his model may relate to a physical distribution rather than a budgetary problem.

If your interests are in abstract and/or analytical areas, then the Ph.D. in Operations Research, Econometrics or the like should be your educational objective. The MBA, MSIA, etc., should be more relevant to you, however, if you are really interested in running a business, making profits, and in hands-on management of people.

Incidentally, after 10 or more years past graduation, more OR Ph.D.'s will be working for MBA's, than the other way around.

It is critical for your own future that you not be steered into a career path merely because you think more money can be made in B, say, rather than A. Spending a lifetime doing things you don't really enjoy, and are not even especially interested in, is just another form of self-imposed purgatory. Economic necessity may drive you in that direction, but why choose this to begin with? Furthermore, if you don't like it in the first place you may not do as well economically as others in the profession who enjoy what they are doing. While psychiatrists have indicated that some substantial preoccupation with success is needed as a driving force to 'get ahead', this preoccupation with success need not be exclusively equated with money, and certainly shouldn't lead you to make decisions with which you are basically uncomfortable. Not so incidentally, if you do make such decisions, your chances of success, by whatever measurement, are slight.

Now let me try to answer some of your specific questions.

First of all, reports to the contrary notwithstanding, there are now, and will continue to be, substantial employment opportunities in the management sciences. Certainly the numbers of persons sought are not as great today as, say, in 1968 or 1969, particularly in relation to the number of available professionals, but the ever-increasing slope of industry's demand curve had to modify itself some time—and a period of economic decline (my hindsight is fabulous) is the logical time. Add to this the fact that, whereas previously OR/MS students could enter defense, aerospace, consulting and government (city, state, regional, federal) in addition to commerce and industry, now the personnel requirements are almost exclusively in industry and only somewhat in government. Also, as a result of the previous very substantial demand for OR/MS practitioners, many students, lured by the high salaries, the attractive ratios of jobs to candidates, and the much heralded exciting nature of the jobs themselves, went into what have been labelled OR or MS programs in colleges and universities, thus increasing the supply of such persons appreciably.

Employers are now, then, in the, for them, enviable position of being able to be more selective concerning their personnel specifications than ever before, and accordingly are making drastic changes in terms of the types of people they seek. These changes have actually been developing over the past several years, but in the current buyers' market they can be insistent on these requirements.

The requirements for OR Ph.D.'s in industry have two different demands, depending on the total package of education leading up to and including the Ph.D. Demand A is relatively constant and this is for those Ph.D.'s where the graduate education is almost exclusively in quantitative technique but where there is little training in or attention paid to the managerial aspects of business and the implications to management, both organizational and political, of the changes sought and recommendations made by management scientists. Demand B is for those who have undergraduate education in engineering, business administration or what have you, so that they have a greater understanding of, and appreciation of, the totality of the problems faced by management rather than the almost exclusively technical/methodological. Demand B has been and will continue to be greater than Demand A. We also anticipate that Demand B will increase at a greater rate than Demand A. More and more companies have commented that many of those with doctorates in Group A are more interested in research and in methodological and technological advancement of the state of the art, rather than in working on those problems management is interested in. In an increasing number of cases we are even being told by firms that they don't want Masters degrees in OR per se, unless the degree is from certain schools. More seek those whose graduate degrees are in business administration or industrial management with a management science option—or major. Amongst the reasons for this is the fact that too many who have degrees in so-called operations research really have degrees in applied mathematics and statistics, with absolutely no appreciation of the implications to management of the managerial and political impact of their technical recommendations on functional activities.

Many other clients prefer a physical science (physics particularly), engineering, accounting, or other functional undergraduate degree, plus an MSOR, to assure themselves that a candidate has some understanding of at least some of the functional parts of their problem.

If I were to advise someone who is interested in operations research as a career, just starting college, I would suggest as follows:

Undergraduate degree	Graduate degree
Engineering (Chem., Mech., Ind.)	MBA or MSIA - (OR major)
Business Administration, Finance, Accounting	MBA - MSIA (OR major) or MSOR
Physics	MBA or MSIA (OR major)

Incidentally, as another indication of the reasons for preferring certain undergraduate degrees to others, note that, while physics is preferred, mathematics is not. Now, in any event, a good mathematical underpinning is certainly necessary for success in OR. However, in studying physics you get a better appreciation for collecting and evaluating data, and making decisions on less than the complete data desirable. Also, you begin to appreciate the techniques and problems of modelling of physical phenomena. Math and statistics, while necessary and substantial components of a good OR background, do not in themselves provide this other training.

In any case, after getting a master's degree, I strongly suggest employment for at least a year or two, and then, if you are sufficiently motivated, go back for your doctorate. The exposure to real world problems will make your doctoral studies much more valuable than could ever be the case otherwise.

Part of the same picture is the fact that there are many complaints from employers concerning the inability (or lack of interest) of the OR practitioner to communicate effectively with others besides his technical compatriots. Those trained almost exclusively in mathematics and statistics are too often inadequately trained when it comes to selling their ideas, their worth, or themselves. Also, since they seldom have the same values as, say, MBA's, or those who run companies, there is often a dichotomy when an attempt is made to evaluate the contributions made to a company. The only problem is that if OR is not sold effectively, i.e. if management is not constantly made appreciative of the contributions of the OR group to the company (What have you done for me lately?), survival prospects diminish, particularly in poor economic times.

The overwhelming majority of OR/MS practitioners we have seen are bright, highly intelligent and distinctly above average in analytical capability. Unfortunately, these characteristics are not enough for success in industry, particularly as you get further away from 'pure' research and move into the applied areas of business. Businesses are not run by numbers, but by people. Theory is important, but the 'art' of management is to get things done through people. Accordingly, if you want to get ahead in management and management science in industry, analytic education and a high level of intelligence must be part of the total package that is you. The other parts, which must be visible incidentally, must include an ability to relate to others, to communicate so that you hear and understand 'them' as well as getting ideas across to others in such a manner that they understand you.

The worst thing you can do is look down upon those who are not trained in your own areas, and act superciliously to others, particularly if and when they are not familiar with your special jargon or nomenclature. It is unrealistic to expect the executives in finance, accounting, manufacturing distribution, marketing and purchasing, amongst others, to speak 'your language'. For your own good, get some training in, and understanding of, these functional activities if you plan to enter industry. A good course in cost accounting will be much more valuable than the umpteenth course in integer programming.

There is, incidentally, a very significant adjustment that you have to make if you do go into industry, which could make the difference between success or failure. By the way, 'success or failure' can make the difference between working or being fired, it is not merely the difference between a grade or a slightly higher or lower grade. Avoid like the plague the 20% syndrome, for, like many a plague, it will surely destroy you. Many of our company clients indicate that this is one of the most exasperating and infuriating problems which they have with Operations Research personnel. These same money-mad controllers feel that the greatest payoff in the use of Operations Research methodology comes in the first 80% of the solution, that the leverage they get in terms of benefit versus expense and time is greatest generally at about this point, although it can vary and, just for discussion, say that the payoff is 5-to-1. On the other hand, the payoff leverage in going from 80% of the solution to the final 100% of the solution may be, at best, 1-to-1. Accordingly, since most business executives are more concerned with making money than coming up with an elegant and complete package, they think that the effort going from 80% to 100% is wasted as compared with investing the same financial and people resources on starting on other problems. One of our clients, as a matter of fact, refuses to hire Ph.D.'s in Operations Research because they insist on working on this last 15% to 20% since it injures their professional and analytical pride to do otherwise. Unfortunately, this same professional pride is not as important to business executives as the bottom line on a balance sheet and, irrespective of what you have learned in school, the bottom line is what it is all about.

To get back to your questions, you ask about average salaries of Ph.D.'s with 15 years' experience. There are none. The first Ph.D.'s in Operations Research were granted in 1957 by the then Case Institute of Technology. Larry Friedman is now at Mathematica and Eliezer Naddor at Johns Hopkins. The statistical sample, even as we went into the early 1960's, is so small as to not be meaningful, especially since the period covered is the beginning of the field from an educational point of view, and a period of tremendous change generally.

Drs. Russ Ackoff (now at the University of Pennsylvania), C. West Churchman (University of California, Berkeley), and Len Arnoff (Ernst and Ernst, Cleveland), having been the founders of the OR program there, would be good sources of additional information concerning the financial progress of the first OR Ph.D.'s.

Current entry level salaries for Ph.D.'s in OR range from about $15,000 to $18,000 with a few exceptions at $20,000. These latter exceptions generally depend on the individual's other experience or training, possibly a brilliant record, etc. Some consulting firms may have paid higher in the past, but generally for exceptional MBA's, rather than Ph.D.'s. I seriously doubt that they are going to pay more in the immediate future or for some time, except as inflation raises salaries generally.

Rapidity of salary advancement is even more difficult to answer. In the past, salaries were increasing rapidly because, amongst other things, of the extreme shortage of trained professionals plus continually increasing demand. Salaries then went up more rapidly than in other fields and from a higher starting base as well. This is beginning to subside and will be brought more in line with other fields as supply catches up with demand.

It would be nice to give you a mechanistic answer, a formula with which you can assay your future. This isn't the way of life, however, and much more will depend on the ability of an individual to contribute to his company, to accomplish things meaningful to his employer, and not just only to himself. Be guided thusly and your income will be well taken care of, rather than preoccupy yourself with charts or progressions based on anticipated rates of increase.

I am enclosing our latest job opportunities listings but am certain that the job market, already beginning to open up somewhat, will be much brighter by the time you are available.

You might wish to show this to some of the professors at the Harvard Business School and get their reactions as well.

I hope the above is of use to you, and I will be interested in your decisions and in what you do in the future. Please keep in touch with me and if I can ever be of assistance, please feel free to call upon me.

Sincerely,
Herbert Halbrecht

P.S. Incidentally, I have shown copies of this letter to a number of friends of mine who have Ph.D.'s in Operations Research and, with almost no exceptions, they indicate that, had they to do it over again, they would take more courses in the financial and business management areas and less in some of the exotic theoretical areas. Accordingly, you yourself may wish to confirm some of what I am saying by finding out the identities of people who have gotten Operations Research Ph.D.'s and who are now successful in industry. Ask them what they think of the Operations Research doctoral education program.

I recognize that this may be a lot of hard work, but your whole life is ahead of you and a very small amount of effort now may be of tremendous benefit to you in terms of your future.

Mr. Herbert Halbrecht
Halbrecht Associates, Inc.
2 Greenwich Plaza
Greenwich, Connecticut 06830

Dear Mr. Halbrecht:

I want to thank you very much for your detailed reply to all of my questions about future opportunities in Operations Research and related fields. Your answers were most informative and certainly were of great assistance in helping me decide on my future career plans.

Incidentally, I have decided to pursue graduate study in Economics with a minor in Operations Research.

Once again, many thanks for your interest and valuable advice.

Sincerely,
Samuel Student

    This article raises interesting and difficult questions concerning the future role of industrial engineers, management scientists, operations researchers, and others in our society. More will be said about these questions in the final chapter.

REFERENCES

1   Bellman, Richard: *Dynamic Programming,* Princeton University Press, Princeton, N.J., 1957.
2   Brockmeyer, E., H. L. Halstrom, and Arne Jensen: "The Life and Works of A. K. Erlang," *Trans. Dan. Acad. Tech. Sci. No. 2,* Copenhagen, 1948.
3   Dantzig, George B.: "Maximization of a Linear Function of Variables Subject to Linear Inequalities," in *Activity Analysis of Production and Allocation,* chap. 21, Monograph no. 13, John Wiley & Sons, Inc., New York, 1951.
4   "General Purpose Simulation System/360, Introductory User's Manual," H20-0304-3, Technical Publications Department, IBM Corporation, White Plains, N.Y., 1968.
5   Hadley, G.: *Linear Programming,* Addison-Wesley Publishing Co., Reading, Mass., 1962.
6   Halbrecht, Herbert: "So Your Mother Wanted You to Be a Doctor but a Ph.D. in OR?", *Interfaces,* vol. 2, no. 2, pp. 44–49, February, 1972.
7   Hillier, Frederick S.: "The Application of Waiting Line Theory to Industrial Problems," *J. Ind. Eng.,* vol. 15, no. 1, January, 1964.
8   Machol, Robert: "The Screwdriver Syndrome," *Interfaces,* vol. 4, no. 3, p. 26, May, 1974.
9   "Mathematical Programming System/360 (360A-CO-14X) Linear and Separable Programming—User's Manual," H20-0476-1, Technical Publications Department, IBM Corporation, White Plains, N.Y., 1968.
10 Moore, James M.: "Optimal Locations for Multiple Machines," *J. Ind. Eng.,* vol. 12, no. 5, September-October, 1961.
11 Taha, Hamdy: *Operations Research: An Introduction,* The Macmillan Company, New York, 1971.

REVIEW QUESTIONS AND PROBLEMS

1   A nurse must assign five nurse's aides to one of each of five tasks. If the relative task times of each aide on each task with equal effectiveness are as given below, which task should be assigned each of the five aides to minimize the sum of task times?

Tasks

		I	II	III	IV	V
	A	3	2	4	6	7
	B	1	2	4	3	5
Aides	C	5	6	3	8	9
	D	2	3	5	7	1
	E	5	5	6	3	8

2 The costs for having five fire engines respond to five different locations simultaneously are given below. If the least-cost response indicates least cost to the property owner and less risk, which fire engine should be sent to which location to minimize losses?

Locations

		I	II	III	IV	V
	A	24	25	23	22	26
	B	26	25	24	23	23
Fire engines	C	23	22	22	27	26
	D	24	25	26	27	30
	E	24	25	25	26	23

3 If an auctioneer can sell only one item to each person, and the prices five customers are willing to pay for each item are as indicated below, to which individuals should the auctioneer sell which items?

Customers

		A	B	C	D	E
	I	6	4	2	3	4
	II	8	7	5	3	4
Items	III	3	7	5	4	4
	IV	6	2	3	7	7
	V	3	2	5	4	0

4 A sawmill company has three mills supplying four retail lumberyards. The costs to ship lumber in multiples of $1,000 per 10,000 board feet of lumber from each mill to each retailer are given below. Multiples of 10,000 board feet of lumber per month required by each retailer and available from each mill are also given. How many 10,000-board foot loads should be shipped from each mill to each retailer to minimize shipping costs for the month?

Retailers

		A	B	C	D		
	1	$3	5	5	7	40	Mill capacities
Mills	2	5	8	3	6	30	(10,000 board-
	3	4	9	3	2	100	foot loads)
		50	60	40	20	170	

Retailer requirements
(10,000 board-foot loads)

5 A commercial painting company has three paint crews. One crew uses an old sprayer (OS), another uses a new sprayer (NS), and the third paints by hand (H). The profits per 100 square feet painted by the three crews for four different buildings are as follows:

		Buildings			
		I	II	III	IV
	OS	$6	5	5	4
Crews	NS	7	7	8	2
	H	3	3	3	3

Buildings I, II, III, and IV are 60,000; 30,000; 20,000; and 40,000 square feet, respectively. The old sprayer and new sprayer can paint 40,000 and 60,000 square feet, respectively. The owner can employ as much labor as he needs to hand-paint up to 70,000 square feet per month. Which buildings should be painted by which crews to maximize profit? Assume that any combination of crews can be used on the buildings and that crews paint in increments of 1,000 square feet.

6 Assume that the following specify a generalized linear programming problem:

Maximize $f(x) = 2x_1 + x_2$

Subject to:

$$x_1 + x_2 \leqslant 6 \tag{1}$$
$$x_1 \leqslant 3 \tag{2}$$
$$2x_1 + x_2 \geqslant 4 \tag{3}$$
$$x_1, x_2 \geqslant 0$$

Graph this linear programming problem, identifying the three constraint equation lines and the feasible zone common to all of them. Plot dotted lines for values of 3, 6, 9, and 12 for the objective function $f(x)$. What appears to be the highest feasible value of $f(x)$, and for what values of x_1 and x_2 does it occur?

7 Assume that the mean arrival rate of moviegoers to a ticket window is three per minute and is Poisson-distributed. Assume also that the mean service rate of the ticket clerk is four per minute and that service times are exponentially distributed. After the system has reached steady state (e.g., at 11:15 A.M.):

a What is the probability that the ticket clerk has no one to serve?

b Is it more likely that one customer is buying a ticket with no one waiting in line, or that one person is buying a ticket and one person is waiting in line?

c What is the probability that there are at least two people waiting in line?

d What is the expected time an arriving customer will have to wait in line before being served?

e What is the expected number of customers waiting in line to be served?

8 Cars arrive at a state inspection center at a mean rate of 15 per hour. Assume that the arrival rate is Poisson-distributed. The inspection center has three parallel lines, and the mean service time in each line is 10 minutes. Assume service times are exponentially distributed. If cars form a single line while waiting, and cars enter any one of the three service cells when one becomes available on a first-come, first-served basis:

a Would this system be expected to exhibit steady-state characteristics?

b What is the probability that at 3:12 P.M. there are no cars in line or being inspected?

c What is the expected time a car will spend at the inspection station if it enters the system after steady state has been reached?

d Someone suggested that if the three service lines were distributed as individual lines about the city, each serving one third of the cars, the time a car would spend at an inspection station would be less. Do you agree, and why?

9 The figure below represents a cable design. The number on each path segment

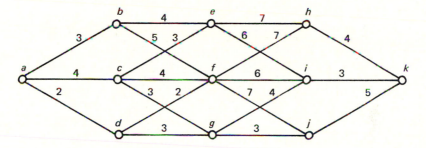

identifies the length of cable in feet needed to connect the two points in the circuit. If it is necessary to run an additional wire from point *a* in the design to point *k* along existing cable paths, what route should the new cable take to minimize the additional wire needed? Use dynamic programming.

10 Assume that in the cable design in Prob. 9, the nodes are towns and the numbers on path segments represent expected profit for a vendor selling vegetables house to house between towns. Assume also that the vendor wishes to go from town *a* to town *k* and make a maximum profit in doing so. Which route should the vendor take?

11 A space capsule needs to have some combination of components to supply oxygen to the cabin. Oxygen system data are given below.

Component	Oxygen (ft^3/min)	Component weight (lb)
A	2	1
B	5	3
C	7	4

Using dynamic programming, determine the combination of oxygen components that produces the greatest supply of oxygen without exceeding the weight capacity for the oxygen system. The maximum system weight is 8 pounds.

Chapter 7

Systems

All the flowers of all the tomorrows are in the seeds of today.

Chinese Proverb

The word "system" means vastly different things to different people. To an IBM systems engineer a system is a computer system. To an electrical engineer a system may be a collection of electromechanical components, such as the flight control or fire control system in an aircraft. For a mathematics professor teaching a course in matrix theory, a set of simultaneous equations may be referred to as a "system of equations."

Another sometimes confusing aspect of systems is that all systems, at least potentially, are part of a larger system. The fire control computer in an airplane is part of the offensive and defensive system of an airplane. This airplane is also part of the defensive or offensive striking power of some segment of a specific military service. This service is also part of our national defense system, which is a major component in our national economy. Our economy is a significant component of the world economy, and world governmental action may well determine the future of our planet as a component of our universe. How many universes are there as components of larger universes or cosmoses or whatever words one chooses to use to describe conceivably larger systems? The film "Orders of Magnitude" (23) dramatizes this limitless range of systems within systems. The problem, familiar to systems engineers, is the need to carefully establish the boundary of a system; otherwise, one must seemingly model the world to consider all potentially related factors. The establishment of a system boundary is more a necessary simplification than a reality.

Churchman begins his text *The Systems Approach* with the following (6, pp. 3-4):[1]

> Suppose we begin by listing the problems of the world today that "in principle" can be solved by modern technology.
>
> In principle, we have the technological capability of adequately feeding, sheltering and clothing every inhabitant of the world.
>
> In principle, we have the technological capability of providing adequate medical care for every inhabitant of the world.
>
> In principle, we have the technological capability of providing sufficient education for every inhabitant of the world for him to enjoy a mature intellectual life.
>
> In principle, we have the technological capability of outlawing warfare and of instituting social sanctions that will prevent the outbreak of illegal war.
>
> In principle, we have the capability of creating in all societies a freedom of opinion and a freedom of action that will minimize the illegitimate constraints imposed by the society of the individual.
>
> In principle, we have the capability of developing new technologies that will release new sources of energy and power to take care of physical and economic emergencies throughout the world.
>
> In principle, we have the capability of organizing the societies of the world today to bring into existence well-developed plans for solving the problems of poverty, health, education, war, human freedom and the development of new resources.

If we can, why haven't we? Is it because we lack the "technology" to do so? If technology means equipment and production methods, the answer is "no." If technology means "Can we appropriately implement solutions to these problems if we possessed the capital and equipment?" the answer is "maybe." This is what systems is all about. Stated another way, we simply have more gadgets today than we know what to do with. It is not so much a question today of coming up with a better mousetrap; rather, the more relevant question is given a full range of mousetraps when, where and under what conditions should mouse traps be employed to best serve the common good within limited resources available.

The automobile, the airplane, the computer, and the nuclear reactor are typical of systems of things. They represent engineering's best in the first half of the twentieth century. The second half of the twentieth century to date has seen a shift toward an additional interest in human systems. For example, who picks up your garbage? When is it picked up? Why does the garbage collector pick it up? Is it a "cost-effective" system? A study of the solid waste collection system in Cleveland (8) showed that $12

---

[1] Excerpt from *The Systems Approach* by C. West Churchman. Copyright © 1968 by C. West Churchman. Reprinted with the permission of Delacorte Press.

million annually was not necessary, $8 million was sufficient, as will be shown later in this chapter. How many systems are there in your everyday life which for lack of technical analysis have been passed down from one generation to the next with little thorough analysis as to how truly effective they are with respect to those they are intended to serve. Churchman puts it this way, "The optimistic management scientist looks forward to a 'systems era', in which man at last will be able to understand the systems he has created and lives in" (6, p. 43).

INTRODUCTION

The "systems approach" is by no means new; politicians have been promising it for a very long time. Succinctly, Churchman states (6, p. 11):[2]

> The idea of a "systems approach" is both quite popular and quite unpopular. It's popular because it sounds good to say that the whole system is being considered, but it's quite unpopular because it sounds either like a lot of nonsense or else downright dangerous—so much evil can be created under the guise of serving the whole.

For years, the emphasis of traditional industrial engineering was "efficiency." How can we do the same thing faster? What fixtures and tools and what sequence of fundamental hand motions will increase output at a workstation? Only recently has there been a shift toward creative system design; both Nadler's (22) and Krick's (18) texts are typical of this overdue change in emphasis. Of course, that is not to say that efficiency is no longer needed. On the contrary, the recent recession in the United States, in the author's opinion, can be at least partially traced to the abandonment of traditional cost control techniques in an essentially unsuccessful search for "behaviorally" oriented labor control techniques that unfortunately violate fundamental principles of human nature. For example, why work if you don't have to, when faced with the reality of putting nut A on bolt B for 480 minutes every day, day after day. Real output per worker hour in the United States, however, is still fundamentally the reason why each worker (i.e., all of us) in the United States enjoys so much real output. Today, when "big government will take care of you if you qualify," some people seem unwilling to accept the fact that the average worker in the United States should expect to enjoy no more and no less than what an average worker produces in a day. Efficiency is necessary, but not sufficient, in solving the problems posed by Churchman. The purely militaristic and authoritarian labor control techniques of the past are inadequate, as have been the permissive and naive "humanistic" approaches of the very recent past. Between these extremes lies an appropriate compromise.

In the February 1975 issue of *Interfaces,* Chinal (5) offered a summary of five years of experience in founding and heading a systems engineering program at the Ecole Nationale Superieure de l'Aeronautique et de l'Espace in Toulouse, France. The article, although more specifically related to the aircraft industry, provides an

---

[2] Excerpt from *The Systems Approach* by C. West Churchman. Copyright © 1968 by C. West Churchman. Reprinted with the permission of Delacorte Press.

interesting introduction to systems engineering in general. My commentary, which
constitutes the accompanying footnotes, attempts to highlight points raised in the
article. In the interest of brevity, the article abstract, footnotes, and references have
been removed from the article as reprinted here.[3]

The Systems Approach:
A French Experience

Jean P. Chinal

The Systems Approach has become an increasingly familiar way of attacking various
problems in engineering, management and possibly, other areas.[4] As a new viewpoint
it can be detected today in three ways; in its applications, in the disciplines associated
with its development and, in the various sources, some outside engineering and
management, which can be ascribed to it.

Applications where the systems approach has been variously identified, claimed
or advocated in the last decade or two have included three successive waves of
problems with decreasing technological contents.[5] First, the aerospace and defense
industries, during the sixties, were able to implement technical systems of stunning
complexity and hence accredited the idea that their successful completion was due to
some original combination of technological and managerial competence which came to
be known usually as the systems approach. Later, near the end of the sixties it was
frequently suggested that such a systems approach might also be a likely method of
solving other problems with lower technological contents, but a greater societal
component and which would be of at least the same overall complexity as the
aerospace problems. Typically this was claimed in the USA for projects such as public
transportation, urban development and water resources. Lastly, and to this day, the
systems approach appears to be advocated to solve problems with little or no
technological contents, which, hopefully, might yield even better to such a
purportedly novel approach. A growing number of books attests today to such diverse
applications as educational curriculum development, language teaching strategy or
organizational development of the firm.

Several modern disciplines also evidenced a concern with the concepts of systems
and systems approach.[6] This is evidenced in their terminology, research

---

[3] Reprinted from "The Systems Approach: A French Experience," by J. P. Chinal, *Interfaces*,
vol. 5, no. 2, February, 1975, pp. 47–56, published by The Institute of Management Sciences.

[4] There is considerably increased reference to the systems approach in today's technical
literature.

[5] The systems approach progressed successively from (1) systems of things to (2) human
systems to (3) such diverse connotation to the word *system* that it now lacks a specific meaning.

[6] Industrial engineering, and electrical engineering to a lesser degree, have probably been the
most overt disciplines in claiming the study of systems. As mentioned earlier in this text, many
industrial engineering departments have changed their name to industrial and systems engineering.
This may be justified if one considers that the function of industrial engineers has significantly

preoccupations and general goals. Most notable in this respect are (control) system theory, computer science, management theory and operations research in the more analytical sense. There appeared over recent years two tendencies in this respect: one was for each discipline to claim the system concept as part of its field, the other, on the contrary, was that we should be able to identify, extract and reformulate some rather general systems concepts, in a manner relatively invariant with respect to these disciplines. This second attitude has resulted in the creation of separate academic programs, colloquia and publications devoted more explicitly to the systems approach in management or engineering.

Lastly, if one looks for the sources of the systems approach it can be found that some of these lie outside the fields of engineering and management. Some strikingly similar concepts appear to have evolved from the Gestalt psychology, linguistics and anthropology to eventually make up the structuralist philosophy and general systems research. It would be hard to prove that there has been an explicit and conscious transfer of concepts between these various disciplines and those of engineering or management and a slow cultural diffusion may in some instances have been the only process of interaction. But it is hard even nowadays to dissociate the basic principles of systems in these various fields and today the structuralist principles and philosophy no doubt play an active role in some very practical areas such as creativity techniques or language teaching.

This threefold variety of applications, of disciplines and origins poses the problem of the nature and usefulness of such a systems approach. In what follows we describe a French experience to express and teach systems engineering at ENSAe France, since 1970. It is shown how from the experience of aerospace systems projects and from a reflection upon the school's own curricula in related fields of control, computer and electrical engineering, the decision was made to develop a new, separate program, based upon a pragmatic concept of systems, as evidenced in the French aerospace and defense industries. This five-year experience is then commented upon and summary of opinions of students, course alumni and industry systems engineers on the concept of the systems approach and teaching is presented.

Engineers' training in France and the USA presents some significant differences and the following points should be noted concerning ENSAe. This is an aeronautical engineering school, recruiting students through a highly selective nationwide exam requiring two years of preparation. It then trains engineers in an additional three years and produces a yearly output of 100 graduates, half of whom go into aerospace and others into high level technology areas such as automation systems, computers and electronics. Due to this market structure for its graduates and to a very explicit training philosophy of the French aerospace industry, the school for many years has been teaching aerospace engineering in its widest sense, i.e., including in it not only

---

evolved in the direction of designing and controlling productive systems (e.g., corporate and other large-scale systems) as compared to a much narrower function twenty years ago. General systems theory is by no means new, but truly general systems theory is in its infancy. In fact, general systems theory is practically nonexistent.

intrinsically aeronautical disciplines, such as boundary layer theory or flight mechanics, but also courses from all major cognate or supporting disciplines, most notably control systems, computers and electrical engineering. Filtering, dimensioning and integration into several three-year programs of some 140 courses is achieved within the school through a full-time central coordination authority, in contrast with the USA where this is the combined effect of student and adviser's thinking. As a result, students are exposed to a much greater variety of subjects than their U.S. counterparts. Thus the explicit philosophy of the school is to deliver engineers with a definitely generalist outlook.[7] This is based upon the industry's and government agencies' specification that more specialized profiles of engineering competence, while obviously useful, should be found in other schools or universities and that ENSAe's graduates should be prepared early to assume positions involving functions of coordination and integration at both management and technical levels.[8]

Interest in the systems approach for engineers developed about 1968 at ENSAe for three main reasons: The generalist connotation of the systems approach, the growing realization in French industry, on an empirical basis, of the systems aspects of various major aerospace projects and, lastly, the increased importance given to the general systems concepts in various disciplines taught at the school such as control systems, computer science and electrical engineering.

Two classes of questions had to be answered at that time concerning, first, the systems approach and, second, its teaching. With respect to the systems approach:

(i) Is the systems approach in engineering really different from the combined effect of established disciplines such as computer science, control system theory and operations research?[9]

(ii) If at least partly different, which part of it can be really made explicit and applied to future engineering projects?

With respect to teaching:

(i) From current industrial practice and existing technical literature, what can we extract and teach?

(ii) How should teaching be combined with existing curriculum at school, should it be diffused into existing courses, set up as a new, last-year optional program (like those in structures, propulsion or avionics), or should it be a separate, postgraduate program addressed to graduate engineers?

We now turn to these two problems.

---

[7] The generalist philosophy in engineering is finding a rebirth, as did "family practice" in medicine. For many companies coming to campus to hire engineers, the industrial engineer has been this generalist. In this author's opinion, this demand for the technically trained applied generalist is the primary factor for the present and forecast future high demand for the industrial engineer.

[8] It is not uncommon for industrial engineering graduates to enter management positions in technical companies sooner than graduates of more technically specialized engineering disciplines.

[9] This is one of few references to IE/OR/MS in this article.

SYSTEMS: AN EMPIRICAL VIEW

The concept of system underlying expressions such as "systems approach," "systems engineering," "systems design," or "systems test" can be traced to both theoretical and empirical sources. However, as we shall see, the great variety of these sources will eventually favor an inductive and empirical definition.[10]

There is in fact today such a variety of disciplines and applications areas using this terminology that the word system has become frustratingly vague and ambiguous. As a determinative in the above expressions it is still relatively clear, but as a determinate in, for instance, "weapon systems," "hi-fi system," "linear system," it has become hopelessly vague and in many current language uses has come to be not much more than an all-purpose word, a kind of syntactic "marker" or "classifier" evidenced in some languages. There is usually an interesting moment of shock when two specialists of different disciplines such as, say, computer systems programming and airborne navigation systems, compare notes on this.

Actually, systems are defined first in a general way and then each field or specialty adds some features of its own, explicitly or not. In a general sense, a system is a set of elements, organized as a whole, so as to fulfill one or several missions (objectives) and subjected thus to various interactions.[11] In any more specific sense, such as in engineering, the above definition is refined in practice, through either the use of complementary statements or, perhaps more frequently, by unstated but actual practices which should be identified. The general definition also explains the apparent initial commonality of the systems concepts and subsequent discrepancies in their actual use. Attempts can be made to delineate strictly each field and to pinpoint its specific actual use of the systems concepts, but these attempts remain frustrating and their results somewhat hard to communicate owing to the vastness of knowledge to be surveyed.

The view at ENSAe was that disciplines or activities can best be defined by the situations they actually deal with rather than with long-term goals. On the basis of this, the following example may serve as a typical system in the aerospace sense and will introduce the general identification criterion we give subsequently.

Modern aircraft or satellite launchers are systems.[12] For example, a supersonic aircraft is basically comprised of structure (fuselage, wings, etc.), turbojet engines, hydraulic systems and avionics. Avionics itself may include autopilot, air data computer, inertial guidance system, doppler radar, frontal radar, map display, airborne digital computer, input-output unit, etc. All these pieces of avionics equipment are electrically interconnected and together make up larger subsystems such as navigation systems, flight control systems, etc. Their various physical interconnections actually

[10] When the need exists and theory is lacking, the approach of necessity is inductive and empirical. An empirical base provides the source, however, for deductive reasoning and consequent future theory.

[11] Chinal's definition of a system.

[12] Previously referred to as systems of things, such systems are also commonly referred to as "engineering systems," "technical systems," or "hard systems." "Soft systems," those including humans as a sensitive or critical element, have been referred to as "human systems."

materialize the functional interdependence of the component "black boxes." For example the Mach number computed by the air data computer is not only displayed to the pilot, but also simultaneously fed to the autopilot, to the air intake mechanism and to the central digital computer. Similarly, the attitude (yaw, pitch, roll) delivered by the inertial guidance unit is used as a reference for the radar, the doppler, the central computer and the autopilot. On a modern aircraft, dozens of other interconnections and interactions of this type could be easily evidenced.

The above is a typical example of systems. Elements, subsystems, systems, and their interactions appear in a technical sense and will determine problems in design and development of the whole system. The system problem will be to break down the overall task in space and time: i.e., on one hand, to decompose respectively the system into subsystems down to a level of "complexity" that can be entrusted to specialists, and on the other hand, to fractionate the project life cycle into phases so as to gradually face the technological and conceptual unknowns in the system. Then, to determine and execute the best structure to conduct this double decomposition and ultimate reassembling without losing track of the overall project and of the overall mission of the system being constructed is the difficult systems engineering problem.[13] Systems may thus, in a technical sense, be said to:[14]

- have sufficient material complexity to require breakdown into physically identifiable subsystems on the basis of conceptual, technical or industrial reasons alone or in combination
- display numerous and/or strong interactions between the parts resulting from the decomposition process
- be so complex as to require, for design and development, the combined efforts of several independent groups (separate departments of a corporation, distinct corporations, disseminated subcontractors)
- be so complex as to require a breakdown in time of the design and construction effort into various phases, separated by major decisions
- pose a non obvious allocation problem of system parameters such as time, funds, reliability, safety, power, to various participants or to various phases
- require a combined application of several or possibly many different technologies.

Systems Approach Originality

Opinions of practicing systems engineers in French aerospace industry and re-examination of our own teaching curricula in electrical engineering, automatic control and computers revealed that the systems approach was generally perceived as different from the doctrine embodied separately by each of these disciplines and others too.[15] This was so because

[13] The "overall mission" aspect is a common requirement in system studies and typically identifies it as a systems study.
[14] This list seems oriented specifically to physical systems; however, it is general enough to include human-machine systems.
[15] There was apparent awareness that electrical engineering, automatic control theory, and computers, although important disciplines, were not sufficient for solving the full gamut of aerospace industry problems.

(i) Awareness for the systems aspects of technical projects was often gained from actual practice. This usually included, first, negative experience from projects having experienced difficulties where the absence of a "global" or "systems" approach could afterwards be proven to have caused problems, and, second, positive experience where successful projects suggest in retrospect (but perhaps do not prove) that the systems approach was used and is likely to have determined success. People with systems experience in aerospace had been trained fifteen years earlier as aeronautical engineers and had not always received extensive formal training in electrical engineering, automatic control or computers. Still they had acquired an empirical notion of systems, had thought it to be useful and had conducted or participated in successful projects.

(ii) Actual practice of electrical engineering, control systems and computer science evidence, beyond a common use of the system terminology and general philosophy, widely varying emphasis at the respective practical levels. Each field has strongly marked preoccupations, skills, and research interests, and specialists exist in each field who do not transfer that easily from one field to the other. A similar case could be made regarding disciplines such as operations research (which then was not taught at the school). This also suggests strongly that the practical common contents of these disciplines is limited if one goes beyond theoretical or philosophical concepts and that otherwise systems approach concepts are either scattered among them or concentrated in one of them. But this last view is clearly untenable: although contributions of each discipline are unequal, the smallest are not inexistent. So that examination of the systems approach in regard to the above "space age disciplines" leads to the view that it is made up of very general "upstream" principles, of specific techniques situated in some of these fields and, possibly, of new concepts outside these. Actually, this statement need hardly be disappointing and such a situation is not specific of systems. General theories of particle physics allow one, in principle, to make a case for the unity of physics and of all disciplines derived from it, but in day-to-day practice for specialized fields with specific methods, this potential unity might not be very readily apparent.

(iii) These aerospace and defense projects of the sixties which had most consistently evidenced and/or claimed a systems approach were strikingly interdisciplinary in nature.

(iv) Various French companies in aerospace or related fields had set up special systems or integrated groups on the basis of an apparently empirical concept of systems, with the aim of better integration of complex projects from design through construction.

(v) A distinct concept of systems and systems approach had emerged in literature and course offerings of universities. (This was mostly a re-inforcing argument. At the time (1968) of the investigation, much useful material had not been diffused in France.)

The Teachable Contents of the Systems Approach

From the above analysis, the teachable contents of the systems approach was seen at three levels of formalization, those of principles, methods and techniques.

(i) *Principles*

- Conduct analysis and design while constantly keeping in view the system as a whole.

- Assume the a priori existence of internal relationships between elements, subsystems, and external relationships with the system environment. Be ready for unexpected or latent relationships, other than those suggested by routine, experience, plain common sense and intuition.
- Give explicit recognition to assumptions or axioms influencing system design. Beware of hidden assumptions left out as a result of mental inertia or blurred on purpose to hide deficiencies. Subject them to mental experiments to avoid omitting important assumptions which would be belatedly revealed by technological or managerial crises.

(ii) *Methods*

Methods or procedures express in relatively normative style the best known rules of the art and are occasionally expressed in manuals issued by private consulting firms or public organizations (e.g., USAF System Command Manuals, NASA or ESRO project management manuals, etc.). They deal in the case of systems with decomposition in time and space and management of the bits and pieces.

(iii) *Techniques*

Those techniques were selected which are most typically systems oriented in that they relate behavior of complex structures to those of the elements and to the existing interactions. Typical in this respect are: scheduling methods (PERT–CPM), computer simulation, reliability, safety and maintainability techniques.

Systems Engineering at ENSAe

A nine-month postgraduate program was set up starting in 1968 and opened in 1970. It was addressed to engineers graduated from the school, or other similar engineering institutions, as a sixth year of study. The program was expected to increase their readiness and awareness for systems work, to shorten the duration of the still necessary on-the-job training and to enable them to generally better integrate all aspects of systems design and development.

Fifty-five engineers from five countries were trained (a figure which compares with or slightly exceeds those for similar programs in control, computer and propulsion at ENSAe).

1 *Acceptance of Systems Concepts by Students*

Students who attend the program appear to be motivated by its management component and its inclusion of techniques not normally taught at length in French engineering curricula (e.g., reliability, safety, maintainability, etc.).[16] They wish to acquire a broader engineering outlook and competence. Also, they hope to graduate faster to positions of responsibility within large scale projects.

In contrast to their previous engineering training they are introduced to problems which are not as well defined as the more restricted analysis or design tasks they know best. They have to get accustomed to the fact that systems have a fuzzy structure and objective at the start, which can be dissipated only gradually through project life. Such

---

[16] A similar interest in management or overall systems design and control courses by engineers trained in technically specialized fields, such as electrical, civil, mechanical, and chemical engineering, is evident in the proportion of such engineers, probably more than half, in graduate-level programs in industrial engineering.

a transition from a more analytical, technique-oriented attitude to a systems viewpoint is noticeable and usually takes two to three months. The initial preference for analytical tools makes room for a concern for methodology. That such a concern may be awakened is, for the school, a positive result; it is felt that the lack of prior systems experience or sensibilization in most of the students can be in part overcome.

2 By Course Alumni

Alumni of the systems engineering program (55 thus far) for the most part seem to view Systems Engineering not as a technical competence but as a combination of both an attitude toward engineering problem-solving and knowledge of explicit methods or procedures for the conduct of large engineering projects.[17] Most of those contacted in 1974 see the program as satisfactory with regard to the build-up of such an attitude but state a yet unsatisfied curiosity for the methods, which they tend to think are not sufficiently developed yet. They also find that the systems engineering concepts are not sufficiently disseminated within industry and that the structure of the firms is not adapted to the inclusion and subsequent practice of such a "speciality."

3 By Industry

On the other side, industry (i.e., here mostly aerospace, electronics, defense and automatic systems sectors) has been associated with the program since its inception through the following actions: preparation of the curriculum, participation in the teaching ($\frac{1}{3}$ of total class hours plus in-house projects), and participation in seminars and surveys on the systems approach in engineering.[18] Opinions received on these various occasions can be summarized as follows, with respect to the systems approach and to the specific systems engineering program set up for its teaching since 1970: Interest appears to focus on the systems design aspect and only to a lesser degree on systems analysis tools. Such an attitude probably reflects only a perception of those areas where the present knowledge is the last satisfactory today and not necessarily an opinion on the respective roles and importance of systems analysis, system design and systems evaluation. Emphasis is on the pragmatic aspects of large systems engineering processes, much less on mathematical techniques. Such an attitude can apparently be ascribed to the following circumstances:

- Systems analysis techniques are already relatively well known.
- Systems analysis plays its most important role at the initial phase of project life, while here it is the whole system project life which is of concern.
- French engineers, who receive, while students, a notoriously abstract and massive mathematical training (even 10 to 15 years ago), in later years develop a keen

---

[17] Systems technology is diffuse at present, developing slowly in numerous disparate disciplines. Few if any sources offer a comprehensive distillation of systems experience, methodology or techniques to date.

[18] Chinal has unintentionally identified one of the few primary weaknesses of the engineering educational systems in the United States. That is, (1) professors are hired in U.S. engineering schools more often than not with no industrial experience behind them, and (2) there is nominal involvement of industry people in teaching and vice versa. European engineering students, as well, have more exposure to the real world during their education.

awareness of the practical limitations of mathematics, occasionally bordering on scepticism.[19]

With regard to the idea of teaching systems engineering, industry in general has reacted very favorably, sometimes enthusiastically. Reservations have usually been of two kinds: first, that existing courses in our school were already well on their way toward that viewpoint and that the systems approach should be retrofitted into existing curricula rather than turned into a new one; second, that the role of experience and ability for human relations and leadership were so vital in systems work that engineers trained to this end in a special program would still have to pass the test later in industry before they could be accepted without reservation by their colleagues.

These objections are incompletely resolved today. Integration of the systems approach into the basic curriculum is certainly worthwhile considering and ENSAe's own solution of a distinct program was dictated in part by some specific and contingent features of the French school systems (loads of programs, incompressible basic disciplines, etc.). The second should be lessened by the planned introduction of courses on human aspects of system development.

There remain three key problem areas: First, the ability to transfer systems experience from one area to another, e.g., from radar systems to civilian aircraft design, appears lower than was expected five years ago.[20] It is harder than was assumed to abstract the particular technology underlying large systems projects, and many typical systems experiences often do not lend themselves to generalization. What often had been found intriguing, challenging or decisive for project success turns out to be heavily contingent upon a particular technology or a particular application. This is a somewhat disappointing aspect and even within the restricted field of aerospace, transferability is still low.[21] We do not have, yet, any strikingly universal methods for systems analysis or design such as was, for instance, the frequency domain approach two years ago in feedback control systems, which could be applied uniformly, regardless of implementation technology (electronic, mechanical, hydraulic) to basically isomorphic problems.

Second, systems engineering is in a sense an attempt at defining a methodology for the conduct of the large-scale system engineering process.[22] It is in part an effort to make more explicit and more rational the design and development activities and

---

[19] As cited in Chap. 6, Halbrecht (10) indicated that far too many operations researchers in the immediate past have failed to gain this "keen awareness of the practical limitations of mathematics."

[20] History has shown that the product of deductive reasoning typically follows considerable empirical study.

[21] Hopes to employ engineering systems technology, such as advanced concepts of feedback control theory, have been relatively unsuccessful to date. The concepts of negative feedback control germane to Forrester's (9) systems approach are, for example, truly fundamental and yet powerful.

[22] Systems engineering is concerned with one of the most fundamental engineering questions of all time; that is, What is it and how do we go about designing and constructing something that best serves the whole with respect to some identified general need? Here lies a hint as to one of the most perplexing aspects of systems engineering—consideration of value and utility.

thus replace intuitive guesses or purely empirical tricks or recipes, perhaps acceptable for smaller projects but no more for large ones. But if we look now at science in general, we can see that, compared to the immense knowledge accumulated through the scientific method, rather little has been said on the mechanism by which such knowledge has been acquired. Also, recorded methods may not, in actual practice, be followed nor be decisive. Methodology of science is mostly implicit or acquired on the job. Thus, it is not quite obvious how much benefit would be gained by such an explicit methodology, unless, and this is important, we consider that science and engineering are significantly different. In this respect there might well be a sharper difference in the French view than in the U.S. one. More than the USA, French engineering education has long separated research and teaching, and engineering from science. This is attested for instance by a long standing division of French higher education in two nearly disjoint sets of educational systems, schools and universities, and by the practice of many engineering schools of having no permanent professors but only part time faculty contributing the results of research done outside the school in industry or research establishments. Although these distinctions are now decreasing, they suggest that Systems Engineering as a design methodology may be an attractive idea in a French context, more so perhaps than in the USA.

Lastly, on the sociological level, acceptance of the systems approach concepts and teaching has some restrictions, typical, probably, of any new discipline. The notion of a generalist associated with the system competence appeals to managers, but perhaps slightly less to senior project engineers. These latter actually wish their young engineers to make their mark in a more established discipline for which references exist, and they may be somehow suspicious of newcomers who by virtue of their new form of training would seem to evade that test. Moreover, due to the generalist connotation of Systems Engineering these senior engineers want to be quite sure that such engineers will actually be able to do the nitty-gritty work and not only deal with comfortably general analyses. Due to these restrictions it is hard to imagine Systems Engineers who would be just that from the start, and to be acceptable today, such competence needs to be combined with some professional achievement in a known field.

Conclusion

It appears after six years of experience that the systems approach in engineering is a concept which, in a design perspective, still retains some vagueness.[23] As a diffuse notion, it has permeated many minds and disciplines in the last decade or two and thus often seems familiar. But as an explicit body of knowledge clearly visible within existing disciplines, or existing autonomously beyond them, it has developed slowly and progress is probably not at par with the popular and sometimes enthusiastic acceptance of the idea. This is due to the difficulty of extracting non-philosophical and general concepts from current systems problems, and to sociological factors relative to the practice of science and engineering.

---

[23] The history of science indicates that major findings do not come easy. Occasionally there is a rare exception, such as Goodyear's spilling a mixture of natural rubber and sulfur onto a hot stove, accidently inventing synthetic rubber.

It has been shown at ENSAe to be convenient and feasible to express, codify and teach systems on the basis of an empirically defined, design-oriented contents and educational objective, with courses relying on both theory and present-day industrial practice. As a concept and a training program, it has received mostly positive acceptance from students and industry. As a result, the particular spectrum of competences implemented will be perfected. Indeed, one major aerospace firm in answer to a school survey, recently wrote: "Most of the courses given are well adapted to the problems our engineers have to treat in our firm. We definitely wish to cooperate with you for the extension and improvement of this form of education."

Appendix: Systems Engineering Program at ENSAe

The first six months are divided in modules whose cumulated duration amounts to one week's work full time. The list of modules is given below; the structuration of modules is not indicated here.[24]

(I) *Theory*

- Algebraic techniques
- Estimator theory and test of hypotheses
- Linear and non-linear programming
- Graphs and scheduling
- Dynamic programming, notions on games and queueing
- Multivariable systems
- Control of multivariable systems
- Mathematic theory of reliability

(II) *Methodology*

- Computer simulation
- Methodology and management of reliability
- Systems life and maintainability
- Creativity
- Large-scale project management
- Multilevel systems
- Diagnosis of failures in systems
- Computer interfaces
- Systems safety
- Decision analysis
- Real time computers

(III) *Examples and Case Studies*

- Cost effectiveness study
- Warning and defense system
- Air transportation
- Satellite system design

---

[24] With little exception, the topics suggested in all three categories fall within the domain of IE/OR/MS, statistics, computer science, and systems analysis.

- Urban dynamics
- Traffic
- Civilian aircraft analysis

SYSTEMS STUDIES

In an effort to give the reader an opportunity to deduce what systems engineering is today, three articles were selected and are reprinted below. They are "A Perspective of the Industrial Engineer's Role in Urban Mass Public Transportation" by Lewis (19); "State Criminal Court Systems" by Brach and Daschbach (2); and "Blood Bank Inventory Control" by Jennings (16). All three articles have a lower than average mathematical content and are more descriptive than most systems studies. It is hoped, however, that they will provide some insight into the nature and diversity of systems studies presently under way and being reported in the technical literature. Some editorial license has been employed in the interest of brevity in omitting references, nonessential abstracts, and footnotes.

A Perspective of the Industrial Engineer's Role in Urban Mass Public Transportation

H. A. Lewis

INTRODUCTION

The role of UMPT is to provide for the movement of people in our urban environment. The major system requirements include: (1) minimizing air and noise pollution, (2) eliminating fatalities, (3) conservation of natural land resources, (4) minimizing consumption of energy resources, (5) relieving automobile congestion, and (6) providing mobility for the 25 to 50% of Americans who are dependent on urban mass public transportation.

Some of the reasons supporting the type of system requirements defined above are becoming very obvious in American society. For example, it has been estimated that 80 to 90 percent of the air and noise pollution in Los Angeles is contributed by the automobile. Also it is estimated that on a national level the automobile accounted for approximately 55,000 traffic deaths in 1971. In addition, there is increasing concern for the preservation and efficient use of our natural land resources. The energy crisis related to shortages of petroleum supplies has become very acute, and there is a very urgent need to find an alternate source of energy or means of transportation within our urban environment. Also, it is becoming more and more apparent that a large percentage of Americans live in our urban areas, and that the increased use of the automobile (approximately 100,000,000 on our urban streets) is creating congestion. Further, estimates indicate that as many as one-half of all Americans can be included in the category of being dependent upon UMPT.

For the past 30 years urban transportation has declined introducing a degenerating cycle leading to a smaller demand for mass transportation which has fostered inadequate service and further reduced demand. The small demand then leads to low revenues. Under these conditions, urban mass transportation managers are unable to justify improvements in facilities based on profits not now experienced nor foreseen in the future. Such conditions may force many private transit companies to discontinue service if the operation must realize a profit.

There are many factors that have and are still adversely affecting urban mass public transportation. These factors include the growing propensity for the private automobile, the existence of poor operating equipment, lack of advanced technology, inadequate service, physical barriers of residential development, peak hour demands, and personal attitudes toward mass public transportation. Changing the adverse effects of these factors will require a major effort.

Consider briefly some of the effects of discontinuing urban mass transportation. First of all a large segment of the urban population within our country who are dependent upon UMPT would be without inexpensive transportation. This condition readily impacts social and economic benefits and freedoms that all individuals should have within the urban community. On a larger scale, the discontinuance of UMPT eliminates a link in our nationally integrated transportation network.

Developments over the past several years, however, have created a more favorable environment for the development of new, efficient urban mass transportation systems. The first major development was the creation of the Urban Mass Transportation Administration (UMTA) in 1968. The major role of the UMTA is to provide a coordinated solution to the transportation needs in our urban environments. Secondly, the passage of the Urban Transportation Act of 1970, which authorized UMTA to spend $3.1 billion over 5 years and $10 billion over 12 years, has provided funding to improve UMPT. Thirdly, the 1970 Highway Act authorized the use of Highway Trust Fund revenues for exclusive busways, passenger loading facilities, bus shelters, and fringe and corridor parking facilities to serve *any* type of public mass transportation. The Federal Aid to Highway Act of 1973 further broadened the uses of the Highway Trust Fund for urban transportation. This legislation allows local agencies to divert Highway Trust Fund monies to urban transportation. Another positive change leading to improvements in urban transportation is the emergence of transportation departments within individual states. This evolutionary process had led to the creation of 19 State Departments of Transportation. Lastly, an improved urban transit environment is being stimulated by large industrial firms and organizations, such as Ford, Mobil Oil, the American Road Builders Association, and the United Auto Workers. They agree that adequate financing of public transportation is needed and others are actively engaged in encouraging the development and marketing of advanced urban transit systems which utilize new technology.

SOME OPTIONS FOR UMPT

Having presented an encouraging outlook for the development of new and improved urban mass public transportation systems, consideration will now be given to some of the options available to urban public transportation planners. A few of these systems will be examined to gain a better understanding of the types of systems available for implementation in our urban environment. These systems may be broken down into three categories: buses, automatic transit systems (loop rapid transit, group rapid transit, and personal rapid transit) and rapid rail transit systems. The following paragraphs review these systems to gain some understanding of the advantages and disadvantages of each.

Bus Systems

Bus systems can be defined in two categories: (1) a fixed route system and (2) a demand response system, or dial-a-bus system. Fixed route bus systems operate on pre-established routes for pick-up and delivery of patrons. The vehicle capacity is between 35 and 50 persons. The main advantages of the fixed route systems are that their operating costs can be accurately established in advance; scheduling is relatively simple and easy to establish; and fixed route systems use larger buses which carry more people per vehicle. The disadvantages, however, are that the larger buses generally do not meet demands in low densely populated areas. These areas require extensive route networks to cover the low-density urban area; and thus, increase passenger travel time. Further, buses experience the same congestion as automobiles when operating during peak hours. Presently, UMTA is sponsoring a R&D program, TRANSBUS, to develop new designs and standards with new emphasis on service to the elderly and handicapped.

The dial-a-bus concept is a door-to-designation service, similar to limousines and taxis, which responds to individual requests for transportation. However, because the dial-a-bus concept utilizes a 10 to 15 passenger vehicle it can transport more passengers per hour and thus potentially reduce the fares. (Taxis and limousines have a lower productibility and the fares are high in comparison with UMPT services.) Most existing dial-a-bus systems are tailored to needs of the elderly and handicapped. An UMTA sponsored system in Haddonfield, N.J. called Dial-a-Ride serves mainly commuter needs to the Lindenwold rapid transit line. The main advantages here are that the demand response system can more effectively meet demands in low density areas than a fixed route system; it more closely approximates the convenience of the automobile in serving the elderly and handicapped. Some disadvantages of this system are its present inability to cover the cost of operation, difficulty in scheduling for random demands on the system; and large-scale systems require a complex computer information network for dispatching and storing data.

There are some distinct advantages in using mass public bus transportation. For example, presently buses are the only form of mass public transit in most of our cities. Secondly, the service capability might easily be extended in a short time. Buses also provide an interconnecting link to other modes of transit such as rapid rail systems, airports, and other transportation centers. Buses also have the best potential for moving large numbers of people economically. Since buses operate on present networks of streets and highways, they are versatile and flexible in their routing and experience lower initial capital cost. In addition, buses can decrease traffic congestion by replacing many automobiles, and thus, decrease air pollution in urban areas. Mr. Frank Herringer, Administrator for UMTA, states that "the bus will remain the mainstay of transit systems in this country for the foreseeable future."

Automatic Transit Systems

A second form of UMPT that applies new technology to urban transit needs is the automatic transit system. Systems under this category include loop rapid transit (LRT), group rapid transit (GRT), and personal rapid transit (PRT).

Loop rapid transit (LRT) systems use automatically controlled vehicles or trains operating on paths or simple service loops. The vehicles or trains follow one another in an unvarying circuit such that the networks are not coupled at intersections, thus preventing switching from one line to another. Headways are greater than 60 seconds. Low-speed systems (less than 35 mph) make short trips and serve major activity centers such as large air terminals, central business districts, universities, shopping centers, and recreation areas. Examples of such systems included the Westinghouse Transit Expressway at Tampa Airport, Florida; the Seattle-Tacoma Airport, Washington; and the Monorail System Rohr at the Houston Texas Airport. Higher speed systems (up to 70 mph) would be desirable for wide area service for large cities. For higher speeds (up to 150 mph), Rohr is developing an LRT system called Urban Tracked Air Cushion Vehicle (UTACV). This demonstration system incorporates advanced state-of-the-art components like air cushions for suspension and linear inductor motors for propulsion and braking. The major advantage of LRT systems is their automatic control to reduce labor cost. The main disadvantages include its inflexible routing because of the fixed guideways that do not permit switching to other networks. Also, the initial investment cost is quite high. While these systems are faster than normal walking rates, they do not compete with automobiles or buses having average speeds of 10 to 20 mph, and are thus not adequate for transit across urban areas.

Group rapid transit (GRT) systems advance the automatic systems by incorporating switches for off-line stations. This allows passing some stations in route to another destination, thus headways can be reduced (5 to 6 seconds) on the main line. This improvement, combined with use of a vehicle which carries 10 to 50 passengers can provide flow rate capacities as high as 30,000 persons per hour. Present systems exhibit speeds below 30 mph with stations 3 to 5 miles apart. Representative systems of the GRT class include Airtrans by LTV at the Dallas-Fort Worth Airport, Texas; the Morgantown system at the University of West Virginia by Boeing and Alden; the Transportation Technology Inc. system by Otis Elevator Co.; the Monocab system by Rohr; and the Dashaveyor by Bendix; the latter three were displayed at TRANSPO-72. The major advantages of this system other than its automatic operation are the higher carrying capacity and the ability to switch to off-line stations. Velocities of these systems still do not compete with the automobile and thus, would not serve low-density population areas. Other disadvantages include its high initial cost of investment, the considerable amount of right-of-way required, and little routing flexibility.

Personal rapid transit (PRT) systems represent a very advanced concept for automatic urban transit systems. The concept for PRT is a personalized vehicle (one to four persons) which provides non-stop service from any origin station to any destination in a network covering an entire urban area. PRT is a design concept to provide automobile-like service, or personal transit when and where desired. To do so, area-wide networks would be required with hundreds of nodes that allow vehicles to turn right or left or continue straight. The computer providing automatic control is programmed to optimize the route from origin to destination and to account for all vehicles at all times. Speeds up to 60 mph are envisioned, but the short headways desired for high capacity flow (in the order of tenths of a second) become very difficult to maintain with high reliability. The main advantage of the PRT system is that its service closely approximates that of the personal automobile. Major disadvantages include the requirement to overlay a complex network of guideways on an urban area, the high initial cost, the unproven technology in computer communication and control, linear inductor motors for propulsion, and the inability to accurately evaluate the long term cost.

Rapid Rail Transit

Rapid rail transit systems (RRT's) are characterized by coupled passenger cars with capacities of 30 to 50 people operating on fixed rail lines between major activity points. The major advantage of this system is its large carrying capacity. However, it does not adequately serve the low-density population areas which characterize most of our urban environments. The major disadvantages of this system are high cost of operation and high initial cost of investment. Also, a considerable amount of right of way must be acquired in order to establish an adequate network to serve low-density urban areas. San Francisco's Bay Area Rapid Transit System (BART) represents an example of new technology in rapid rail system design. RRT systems are also being designed for Atlanta, Ga.; Baltimore, Md.; Washington, D.C.; Buffalo, N.Y.; and Miami, Fla. (Dade County).

BARRIERS TO IMPROVED URBAN TRANSIT

Having introduced some of the potential alternatives for urban mass public transportation, the barriers that impede improvement of urban mass transportation in our urban environment will be briefly considered. One of the main problems is that people attempting to employ new technology do not understand the governmental process operating within our urban environments. This may be more likely represented as a problem of technology transfer which is being addressed by such large aerospace operations as the National Aeronautics and Space Administration (NASA) and the Department of Defense (DOD). Also, there is no well defined market potential to justify large investments by industry, and local governments are unwilling to take technical risks on "iffy" solutions. For the most part officials within urban communities will not implement advanced technological concepts; they desire proven products before they are willing to extend their effort to secure Department of Transportation (DOT) funding for transit operations. Generally, cities do not recognize transit as a service to be provided by a city. Therefore, funds are generally lacking for matching UMTA assistance programs.

Another barrier to the implementation of new urban transit systems is the fact that industry tries to sell ready-made solutions with little flexibility to meet the varying needs of the communities across the country. Associated with this problem is the lack of consistent procurement rules within the urban environment. In general, there is a need for a systems architect. It has become necessary that public agencies and consultants develop a new level of sophistication for design and implementation of urban transit systems in a complex economic, social, and political environment.

INDUSTRIAL ENGINEER'S ROLE IN
URBAN TRANSIT RESEARCH

Now that the basic problem of UMPT has been presented, and the alternatives and barriers to urban mass transportation have been considered, some of the needs in urban transit research and the role of the industrial engineer in this research will be identified. For purposes of discussion, urban transit research is divided into the four categories: (1) transportation planning, (2) system definition and implementation, (3) operating system, and (4) rider public. The following paragraphs discuss some of the needed research activity and the Industrial Engineer's role in helping support the research activity.

Research in Transportation Planning

Transportation planning is a difficult task because transportation is related to almost every aspect of urban activity. One important research need is to develop techniques for using census data for transportation planning. Since census data represents the most extensive urban and regional information system, it is imperative that techniques be defined which can optimize the use of this data for transit planning. Techniques for extracting land use and population distribution data and identifying potential transit demand and population shifts, would be of valuable use in transit planning. Industrial engineering skills applicable to this research include statistical analysis, data reduction and analysis techniques, and computer programming.

Another research need is to identify the factors underlying the choice of transportation modes. This includes identifying criteria that determine rider preferences, identifying value scales related to transit choice and formalizing methods of measuring preferences. Industrial Engineering skills applicable to this area include human factors, human motivation psychology, and design of experiments.

A major contribution to transportation planning lies in research efforts to design models for economic evaluation of transit efficiency. This task includes the simulation of the interrelatedness of the structural characteristics within the economy of transportation. Methods for evaluating proposed transit operations would be formulated by integrating the factors of transit demand, motivational response, operational functions, and cost. Skills in economics, computer simulation, computer programming, engineering economy, and system analysis would be applicable to this effort.

Another research need is the evaluation of the economic consequences of urban transit to the transit user. This effort would attempt to identify the criteria that users employ in evaluating urban transit. A follow-on to this study would be to relate the criteria to the economics of urban transit to help set the requirements for system definition. The Industrial Engineering skills of economic analysis and human motivation are applicable here.

Associated with economics is the need to evaluate the effort of transit investment on local economics. The emphasis would be to identify changes in land values, land use, and business activity and to develop the methodology for predicting these influences. Skills in economic analysis, mathematical analysis, and cost and benefit analysis would be applicable.

One final need in transportation planning research is to identify the underlying factors in urban transit analysis. The effort would be directed to defining critical variables underlying the urban transit requirements and to developing methods for their reliable measurement. An analytical expression of the functions which determine the needs, performance, and constraints on movement should also be developed. The general concepts of systems analysis are applicable here along with the techniques of experiment design and mathematical analysis.

Research in Transit System Definition and Implementation

One of the weakest links in the chain for realization of an effective urban transit system is the definition and implementation activity. Considerable effort is needed to assure the transformation of concepts and planning factors into the effective, efficient transit operation within our urban environment. Some of the areas that need to be researched in order to more effectively bridge the gap between concept and deployment in the urban environment are outlined below.

One of the areas, closely associated with transportation planning, is the research in market analysis. This entails formulating procedures and performing activities to define transit needs of individuals and whole communities. Involved in this effort is the research of methodologies for forecasting transit resources, economics, and defining population shifts. Identifying transit accessibility measures to help evaluate transit needs would be included. Skills associated with survey techniques, data analysis, and mathematical analysis are applicable to this effort.

A major effort in transportation system definition is establishing the requirements for various urban environments. Research is needed to develop the methodology for formulating the transportation requirements which lead to specifying the type of urban transit to be implemented. This research should formulate techniques which would aid planners in specifying urban transit systems from the new evolving transit alternatives such as PRT, GRT, and bus systems. Industrial Engineer's skills related to this effort include engineering analysis and mathematical programming.

Funding of urban transit systems through subsidy is becoming more and more an accepted policy; however, the source of subsidy is not in universal agreement. Considerable research is needed to clarify the advantages and disadvantages of alternate subsidy sources such as highway trust funds, gasoline sales tax, bonds, general funds, and property tax. The question as to the equitable, proportional breakdown of the applicable alternatives should also be addressed. Further, analysis is needed to address the question of subsidizing operating expenses as well as capital acquisition of urban transit systems, since many transit systems are unable to realize revenues that cover operating expenses. Presently, a free transit experiment is being conducted in Seattle, Washington, where free bus service is provided in the downtown area. Free service may cause many automobile riders to leave their cars in peripheral parking lots and use public transit to travel downtown. Study should be directed on the next higher level to address the question of equitable funding between national, state, regional, and local levels of government. The Industrial Engineer's skills applicable to this area of research include economics and management science.

The long-term effectiveness of urban transportation is greatly dependent on the effective design of the organizational structure. The institutional aspects of transit operations are most important if new innovations in modern system operations are to be used advantageously. Of equal importance is the design of the legal structure which establishes the authority levels. It is imperative that the authority for establishing and implementing policy be defined so that an effective operating system exists between the local, regional, state, and national levels of government. Industrial engineering skills that are applicable to research in this area include management science and organizational design.

In view of the modern urban transit concepts that exist, it is becoming imperative that accurate methods of estimating cost be applied. To this extent, research is needed in determining the life cycle costs for a given system configuration. Large government contracts use this approach to account for the long-term operation of a system. The life cycle costing approach is most applicable to the new urban transit concepts like PRT, GRT, and demand-response systems. A major element in the research, is the aspect of breakpoint analysis. The industrial engineering skills applicable to this research area are cost analysis and engineering economy.

Research in Transit Operating System

It has been noted that many urban transit systems have been taken over by public ownership. One consequence of this evolution is that urban transit operations have not kept pace with modern techniques supporting large system operations. Also, the existence of a small demand for urban transit has not provided incentive for advanced developments in transit equipment. The following paragraphs outline needed research area for improving the operating aspect of urban transit.

One area of research greatly impacting the operation of urban transit systems is computerized management information systems. Specifically, research effort is needed in the areas of (1) inventory and parts ordering, (2) maintenance scheduling, (3) dynamic vehicle scheduling, assignment and routing models, and (4) cost performance evaluation and effectiveness tools. Emphasis should be placed on designing tools for use by the individual transit systems so as to improve their economic efficiency and service effectiveness. Industrial engineering techniques embodied in the field of operations research are applicable to this effort. Such skills would include inventory models (economic order quantity), assignment models, network analysis, scheduling, simulation and effectiveness analysis.

Recognizing that advanced developments in transit equipment have not been deployed extensively in our urban communities, it follows that research is needed in vehicle design for comfort, safety, pollution control, and aesthetics. Of equal importance, is the need to formulate some form of standardization of urban transit equipment. This factor was noted as a barrier to improvement of urban transportation in our communities. Industrial engineering skills applicable to this need include human factors engineering, material processing, manufacturing processes and safety engineering.

Considering new technology systems, research is needed to establish maintenance policy as related to component reliability. This would lead to developing methods and standards for all aspects of maintenance. Industrial engineering skills applicable to this effort include maintenance engineering, quality control, and reliability.

A need for research in transit facility layout is also needed to complement the design of an advanced organizational structure. The emphasis here is directed toward formulating methods to optimize the facility environment. An important element in this effort is to design for compliance with the guidelines of the Occupational Safety and Health Administration. Skills associated with this effort include facility layout and design, safety engineering and methods analysis.

One of the major complaints of the present and potential rider public is the slow service provided by public urban transit systems. Research is needed to define the constraints on service times and to formulate methods for increasing the operating speeds of public transit operations. This effort would include investigating fare collection systems; movement of people into, within, and out of vehicles; and the movement of vehicles through traffic. These factors are closely associated with improving the convenience factor of transit. Industrial engineering techniques that apply to this area include systems analysis and human factors engineering.

Communication and control technology represents one of the major advances being applied to urban transit systems. This is especially true for the automatic transit systems like LRT, GRT and PRT, but also important in the design of advanced concepts of demand response systems. Research in the area of digital computer control and sensing techniques is required. Skills in computer software design, computer programming, and systems analysis are applicable.

Research Related to Rider Public

Earlier it was noted that the present public attitude adversely affects the improvement of public urban transportation. Recognition of this factor leads to the increased need for research related to the rider public. A considerable effort is needed in defining methods to create a positive public attitude. Obviously, the public attitude is linked to improvements in transit operations, but it is necessary to ensure that the changes in improved transit operations are followed by positive changes in public attitudes. This effort would include devising advertising strategies which incorporate the positive factors and benefits to urban public transit. Skills applicable to this area include human psychology and social science.

Associated with the need for creating positive public attitudes is the need to understand changing public attitudes. This would involve identifying social changes and causes and the environmental factors influencing the public attitude. The skills in social science and system analysis are also applicable to this research area.

Another factor related to the rider public is fare structuring. If consideration is to be given to total subsidy (capital acquisition and operating expenses), it is of interest to understand the effect that a "no fare" policy has on public attitudes. Associated with this area is the need to identify the populace reaction to public subsidy of urban transit. Skills in economics, human motivation, and psychology are applicable to this area of research.

Research in organizing advanced methods in personnel training for new urban transit operations is also needed. Due to the unprofitable nature of present transit operations, it is imperative that means be found to increase the productivity of transit personnel at all levels. Skills in labor management and human motivation are most important contributors to this area research.

The area of rider safety is very closely associated with creating a positive public attitude. Of interest here is determining ways to minimize the effects of vadalism on transit equipment and to optimize passenger security at all transit stations and while in route. Industrial engineering skills in safety engineering, human factors and facility design are applicable to this area of research.

Research into the methods of presenting information to the transit rider is also needed. This effort should emphasize development of effective, inexpensive information displays for transit riders. Also attention should be given to the development of systematic and easily changeable formating of displays. The importance of these areas of research can be realized when considering the magnitude of display equipment and the flexibility required to accommodate the changing patterns of urban transit demands. Skills related to this area of research include design of displays, graphic techniques, and communication techniques.

SUMMARY

Urban mass public transportation is getting a "shot in the arm" at the national level. This is exemplified through the creation of UMTA and the legislated increases in funding. Many options, incorporating new technology, exist for UMPT, and some are available now to meet the mobility requirements within our urban areas. However, improvements in urban mass transportation greatly rely on the desires and aggressiveness of local and regional communities. The U.S. Department of Transportation is there to serve the needs and demands required by U.S. citizens. If local communities fail to take the initiative to seek improved systems and organize the necessary legal structure at the local level, the needed new improvements in UMPT will not come about.

The Industrial Engineer's role in supporting the transformation from the concept to the reality of improved UMPT has been presented. It is significant that the Industrial Engineer's training emphasizes the systematic approach of integrating systems of men, materials, and equipment. In particular, the Industrial Engineer's training in human engineering concepts highly qualify him to be a major contributor in the development of efficient, effective UMPT systems to serve human needs in our urban environments.

State Criminal Court Systems

Raymond M. Brach
James M. Daschbach

PART I

Several years ago, from a setting very similar to this one, an Air Force General and our own, now deceased, Wilson Bentley spoke on human resources during a conference in San Francisco entitled "The Man in Manpower and Management." We in the Industrial Engineering discipline in the recent past have seen an increasing emphasis on the attempts in our society to engineer the man. Materials, machines, and even money can be engineered to a degree. Phases I and II of President Nixon's recent economic program are good examples of attempts to "engineer" the money in our economic system. However, the homo sapiens has to a large degree evaded the attempts to engineer a more effective man. This evasion is largely due perhaps to the complexity of the unround peg to be fitted into the nonsquare hole. My colleague and I will expose you today to a project in using a team of engineers and lawyers to "engineer" improvements into the criminal justice system. In society's structure to process felonious crimes, tradition has dictated that basically the structures shall be made up of men and there has been little or no thought given to machinery or technology to assist in making the system more effective and efficient. We have been asked by your conference chairman to emphasize the unique qualities of this study—both problems and successes—and to deemphasize the mathematical intricacies in order to better bring out the potential engineer's contribution in "solving" social problems.

An overall view of this project can be put into three basic technical categories

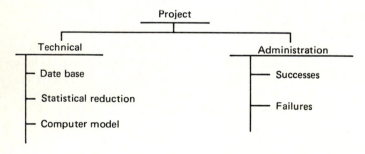

Figure 1 Overview.

plus two administrative ones. These technical categories are the data base, statistical reduction of data and the computer model. The administrative categories I will broadly signify as successes and failures.

We are pleased to be here today to relate to you what we think are the unique attributes of a project wherein professionals from both Engineering and Legal disciplines are joined together on a common project.

In mid 1970, the Law Enforcement Assistance Administration, a portion of the Justice Department, made a grant of some $192,000 to a College of Engineering and Law School joint team at the University of Notre Dame for the purpose of studying the state of Indiana criminal justice system. The project was to be one year in length. We have entitled the project "LEADICS" and the acronym stands for

LAW-ENGINEERING
ANALYSIS OF DELAY
IN COURT SYSTEMS

Figure 2 LEADICS

Law-Engineering Analysis of Delay in Court Systems. It is interesting to note that the combination of lawyers and systems engineers is rather an unusual one. The first time apparently this combination or joint team effort was suggested was during the late 1960's when "The Challenge of Crime in a Free Society," a report of the Crime Commission under President Lyndon Johnson, was submitted to the public. Since that time, there have been several attempts to combine lawyer and engineer talents. It is surprising, however, how the joint efforts seemed to result in the personnel from each of the diverse disciplines concentrating on their own forte—the qualities of law or the hardware and technology.

In a presentation to the Notre Dame Law School Alumni, just prior to the initiation of this study, I was impressed by the comments from this distinguished group of judges and other lawyers from the Midwest States that the major problem to be encountered was that of communication. This was to be a very difficult one to overcome as the reference frame and the vocabulary as used by the two disciplines are so diverse. The prophecy turned out to be all too true. There is a very strong difference between the two disciplines. The frame of reference which we engineers use normally involves statistics, quantitative aspects, diagrams, machinery, and facts in attempting to fit elements into a system. Lawyers, however, do not focus on system functions. They normally look at single cases as unique and isolated experiences for which their services are required even when there is a team effort. The lawyer work habits are usually rather self-centered and the individual sets his own pace and deadlines for the completion of various elements of the work. In another facet, communicating by the written word in order to account for his time, his money, and/or his efforts in precise factual terms is not a habit with lawyers except to charge a client—oft times a very subjective function. The language of the lawyer is equivocation and qualification of a phrase and these attributes become inherent in a lawyer's manner in attempting to fit a situation within an interpretation of the law or to make a reinterpretation of the law. These are then some of the qualitative problems which had to be solved or at the very least satisfactorily compromised in order to provide a team effort.

Organization of the Project

For definitional purposes there are two major aspects requiring a bit of outline. The structure upon which the study was made is defined as the criminal justice system. In

A Police department(s)
B Prosecutor's office
C Court
   1 Municipal
   2 Superior
   3 Circuit
   4 Appellate
D Defense
   1 Lawyer
   2 Accused

Figure 3 LEADICS: The criminal justice system.

the State of Indiana this is composed of four functions: (a) the police department, (b) the Prosecutor's office, (c) the court itself, and of course, (d) the defense functioning as a part of the defendant group. These four major functions form the process to judge an accused. The emphasis of governmental and judicial concern has been on the delay in the court system and has not included the rehabilitation or penal system as a part of the technological research function.

The organization of the project personnel to perform this research, I think, is of some interest to you and this chart then shows how we were organized along

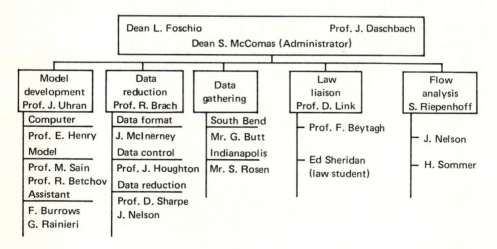

Figure 4 Department of Justice research systems study in court delay.

functional lines. There were codirectors, assisted by an administrator, and the Model Development group on the left was made up of those skilled in systems theory. The Data Reduction group, headed by Professor Raymond Brach, who will be your next speaker, consisted of a group familiar with and capable in statistical reduction of data. The third group completed the data mainly using law students; however, this group with engineering graduate students was the actual interface with the public records. The Law Analysis group provided an analysis of the Indiana state laws, rules and regulations governing the entire legal system. This was a very difficult and highly overlapping function throughout the entire project. Finally, the Flow Analysis group which assembled the flow analysis charts became an extremely important part of the interfacing between the lawyers and engineers.

There was a very high integration of activities across the breadth and scope of this entire project team. The data analysis group or those who were actually gathering the data had to know in what form the modeling group could best use their work and in addition how to present the format for best statistical use. The lawyers had to decide upon and inform the engineers what data was actually of great importance to them and what was of secondary importance in order to establish the priority of the data to be gathered. Thus, there was a very great integration of the activities and a close working relationship was eventually established between the lawyers and the engineers.

The Data

The original concept of data gathering in this particular project was that law students or someone with a background in legal systems would gather the data in handwritten form developed from the public records. A cursory survey showed insufficient computer-oriented personnel with legal knowledge available for this work. The written data would be passed to an encoder who would transform the information into a digital format. The numerical data would then be given to keypunchers who would place the data on computer tape which would then be made available to the statistical reduction group and the modeling group. As can be visualized this design scheme presented several areas of potential error. As many of you are no doubt aware, with increasing numbers of personnel further and further removed from the actual data gathering element, the probability for increased frequency of error goes up rapidly. This became a serious problem in trial tests and was the key in developing the finally accepted scheme.

The final concept of the design for data gathering utilized trained personnel in the use of data sheets which are "mark sensitive." A two-day training course was established for the law students. Thus, the information from the public records was transformed into digitalized format immediately upon being gathered by the law students. The forms were then sent to an optical scanning machine which optically "looked" at the data and placed it on tape. Our human errors were thus minimized insofar as possible even though machine problems were encountered and perhaps, in retrospect, required almost as much time to correct as the human errors which might have been encountered under the original design.

A brief look at the dimensions of our data I think would be of some interest (see Fig. 5, p. 218). The criminal justice system in the State of Indiana is arranged on a county-wide type basis. We studied two counties—Marion county and St. Joseph county—in the state.

There were two types of courts, the criminal court and the circuit court each having felony jurisdiction in this state. Each of these, however, breaks into a further delineation and is thus another dimension for the study's complexity. There are six police departments, seven judge categories, and nine different crime categories within the study framework. In order to provide a proper statistical base, a long period of time was required and finally it was decided that a four year interval for an adequate

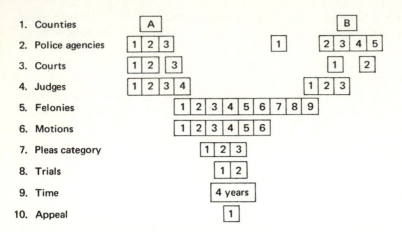

Figure 5 LEADICS: Data complexity.

sample was required. Because of the infrequent occurrence of cases in several of the crime categories within the St. Joseph county area, such as murder, the sampling of cases from among those available amounted to some 100% of the possibilities. Each crime category has its own particular complexity and within a county a case may have one or more Change of Judge motion, change of venue motion or other motions which further complicate the analysis. Within the court system itself, you have two paths to the trial—affidavit or Grand Jury. There is also an appeal from a lower court such as a JP or Municipal court, but this is only a variation on the affidavit system. Within law enforcement agencies, there are jealousies between and among the sheriff, the city police, the state police and the county police. Finally, within just the time category, felonies may have long cases with continuances, the accused may be on bail which is jumped, and various other qualifications or complications of the time factor must be accounted for in each case. These few examples give a glimpse of the date base complexity.

Each case for which data was gathered is traced from the commission of the crime itself through to a dismissal, the sentence of the defendant or an appeal to a higher court. The appeals process, though affecting only 2% of the cases, is important from several aspects including the rights of both the accused and Society. Overall some 2600 cases were analyzed. Seven hundred and sixty of these are from St. Joseph

	Analyzed	Accepted	Useful
County A	783	754	738
County B	1868	1424	1276
Total	2651	2178	2014

Figure 6 LEADICS: Data totals.

county and some 1800 or more from Marion county. Those of you familiar with sampling will, I am sure, wonder how many of these cases had to be dropped due to errors. Surprisingly enough some 740 cases in St. Joseph county and over 1200 in Marion county are of use. Thus, just over 600 cases had to be dropped out as bad samples.

Error Correction

As anyone familiar with data gathering knows, there are still errors no matter how much quality control you put into a system for correction of data. The validation of the data was accomplished by computer as much as possible. A negative time factor check was developed by sequencing the blocks and analyzing the data tapes through use of the computer. For example, if the crime data occurred after the arrest and/or after the trial, the data, of course, had to be corrected. By selecting one of the various network paths as shown on the flow chart and aligning the dates in a sequential sense we were able to correct many of the errors as originally gathered in the data. The Statistical Reduction group, of course, added other control elements such as certain alpha or numeric terms required in the data format for each case.

Upon receipt of the error printouts from the statistical reduction group, the data gatherers were sent back to the public records to correct their data. The time and effort spent in this mundane task of going back to the files, noting that errors existed even in the files themselves and thus having to return to the prosecutor, the defense, or the court personnel was a very time consuming and expensive part of this project. Thus, quality control of data is an extremely difficult area in a project where people are the only working element. The social problems facing our society are, of course, in this category—and we on the "LEADICS" team have, I feel, a much greater appreciation for the complexity of the analyses—much less the solutions—for our social structure problems.

Eventually, however, two sufficiently accurate tapes with the data from all 2600 cases were presented to the statistical reduction and the modeling groups.

Conclusions

For the first time then, apparently in the history of legal analysis or legal research recently fostered, a definitive base for data has been provided for the criminal court system. In the course of this work we have noted other studies using questionnaires as the basis for their data. This data system is, naturally, error prone to an unmeasurable degree and is therefore unacceptable in the engineering analysis. In other areas, annual reports of police, courts and so forth have been developed into data banks. Too often these reports are self-justifying and highly biased in favor of the authors or the authoring agency. It must be recognized that the data base becomes a most difficult problem for any project of this size—its very size causes mechanical problems and the control of accuracy is quite formidable.

The immediately applicable results focus upon administrative practices of the court. The data show conclusively that pre-trial activities such as lawyer meetings to specify charges, bargaining for an accused's plea, and filing motions for various types of continuances are the major time consuming elements in the process. Many of these results show sharp contrast to what legal authorities and the many periodicals have been specifying as the major delay factors in the system. The overall conclusion is that the power to improve presently exists within the process as now constituted. These improvements can be attained with strict adherence to existing rules, procedures and time restrictions plus a change of attitude on the part of a portion of the legal profession.

Finally, we believe that this project has shown potentially great advantages in joint Engineering-Lawyer cooperation. More particularly, the contribution of the engineering disciplines to many of our social research projects may very well be a vital key to the analysis—if not the solution.

I would like now to introduce my colleague, Dr. Raymond Brach, who will give you some insight on the analysis by the other technical groups.

PART II

As you have seen, the early effort in this project consisted of two steps. The first was for the legal analysts to characterize the important parts of the legal process starting from the initiation of a case to the termination of the case, that is when the disposition is reached. The second step was for the data gatherers to collect information concerning many individual cases in the two jurisdictions. The data for each individual case can be broken down into two types, the first being attribute data such as the type of crime, the sex of the defendant, the jurisdiction in which the crime took place, and so on. The other type of data is quantitative such as the number of days between important steps of the process, the number of motions filed in an individual case, and so on. To give you an idea of the magnitude of the data assembled consider that of the 2600 cases each had approximately one hundred dates recorded. These corresponded to events such as the date of crime, date of arrest, date of arraignment, date of filing of motions, etc. As you can well imagine the total amount of data was overwhelming. The function of the technical groups in this project was to obtain some meaningful insight into the criminal court system by analyzing these data. This task was broken down into two main efforts; the first was that of a statistics group which used classical methods of statistical analysis to obtain information concerning variable relationships. The second effort was by a systems analysis group which set up a computer model of the system. In the following I will give you an idea of the way in which these two groups accomplished their respective tasks.

The first task of the statistical analysis group was to review the type of data available and to define the ways in which statistical inference could be used to obtain some meaningful information from these data. This amounted to a review of the available, classical statistical methods. For example, consideration was given to the possibility of using factorial design and then following this with an analysis of variance. This approach was not followed because a reasonable number of clear cut variables could not be defined which could then be further sampled on a designed basis. In other words, one could not go into the system and cause it to operate in a predetermined fashion such as to require that the system have a certain number of each type of felony, a certain number of cases in each jurisdiction, a certain number of pretrial motions, etc. The system could only be observed, not manipulated. Another classical technique was that of contingency tables normally used to analyze independence between variables. This method had some potential, however it was felt not to be generally applicable and thus not pursued.

Since there were a large number of variables associated with the felony process, it was realized that the problem fell into the area of multivariate analysis. Many of the questions asked by the legal analysis seemed to refer to the way in which many of the variables were interrelated. This led to the computation of correlation matrices in order to get some idea of the degree of relationship between these variables. For example, at one point in the study a list of 43 variables was made, and a correlation matrix computed. Inspection of a matrix of this size led to many interesting conclusions; however, the large number of correlation coefficients still taxed the individual in the effort to draw meaningful conclusions. At this point it was decided to apply the technique of factor analysis. Factor analysis is a name given to certain techniques which fall into the category of multivariate statistical analysis of correlation matrices. These techniques begin with correlation matrices and by performing mathematical operations attempts to uncover dependency structures among the variables. In particular the technique used in this project is called the method of Principal Components.

Historically the beginning of factor analysis was in the early 1900's when Spearmen hypothesized that the correlations among a set of intelligence test scores could be explained by a single factor of general intellective ability for an individual and other factors reflecting the qualities of the type of test administered. Application here to the criminal court system is illustrated by means of an example.

Suppose that eight variables are defined as in Figure 7. All of the intervals are

List of variables:
1   Interval from arrest to disposition
2   Interval from crime to arraignment
3   Number of pretrial motions
4   Number of previous arrests
5   Interval from crime to arrest
6   Interval from arrest to arraignment
7   Interval from arraignment to disposition
8   Interval from crime to disposition

Figure 7   Factor analysis: Principal components.

measured in days and the order of the above variables is irrelevant. What factor analysis does is to perform a linear transformation on the matrix of correlation coefficients of the above variables. This linear transformation accomplishes the following purpose. It assembles the variables into a few groups of highly related variables where each group is called a factor. These factors are arranged in order such that the first factor accounts for the largest amount of variance of the correlation matrix. The next factor has the next highest variance, and so on. In this particular example the factor matrix, which is the result of the linearly transformed correlation matrix, resulted in three factors from the original eight variables. Thus the complexity of the analysis is now reduced from a level of 8 quantities to a level of 3 quantities. (See Fig. 8.) It is the function of the analyst at this point to determine the

A. Retained counsel (114 cases)

Variable means
285.3 94.32 2.991 3.623 31.50 62.82 222.5 316.8

Standard deviations
291.2 183.4 2.196 4.069 92.04 160.7 260.0 302.9

Correlation coefficients

1.000	.3888	.4807	.0643	−.0279	.4597	.8358	.9528
	1.000	.0328	−.0073	.4824	.8650	−.0993	.5203
		1.000	−.0350	−.0037	.0396	.5139	.4609
			1.000	.1246	−.0796	.1212	.0996
				1.000	−.0221	−.0176	.2770
					1.000	−.1033	.4352
						1.000	.7981
							1.000

Rotated factor matrix

	Factor 1	Factor 2	Factor 3
Variable 1:	.9099	.3637	.0192
Variable 2:	.0617	.9657	−.2411
Variable 3:	.6934	−.0446	.0824
Variable 4:	.0949	−.1791	−.6755
Variable 5:	−.0638	.2926	−.8011
Variable 6:	.1069	.9346	.1837
Variable 7:	.9530	−.1705	−.0921
Variable 8:	.8552	.4385	−.2250

Percentage of total variance (3 factors): 81.4

B. Appointed counsel (162 cases)

Variable means
139.7 41.30 2.333 2.079 5.105 22.96 18.34 121.4 162.7

Standard deviations
162.5 78.35 2.079 5.373 74.59 27.88 159.5 186.0

Correlation coefficients

1.000	.1715	.6983	.0316	.1087	.1910	.9852	.9171
	1.000	.1682	−.0330	.9347	.3095	.1206	.5246
		1.000	−.0070	.1051	.1914	.6778	.6521
			1.000	−.0082	−.0709	.0446	.0243
				1.000	−.0487	.1193	.4960
					1.000	.0198	.1473
						1.000	.9084
							1.000

Rotated factor matrix

	Factor 1	Factor 2	Factor 3
Variable 1:	.9824	.0424	−.0406
Variable 2:	.1275	.9688	−.1964
Variable 3:	.8072	.0287	−.1419
Variable 4:	.0634	.0176	.6155
Variable 5:	.0809	.9834	.0998
Variable 6:	.1417	.0917	−.8189
Variable 7:	.9758	.0271	.1017
Variable 8:	.8905	.4314	.0045

Percentage of total variance (3 factors): 85.0

Figure 8 Factor analysis—St. Joseph county.

meaning of each of these factors. It can be seen from the factor matrix corresponding to these 8 variables that factor one is associated with the overall time through the process since 3 of the four significant variables in factor one are intervals which terminate at disposition. In all of the variables associated with factor two the intervals begin with either the crime or the arrest dates. As a result factor two seems to be associated with the early intervals of the criminal process. Factor three always contains the variable 4, number of previous arrests, which is closely associated with the police-defendant interaction. Thus these three factors could be named the overall process time factor, the prearraignment factor, and the prearrest or predefendant factor, respectively.

Further use of the method of principal components can be made by repeating the same factor analysis for different categories of some of the attributes associated with the process under study. For example, the cases can be segregated into different categories such as those which had a retained counsel or those which had an appointed counsel or public defender. Furthermore, the cases can be kept in the jurisdiction from which taken as data. The resulting factors between each of these categories can then be compared.

This was done and resulted in many interesting conclusions. For example, it was noted that the differences in the factors from the two counties was very slight. This indicated that the systems operate similarly. Furthermore, there were differences in the results obtained between the different types of attorneys associated with each case. This led to the conclusion that retained attorneys handled their cases in a different fashion from appointed attorneys. Specifically, it was found that retained attorneys exert much more influence in the crime to arraignment interval. This was further supported by examination of the averages of some of these intervals. For example the average interval from the time between crime to arrest for the category of retained counsel in St. Joseph county was approximately 27 days; the average interval between arrest to arraignment is 53 days. For appointed attorneys however, the average number of days between crime to arrest was 19 days and the average interval between arrest to arraignment was 19 days. These and other worthwhile conclusions were drawn from the result of the factor analysis.

At this point I would like to turn to the work done by the systems modeling group. The task of this group was to set up a block diagram model of the criminal court system with as much generality as possible. The generality of the model was not to be so large however, that its mathematical description and computer simulation would be too complex. Early in the project Industrial Engineering students went into the court systems and developed some rather extensive flow charts which indicated all of the possible paths through which a defendant could travel while his case was in the system. These charts contained hundreds of alternative paths; however, it was realized from the data that many of these paths were used infrequently. As a result it was possible to reduce the complexity of this flow through the system to a smaller but meaningful flow chart and still retain the essential characteristics of the criminal system. Once the flow through the system was characterized, the next step was to develop a mathematical technique to simulate the operation of this system. Following this the mathematical model was put on a computer and then finally exercised.

The systems group investigated all of the applicable simulation techniques. For example, such classical techniques such as GPSS, GASP, and GERTS, were considered. For various reasons such as complexity or difficulty in adapting these techniques to existing computer facilities, these methods were not adapted. Instead a new approach was developed at Notre Dame to simulate systems of this nature. Basically the procedure was to define a step of the process to be an interval between identifiable and measurable events in the process. Each elementary step was then placed into a block diagram. Throughout this block diagram time is treated as an independent random variable. One of the novel features of the approach developed was to use the data describing each of the various intervals to develop a smooth rational approximation to the probability density function for each of the elementary steps or blocks. The approximation is developed from the statistical moments of the data. For example, one elementary step would be the time between the date of the crime and the date of the arrest and booking measured in days. The data collected earlier on the project allowed the computation of the mean, standard deviation and other higher statistical moments. Thus, the flow chart was set up as a group of blocks connected in series and/or parallel, each with an associated probability density function. A few "feed back" paths were also represented. The simulation was completed by using fast Fourier transforms to convolve combinations of these blocks and furnish probabilistic estimates of the system operation for any desired combination of paths through the system.

For example, we can look at the block diagram of the model of the St. Joseph county criminal court system. (See Figure 9.) The St. Joseph county system contains two courts, the Superior Court and the Circuit Court, each of which in its normal routine handles felonies. Thus some of the path junctions between the various blocks in the diagram require a decision which in the model can be based on the statistics gathered from the actual court operation. A defendant can traverse through the system from the commission of a crime to arrest and booking through the prosecutors office and then follow a path through to the Superior Court. Or the affidavit may take him to the City Court with a warrant, a preliminary hearing, a grand jury session and then possibly on to the Circuit Court. Following this, in either of the courts a series of decisions is made at and following the arraignment. If there is a trial, it could be either a bench trial or jury trial. Throughout the system the person may be declared innocent or guilty; the guilty person then goes into a pre-sentence investigation and finally sentencing. At many of the points in the system the case may be either dismissed or the person judged innocent in which case he leaves the system as indicated by the dark arrows.

This and other models (see Figure 10) were simulated on the computer with interactive programs. This allowed the user to control the input and operation of the system from the console of the computer. It should be mentioned that one of the novel features of this simulation was that it was done on a small computer namely an IBM 1130 with a disc drive and an eight-thousand word memory.

The various models developed in this study were exercised on the computer in order to determine results as to system operation under various conditions. Some of the runs were conducted where various blocks were combined in order to comprise a

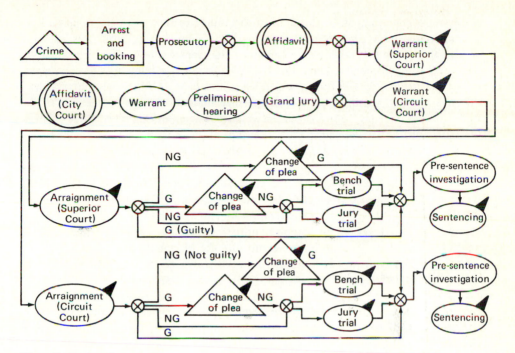

Figure 9 St. Joseph county.

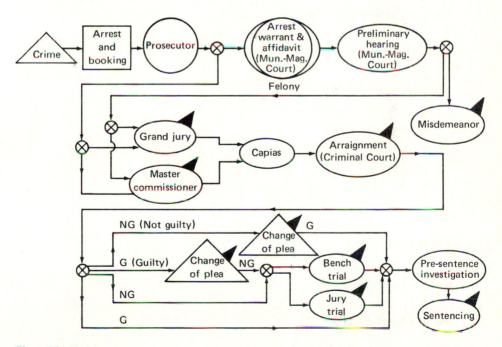

Figure 10 Marion county.

new system. For example, in the St. Joseph county system, both the Superior and Circuit Courts were combined to form one court for this county. This exercise revealed that if the two courts were combined the average time between various steps in the process could be shortened. For example, the average number of days between arrest to sentencing in the Superior Court was 191 days, in Circuit this interval was 240 days. In the "combined" court the average number of days between arrest to sentencing was 195 days. Another use to which this computer simulation was put was to modify some of the individual blocks or steps in the process. This can be done on a hypothetical basis; for example, one could arbitrarily reduce the average time through any portion of the system by some factor, say $\frac{1}{2}$. By operating the system on the computer one could then determine the effects in the overall operation of the system of this modification of one portion. Another use to which this simulation was made was to actually modify the system operation by eliminating some of the blocks. For example, the court system in Marion county traditionally makes much use of the grand jury process. It was interesting to simulate the Marion county system with the grand jury eliminated. It was found here that the average number of days between arrest to the date of issuance of a warrant was 97 days in cases in Marion county which used the grand jury. By eliminating the grand jury and running the computer simulation it was found that the average number of days between arrest to issuance of the warrant was 58 days. Other exercises of the simulation were made and much of the information was used by the legal analyst to draw various conclusions concerning more efficient operation of the system.

In conclusion we would like to state that many things were learned from this study. The interaction between people of different fields namely Engineering and Law could be made to work effectively and result in a good approach to a rather complex problem. Furthermore, the statistical analysis of the data using classical techniques and the computer simulation were complementary. The results of these analyses furnished useful information to the legal analyst by which he could intelligently discuss the modification of the system in order to reduce the overall problem of delay in the courts.

Blood Bank Inventory Control[25]

John B. Jennings

I. INTRODUCTION

One of the important medical resources of any community is its system of blood banking facilities. It is through such systems that blood is collected from human donors at one time and place, processed, stored, and ultimately provided for transfusion to hospital patients at some other time and place. In most areas, blood banks are organized into loose regional systems, each composed of anywhere from 20

[25] Reprinted from "Blood Bank Inventory Control," by John B. Jennings, *Management Science*, vol. 19, no. 6, February, 1973, pp. 637–645, published by The Institute of Management Sciences.

to 200 hospital blood banks located in some geographically or politically defined area (e.g., a city, a portion of a state). While the hospitals in such a system generally acquire a portion of their blood supply from one or more common central blood banks and one or more donor services, the hospitals typically interact with each other only infrequently in times of emergency. In most cases these systems have developed without the aid of central coordination and presently face a variety of problems which result from ineffective and inefficient modes of operation.

Three of the most common and pressing problems are the following:

A chronically short supply at the same time that as much as 15 to 25 percent of the available supply is lost through outdating (the shelf-life of whole blood is limited by law to 21 days in most areas). This condition results both from a maldistribution of blood among blood banks and from a system-wide deficient supply.

Susceptibility to sudden stockouts resulting from unpredicted large demands at one or more hospitals.

High operating costs—in particular, large expediting costs and outdating losses.

These problems reflect the importance of the inventory function in blood banks and call for improvements in the control of blood inventories. However, four characteristics of blood banking inventory systems make this a very complex task:

1 Both supply and demand are probabilistic.
2 Approximately 50 percent of all units of blood requested by physicians, "cross-matched" for compatibility with the blood of prospective patients, and reserved are eventually found not to be required for the patient in question and are returned to the "unassigned" inventory.
3 Blood is perishable, the present legal lifetime being 21 days in most areas.
4 Each blood bank typically interacts with a number of other banks.

Because of these complexities, past work in the field of inventory theory has limited applicability. Most attempts to deal with the blood inventory problem have consisted, on the one hand, of rules of thumb found in practice to provide adequate service, and, on the other hand, of analyses and simulations of highly simplified models of a single hospital blood bank. A good review of these studies is given in Elston. The only previous attempt to consider the interaction of hospital blood banks concentrated on the collection policies of the central blood bank and suffered from an oversimplification of the hospital models.

In the remainder of this paper we shall briefly investigate the potential costs and benefits of improved control of inventories of whole blood both at the individual blood bank and at the regional level.

Before proceeding to the analysis, we must have a set of specific criteria or measures of performance by which to evaluate alternative inventory policies. The three most important measures are the following:

1 Shortage. The amount of blood that is requested by physicians and not immediately available.

2 Outdating. The amount of blood entering a bank and not transfused before the expiration of the shelf-life (21 days).

3 The cost of information and transportation systems needed to support inventory policies.

An additional criterion—the age of blood transfused—was included in the study as an index of quality but will not be discussed here, since it was found to be relatively insensitive to the policies under consideration.

II. INVENTORY CONTROL IN AN INDIVIDUAL HOSPITAL BLOOD BANK

The first phase of a blood inventory control program must focus on the individual hospital. A general model structure for a hospital blood bank, developed on the basis of a survey of the operations of a number of hospitals in several localities, is presented diagrammatically in Figure 1. As shown, the inventory of blood is divided into two portions: the assigned inventory, which consists of blood which has been crossmatched with samples of prospective patients' blood to ensure compatibility and reserved for those patients; and the unassigned inventory, which is available to meet new requests for blood.

The flow of blood through the system is as follows. The unassigned inventory is depleted by physician demands for blood to be crossmatched, by the outdating of blood, and by the shipment of blood to other hospital blood banks. It is replenished with blood ordered from one or more central blood banks, with blood drawn from selected donors (or "ordered" from one or more donor services), with blood drawn from random, or unsolicited, donors, with blood released (unused) from the assigned inventory, and with blood received from other hospital blood banks. The reader should note that shortage, as defined above, applies to the unassigned inventory only. The assigned inventory is supplied with blood demanded by physicians; it is depleted by blood usage (transfusions), by the release of blood demanded but not used, and by outdating. The development of this model is described in detail elsewhere.

Several characteristics of whole blood inventories—particularly the need to

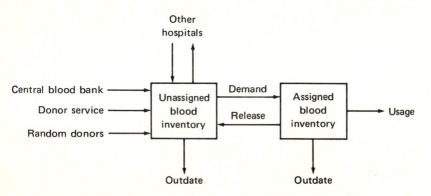

Figure 1 The hospital blood bank whole blood inventory model.

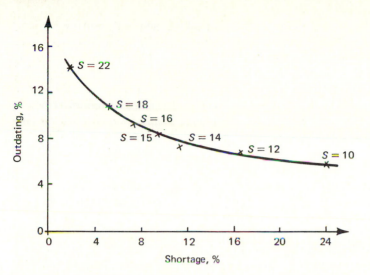

Figure 2 Shortage-outdating operating curve: An independent hospital.

describe a blood inventory as a 21-dimensional variable (one dimension for each possible age)—defeat analytical approaches to the study of practical whole blood inventory problems. Accordingly, the model described above has been utilized to examine a variety of inventory policies with the aid of computer simulation. The necessary data were collected at a large hospital blood bank in Boston (the Peter Bent Brigham Hospital blood bank). It was found that the eight major blood types could be treated independently, and here we shall report the results for a type (B+) which was used at the rate of about 800 units per year. While these results apply in detail only to this particular blood bank, the characteristics of the results are of general validity.

The most basic inventory policy in a hospital blood bank is the establishment of a daily inventory ordering level (a different level for each blood type), which we designate by "S." With such a policy, the *unassigned* inventory is counted each morning, and if it is less than S, the difference is ordered from available sources; if the unassigned inventory is equal to or greater than S, no blood is ordered. Here we shall assume that the bank is always able to begin with at least S units; the impact of randomness in supply is discussed in [Jennings].

Of the various criteria suggested earlier, the two which are significantly affected by this policy are shortage and outdating.

To demonstrate the simultaneous effects of the inventory ordering level on shortage and outdating, we shall represent the model's operating statistics in the form of a graph of attainable combinations of shortage and outdating, both expressed as percentages of the annual number of pints transfused, with the inventory ordering level S as a parameter. This graph, which we shall refer to as the shortage-outdating operating curve, is shown in Figure 2. In other words, for each value of S, the average annual shortage and outdating observed in the simulation of the equivalent of a three-year period of operation of the model are plotted as a single operating point in Figure 2 and labeled with the appropriate value of S. Thus, for example, for an inventory ordering level of 18 units, shortage is 5.8 percent and outdating is 10.3 percent. (The blood bank which served as the source of the data for this model operated at a 15- to 18-unit ordering level.) As S is raised, the operating point moves up and to the left, corresponding to increased outdating and reduced shortage, and vice versa. This curve represents the trade-off between the two most important measures of the blood bank's effectiveness, and the blood bank administration can select the most desirable operating point on this curve simply by specifying the daily inventory ordering level. This decision will depend on the administration's assessment of the relative tangible and intangible costs of shortage and outdating.

This same model may be used to investigate a variety of other inventory operating policies. Several have been investigated and reported elsewhere: the possibility of obtaining all supplies directly from donors with the full shelf-life remaining, rather than from a central blood bank with only a portion of the shelf-life remaining; the possibility of ordering blood more often; and the possibility of ordering whenever the inventory falls below some lower threshold. Of course, certain policy changes—for example, changes in the rules governing the entire crossmatching process—cannot be examined without revising the model.

III. REGIONAL BLOOD INVENTORY CONTROL

At the regional level, interactions between hospital blood banks are presently extremely limited in most areas, and there is a wide range of potential inventory strategies from which to choose. However, those which are capable of contributing to blood banking goals fall into two classes:

1 Shortage-anticipating transfers of blood from one hospital to another. Policies in this class allow a hospital that is experiencing an unexpectedly large demand to avert (or end) a shortage by "borrowing" blood from relatively well-supplied neighbors.

2 Outdating-anticipating transfers. Under this class of policies, a blood bank which is passing through a period of low demand may seek to reduce outdating by "lending" blood to banks that are more likely to use it.

The benefits to be derived from such policies must, of course, be balanced against the costs of the systems required to support them. This trade-off may be explored by investigating the effects of specific policies on a model of a regional system.

For simplicity, we shall concentrate our attention on a model of a hypothetical, homogeneous system composed of a variable number of identical hospital blood banks, all operating under exactly the same policies. This procedure has the advantage of facilitating our understanding of the effects of the policies to be examined. Since we already have a validated model of one hospital blood bank, we shall use it as the archetype. The central blood suppliers are modeled implicitly, just as in the one-hospital model. (If we were modeling a particular operational system in its entirety, it might be preferable to model the suppliers explicitly.) Further details may be found in [Jennings].

A useful point to begin an analysis of regional inventory control policies is with the policies which specify minimum and maximum interaction, respectively. The former has already been examined: it is simply the case of the individual, independent hospital for which the shortage and outdating results are given in Figure 2.

The policy which maximizes interaction in terms of both shortage-anticipation and outdating-anticipation will be called the Common-Inventory Policy. Under this policy, all hospital inventories are, in effect, fully shared by all participating blood banks; the oldest bloods in the system are always crossmatched first, no matter where they might be initially located. The performance of the model under this policy, in terms of shortage and outdating, is the best attainable.

The shortage and outdating results of a number of one-year simulations are shown in Figure 3 as a family of shortage-outdating operating curves. In each case, a particular operating point on one of the curves may be selected by adjusting the inventory ordering level. For example, two hospitals operating under the Common-Inventory Policy described, at the 12-unit inventory ordering level, would be expected to outdate blood at the rate of 4.9 percent of annual usage and experience shortages in the amount of 5.0 percent of average annual usage. Five hospitals operating under these same policies would have 4.2 percent outdating and 0.5 percent shortage.

A useful index of the extent of the inward shift gained by expanding the size of the system may be determined as follows: (1) Identify the most desirable operating point on the curve for the basic model. (2) Draw a straight line from the origin through this operating point. (3) Treat as the approximate most desirable operating point on any inner operating curve that point at which the curve intersects the constructed line. (4) Take, as an index of improvement in shortage and outdating performance under the new policy, the common percentage reduction in *both* shortage and outdating.

Suppose one is interested in points at which the ratio of outdating to shortage is unity. One may then determine from Figure 3 that, with respect to the one-hospital

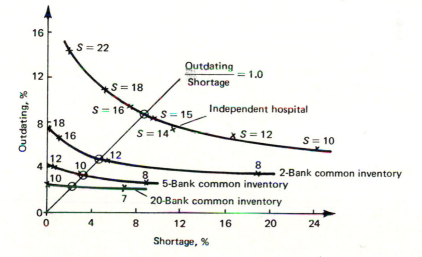

Figure 3 Shortage-outdating operating curves: Common-inventory systems.

starting point, a two-hospital Common-Inventory system reduces both outdating and shortage by about 45 percent, a five-hospital system by about 64 percent, and a 20-hospital system reduces both measures by about 72 percent.

While we have shown the operating curves only for systems up to 20 hospitals, there is clearly a very strong effect of diminishing returns as the size of the system is increased. We may conclude that a further increase of the size of the system to even 50 hospitals would produce little additional gain; and not even in the largest urban centers can one find a concentration of 50 hospitals handling blood in the volume being modeled. Note, however, that further reductions of shortage and outdating can be achieved through changes in internal policies, particularly those which govern the entire crossmatch process.

With regard to the support system requirements of the Common-Inventory Policy, we find, as might have been anticipated, that they are quite heavy. The supporting information system must provide essentially continuously updated information on both the locations of the oldest bloods of each type in the system and the total inventories at those hospitals. The transportation system is found to be required to transfer—one at a time—about 550 units *of this type* per bank per year in a two-bank system (i.e., 1,100 units per year), about 900 units of this type per bank per year in a five-bank system (4,500 units per year), and almost 1,100 per bank per year in a 20-bank system (22,000 units per year).

In view of these "costs" of implementing a Common-Inventory Policy, it behooves one to seek other policies which yield benefits approaching the maximum gains indicated, but at a lower cost.

Clearly, there are many reasonable policies combining the two types of interactive transfers identified earlier. Here we shall illustrate our method of analysis using one policy which was found to be relatively attractive in terms of the gain achieved and the support systems required.

The policy we shall review here will be referred to as a "Threshold Transfer Policy" and is defined as follows: Whenever the inventory level (for the blood type under consideration) at any bank falls below a lower threshold of one unit, a transfer is initiated. Each bank is willing to lend only those units it may have in excess of a retention level of one unit. When more than one bank is "willing" to lend blood, a single lender is selected as follows: Once in the middle of each day the status of the inventory at each bank in the system is determined. When a lender is being sought, the hospitals are "polled," beginning with the one most recently known to have had the largest inventory, until a willing lender is identified. The amount borrowed is five units, oldest first, or less if five would reduce the lender's inventory below the retention level.

Rather than present another graph of the applicable family of shortage-outdating operating curves, we compare in Table 1 the simulation results of this policy with those of the Common-Inventory Policy for operating points at which the ratio of outdating to shortage is unity. As shown, the Threshold Transfer Policy for two hospitals yields an inward shift of the shortage-outdating operating curve about two-thirds as great as that obtained in a Common-Inventory System (29 percent compared to 45 percent). For five- and 20-bank systems, the gains are each about

Table 1   Shortage and Outdating Reductions in the Threshold Transfer and Common-Inventory Policies

Number of participating hospitals	Percent reduction of shortage and outdating at operating points where outdating/shortage = 1.0	
	Common-inventory Policy, %	Threshold transfer Policy, %
1	—	—
2	45	29
5	64	54
20	72	61

five-sixths as great under the Threshold Transfer Policy as they are under the Common-Inventory Policy.

These findings suggest that we have, in fact, found a policy which yields gains approaching the maximum attainable (in the situation modeled). Let us now examine the costs of these two policies. Specifically, Figure 4 presents, for each policy, the relationship between the primary benefit—the reduction of shortage and outdating (along the line representing operating points for which the ratio of outdating to shortage is unity)—and the primary variable cost—the number of inter-hospital shipments (trips) required to execute the policy. Here, for example, the five-bank point on the curve for the Threshold Transfer Policy represents the combination: shortage and outdating reduced by 54 percent; 90 inter-hospital shipments per bank per year (for this blood type).

Clearly, for reductions in shortage and outdating that can be achieved by the Threshold Transfer Policy, this policy requires far fewer inter-hospital shipments than does the Common-Inventory Policy. However, suppose that the size of the system is limited to some number of banks N. In such a case, the maximum reduction of shortage and outdating provided by the Threshold Transfer Policy alone is also limited and can be determined by referring to the appropriate point on the lower curve in Figure 4. If larger reductions are desired, this policy must be replaced by or combined with some other policy which is capable of shifting the N-bank point on the Threshold Transfer curve in Figure 4 toward the N-bank point on the Common-Inventory curve. Note, though, that the ratio of additional shipments to additional gain in such a shift would be extremely large.

Finally, the information requirements of the Threshold Transfer Policy are far less severe than those of the Common-Inventory Policy, and could be very simply met with a manual, telephone-based system requiring only part-time clerical staffing. In comparison, the Common-Inventory system requires continuously updated information on the location of most of the units in the system, pointing to the need for a large-scale automated information system.

In the context of particular blood banking systems, these and other policies can be "costed out" in detail and their total advantages and disadvantages evaluated. Such an extended analysis would require attention to a number of factors we have not dealt

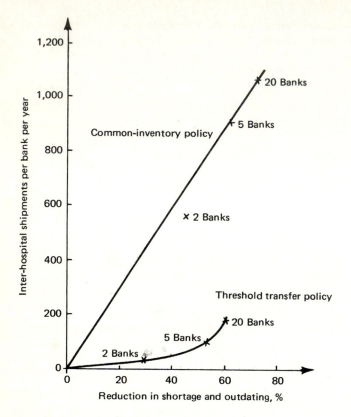

Figure 4 Inter-hospital shipments vs. percent reduction in shortage and outdating: the threshold transfer policy and the common-inventory policy.

with here: for example, extension of the model to include all eight blood types, secondary uses of the information and transportation systems required, distances between hospitals, and so on.

IV. CONCLUSIONS

The accomplishments of this study are threefold. First, the whole blood inventory problem for a single hospital has been structured and several classes of alternative policies identified. Second, a realistic model of the whole blood inventory system in a group of hospitals has been developed for use in the analysis of the operations of such systems. Third, the feasibility of testing the effects of specific policies using computer simulation has been demonstrated, a number of such policies have been examined in detail, and the simulation programs have been documented for use in analyzing specific systems.

The analyses performed should provide useful insights with direct applicability in a variety of blood banking systems. Specifically, we have determined (a) the range of improvements in shortage and outdating performance that can be expected in systems of various sizes, (b) the nature of the corresponding support system capabilities that must be provided, and (c) the fact that a group of only five cooperating blood banks can achieve most of the gains available in larger coordinated systems.

The studies reported here are not the last word in the investigation of blood bank inventory control. They should, however, establish a firm basis and the necessary tools for further analyses of specific operating systems and for the initiation of controlled experiments to test the effects of the most promising policies in the "real world."

These articles are intended to provide a frame of reference for the remainder of this chapter. They indicate the need for extending the use of traditional industrial engineering and management techniques into nontraditional environments. Much of present-day systems engineering work is simply the application of traditional techniques to these previously unanalyzed environments. The remainder of this chapter will highlight nontraditional techniques of value to systems analysis.

SYSTEM OBJECTIVES

Most systems authors seem to agree that the most important step in systems analysis is the determination of objectives. Churchman states, "The objectives of the overall system are a logical place to begin, because, as we have seen, so many mistakes may be made in subsequent thinking about the system once one has ignored the true objectives of the whole" (6, p. 30). It is disheartening to students and practicing engineers and scientists alike to discover that they have an excellent solution to the wrong problem. Hall emphasizes this point by stating, "Yet it is much more important to choose the 'right' objectives than the 'right' system. To choose the wrong objective is to solve the wrong problem; to choose the wrong system is merely to choose an unoptimized system" (11, p. 105).

The fundamental problem with establishing objectives is that objectives implicitly involve value judgments. Hall states, "The operation of setting objectives (alternatively, the design of a value system) is truly the key to good systems engineering; it summarizes all the results of environmental research guides and sets the standards for all future work, provides the means for optimizing systems, and provides the rules for choosing among alternative systems" (11, p. 229).

The determination of objectives is sufficiently difficult that it is far too often almost totally ignored in systems studies. Hall identifies the source of difficulty as follows, "Unfortunately, there is no general theory of value, though many individuals and groups in all walks of life, including science and engineering, pretend to have a philosophy, system, code or a pipeline to God that permits them to make end judgements as simply as they make instrumental judgements" (11, p. 253). The lack of a value theory does represent a significant factor limiting success in systems studies; however, complete avoidance of this aspect of the problem is not the solution.

The first look at a systems problem usually uncovers a short list of primary objectives. It is not uncommon for there to be more than one primary objective.

Additional consideration typically results in the identification of other secondary objectives as well. It is rare, for example, when money is the only dimension to the problem. Churchman refers to typical secondary factors such as aesthetics, recreation, and health. The author recalls some associates instrumental in the establishment of a fried-chicken outlet in Germany. They eventually went out of business, but who is to say that they managed their business badly considering the phenomenal amount of delicious German wine they consumed while they were in business.

Profit motive gives an inadequate measure of success. Individuals responsible for profit typically have aspirations quite different from the assumed corporate goal of maximum profits. Rensis Likert, in *New Patterns of Management*, stresses this point in his "theory of supportive relationships," as follows (20, p. 103):

> The leadership and other processes of the organization must be such as to ensure maximum probability that in all interactions and all relationships with the organization each member will, in the light of his background, values, and expectations, view the experience as supportive and one which builds and maintains his sense of personal worth and importance.

Essentially, the theory suggests that management should attempt to develop management systems such that the aspirations of individuals are compatible with the objectives of the firm. To the extent that incompatibility exists, conflict will diminish the desired outcome. In a more cynical vein, the essence of this theory may be paraphrased as, "It is practically impossible to get anybody to do anything unless they want to." The management approach, then, is not to make people do things, but to provide an environment in which what they want to do is consistent with what you want them to do. Numerous studies have shown that people are concerned with such factors as peer respect, power, and personal satisfaction, with money far from the top of the list. In human-machine systems, then, it is important to optimize some combination of all of these motivating factors.

Concern with needs leads one to needs research. Human systems typically involve satisfaction of some combination of human needs. How does one determine what needs should be satisfied? One solution, not particularly recommended, is to ask a politician or a student activist group. Typically, they honestly believe they know, and who has the time, patience, or resources to say they do not know. It is to be hoped that systems analysis will bring order to chaos in more and more systems areas.

The optimum system often does not result from operating individual components in what appears to be the optimal fashion for each component. As an example, a public utility group funded the Battelle Institute in Columbus, Ohio, to perform a system study (12) of the Susquehanna River basin to determine optimal future development of the basin. Experts were engaged to study various aspects of the problem. The specialties represented included demography (population), water pollution, recreation, economic development, hydroelectric power, and others. Some time later these experts met and found that they had vastly opposing views as to what was best for the river basin. In an effort to progress from this point Pugh and Roberts, members of Forrester's (9) industrial dynamics group, were invited to participate in the study. Their industrial dynamics simulation model became the unifying mechanism for combining the data from the various specialties represented. As is true with many systems studies, the optimum solution for the system was at best a compromise of all interests represented. Industrial dynamics will be discussed in more detail later in this chapter.

One of the key problems that arose in the above study, and that is common to many systems studies, was the need to determine just who were the "customers" in the study. The determination of customers can be sufficiently difficult that some systems are best studied without the study group's making value judgments as to desired outcomes. The author was involved in a systems study in which it was considered appropriate to attempt to link expected outcomes to actions, and let the local political process determine which actions and outcomes, in light of benefits and costs, were most desirable for the common good. The study group, therefore, concentrated on identifying expected outcomes based on assumed inputs to the system, which was a far more manageable problem than valuing outcomes.

In a typical systems study, as in the example above, one must deal with a set of outcomes (e.g., multiple objectives). Figure 7-1 is a detailed list of multiple objectives offered in a policy planning problem in Quade and Boucher (25, p. 399). If a

	Criteria	
1	Destructive potential	How well can the force mix destroy targets?
2	Responsiveness	How rapidly can the force mix be ready for military action?
3	Deployability	How rapidly can the force mix move to different theaters?
4	Mobility	How rapidly can the force mix move in the theaters?
5	Supportability	How effectively can the force mix be supported and maintained?
6	Survivability	How vulnerable is the force mix to enemy actions?
7	Flexibility	How many different postures or capabilities can the force mix employ?
8	Controllability	How responsive is the force mix to command requirements?
9	Complementarity	How well does the force mix complement the forces of our allies?
10	Versatility	How effective is the force mix in a variety of military and politicomilitary situations and crises?
11	Deterrent capability	How much does the force mix contribute to our ability to deter aggression?
12	Expandability	How fast can additional capability be mobilized for the force mix?
13	National acceptability	How readily will the force mix be accepted domestically?
14	International acceptability	How readily will the force mix be accepted by other nations?

Figure 7-1 Example Multiple Criteria. [*From Quade and Boucher (25, p. 399).*]

particular set of actions with respect to some system were to produce various levels of accomplishment in relation to these criteria, there are two immediate problems: (1) How do we measure with respect to these attributes? If we could measure, how would we define the scales upon which we would measure? and (2) How do we rank and combine values obtained with respect to individual criteria to provide some single useful measure of success with respect to the overall mission of the system?

The specific area of decision theory concerned with defining, determining, scaling, measuring, and combining multiple criteria measures is a frontier of technology that, in this writer's opinion, will receive considerable attention in years to come. There is little applicable knowledge in this important area today, and the need for such technology in an ever-increasing number of systems studies will likely greatly expand this complex technical area in the future.

One of the simplest approaches to determining relative values of outcomes, referred to as "Procedure 1" in Churchman et al. (7, pp. 139–140), is given as Table 7-1.

Table 7-1   A Procedure for Determining Relative Values of Outcomes[*]

1   Rank the four outcomes in order of importance. Let O_1 represent the outcome that is judged to be the most important, O_2 the next, O_3 the next, and O_4 the last.

2   Tentatively assign the value 1.00 to the most valued outcome and assign values that initially seem to reflect their relative values to the others. For example, the evaluator might assign 1.00, 0.80, 0.50, and 0.30 to O_1, O_2, O_3, and O_4, respectively. Call these tentative values v_1, v_2, v_3, and v_4, respectively. These are to be considered as first estimates of the "true" values V_1, V_2, V_3, and V_4.

3   Now make the following comparison:

$$O_1 \text{ versus } (O_2 \text{ and } O_3 \text{ and } O_4)$$

i.e., if there was a choice of obtaining O_1 or the combination of O_2, O_3, and O_4, which would the evaluator select? Suppose the evaluator asserts that O_1 is preferable. Then the value of v_1 should be adjusted so that

$$v_1 > v_2 + v_3 + v_4$$

For example: $v_1 = 2.00$, $v_2 = 0.80$, $v_3 = 0.50$, and $v_4 = 0.30$.

Note that the values of O_2, O_3, and O_4 have been retained.

4   Now compare O_2 versus $(O_3 \text{ and } O_4)$. Suppose $(O_3 \text{ and } O_4)$ is preferred. Then further adjustment of the values is necessary. For example: $v_1 = 2.00$, $v_2 = 0.70$, $v_3 = 0.50$, and $v_4 = 0.30$.

Now each value is consistent with all the evaluations.

5   In this case, the evaluations are completed. It may be convenient, however, to "normalize" these values by dividing each by Σv_j, which in this case is 3.50. These standardized values are represented by v_j':

$$v_1' = 2.00/3.50 = 0.57$$
$$v_2' = 0.70/3.50 = 0.20$$
$$v_3' = 0.50/3.50 = 0.14$$
$$v_4' = 0.30/3.50 = \underline{0.09}$$

Total   1.00

[*]From Churchman et al. (7, pp. 139–140).

It is suggestive of the types of procedures or techniques that must be developed, if only empirically, for valuing system outcomes. Theories of value will likely follow.

Computer company systems engineers at least qualitatively offer a modern example of the necessity to view problems in a systems sense. In fact, a computer system salesperson must deal with the reality of a system within a system. The first system is that combination of computer components (i.e., a system of things) which meets the customer's needs. The second system, the human system, must be understood with respect to such factors as: (1) How will the initial investment be recovered? (2) How many people (if any) will it replace? (3) What will become of the people it replaces? (4) Will it be reliable? (5) If the power goes out, where does it leave us?

In sociotechnical human systems there are not only multiple criteria but multiple customers as well. As will be shown later, there is a further complication in the sense that each individual possesses a unique set of "utility functions" with respect to the satisfaction of needs. For example, consider the relative negative value of a $25 traffic ticket in two different situations. Assume in the first situation that a soldier is stopped on Sunday 400 miles from home after driving for a day on a three-day pass in the hope of seeing a favorite friend and has only $18 available. In the second case consider a millionaire stopped for the same violation. The first case may well represent a tragedy, the second a temporary inconvenience.

In any case, multiple outcomes, multiple criteria, multiple customers, and multiple utility functions complicate systems analysis. The following reprinted article by Chaiken and Larson (4)[26] is offered to provide additional understanding of the nature of the sociotechnical systems coming under analysis today. This article should provide a reference for the consideration of such things as "value" and "utility" that will follow. Again, in the interest of brevity, the abstract, footnotes, and references have been omitted.

Methods for Allocating Urban Emergency Units: A Survey

Jan M. Chaiken
Richard C. Larson

I. INTRODUCTION

Urban police and fire departments, emergency ambulance services, and similar urban emergency service systems comprise an important class of governmental service agencies that until recently has not benefited from systematic analyses of operational problems. These systems operate in a complicated environment that includes temporally and spatially varying demand patterns, both explicit and implicit administrative, legal, and political constraints, and often ill-defined mixtures of objectives.

[26] Reprinted from "Methods for Allocating Urban Emergency Units: A Survey," by Jan M. Chaiken and Richard C. Larson, *Management Science,* vol. 19, no. 4, December, 1972, pp. 110–130, published by The Institute of Management Sciences.

Our purpose in this paper is to review those operational problems of these agencies which are related to the deployment of their vehicles and to report current progress on mathematical modeling approaches to these problems. One cannot expect to find universally acceptable solutions—the agencies and their problems differ too much from city to city. Instead, we discuss the methods which are available, the extent of improvement that can be achieved as a result of quantitative study, and the type of solutions that can be obtained. Details of the models described here may be found in the cited references.

Agency administrators faced with the need to provide remedies for service delays and overworked personnel can use the methods described here in two ways. First, they can improve the deployment of their existing personnel. This is desirable even if there is a possibility of hiring more personnel, since the cost of even a single additional emergency unit is usually large enough to justify whatever analysis may be required to bring about the enhanced performance level without the added unit. Second, they can determine the number of men required to meet specified objectives in the future. This application is recommended for budgeting purposes in planning and administration texts since personnel costs constitute as much as 90 to 98 percent of the budget of an emergency service.

We are considering here, as an urban emergency service, any system having the following properties:

Incidents occur throughout the city which give rise to requests or calls for service (e.g., fire alarms, crime victim assists); the times and places at which these incidents occur cannot be specifically predicted in advance.

In response to each call, one or more emergency service units (vehicles) are dispatched to the scene of the incident.

The rapidity with which the units arrive at the scene has some bearing on the actual or perceived quality of the service.

In addition to such examples as fire engine and ladder trucks, police patrol cars, and ambulances, emergency service units include certain tow trucks, bomb disposal units, and emergency repair trucks for gas, electric and water services.

Although all urban emergency service systems share the above characteristics, they may differ in certain significant details:

First, some emergency units are ordinarily found at *fixed locations* at the time of dispatch. Others, such as police patrol cars, are *mobile*. This distinction is important for both administrative and analytical purposes. For instance, in principle it is possible to vary the location, size and shape of police patrol sectors at will, whereas the response areas of fire units must be designed in relation to the (fixed) locations of the fire stations. Also, the dispatch strategy for *mobile* units can often be improved by a variety of location-estimation techniques which are not needed if units are positioned at known locations. For instance, a police dispatcher could improve his decisions by querying the cars as to their locations or using information from an automatic car locator system.

The distinction between mobile and fixed-location units begins to break down during periods of high demand. At such times the units may be dispatched directly from one incident to the next, or they may be dispatched while en route from a previous incident to their home locations. Under such conditions, system operation is not very sensitive to the distribution of initial locations, either fixed or mobile.

Second, emergency services differ in their ability to determine the urgency of their calls in advance. For example, false alarms of fire, while not urgent, cannot often be identified with assurance. On the other hand, a telephone call to the police reporting a past burglary can be identified as not requiring the immediate response of a patrol car. The fraction of nonurgent calls to an emergency service ranges from around 10 percent in the case of ambulances to 40 percent for fire alarms in New York to 75 percent for police departments.

The ability of an emergency service to distinguish the priorities of its calls determines the options open to the dispatcher under near-saturation conditions. If a call can be identified as not urgent, the dispatcher may decide not to send any units, or he may hold the call in queue to await the availability of a unit near the scene; he may even place a call in queue when some units are still available, thereby protecting his ability to dispatch units to future high-priority incidents. However, if the dispatcher cannot recognize the nonurgent calls, he will have to dispatch at least one unit to each incident so that its nature can be determined; no calls can be queued.

Third, for some emergency services the time the units spend between servicing calls is used for another important activity. For example, it is widely believed that routine patrol by police cars acts as a deterrent to certain types of crime.

If police cars spent nearly all their time handling calls for service, the preventive patrol function would suffer. Such an important secondary function is not present in all emergency service systems and should be distinguished from routine internal functions: rest, meals, and training for the men, maintenance of equipment, and preparing written reports.

For units which do have an important secondary function, questions involving the dispatch of units cannot be answered exclusively in terms of the effectiveness of response to emergencies. For example, it may be desirable to place some calls to the police in queue simply to preserve the deterrent patrol. A fire dispatcher would rarely have occasion to make such a decision, since the available fire units are not engaged in any activity which could be judged more important than responding to an alarm.

The *allocation* (or *deployment*) *policy* of an urban emergency service system may be defined as the collection of rules and procedures which determine

The total number of units of each type on duty at any one time. (This may differ by time of day, or day of the week, or season of the year.)

The number of men assigned to each unit.

The location or patrol pattern of each unit.

The priority attached to different types of calls, and the circumstances under which calls are queued.

The number of units of each type dispatched to each reported incident.

The particular units dispatched.

The circumstances under which the assigned locations of units are changed. (This operation is variously called *relocation, move-up, redeployment, repositioning* or *reinforcement*.) When relocations are required: the number of units relocated, the particular units relocated, and their new locations.

In §§II–V which follow, four of these topics will be discussed in some detail. We begin with a review of methods for selecting the number of units to have on duty, continue with an approach to selecting which unit to dispatch, and then discuss location-relocation problems and strategies for preventive patrol. §VI describes other, more general, allocation models and includes some observations about the insights obtained from use of quantitative methods.

II. DETERMINING THE NUMBER OF UNITS TO HAVE ON DUTY IN EACH AREA

The number of units on duty in each area of a city is a major determinant of the performance of the system. We shall discuss two commonly-used methods for selecting the number of units before describing the recently-developed methods based on models.

A. Methods Based on Geography and Land Use

Decisions regarding the total number and locations of a city's fixed facilities (e.g., fire houses and police precinct stations) have usually been made solely on the basis of geography and land use patterns. This reliance on geographical factors has been reinforced by geographical standards and regulations which apply to many cities. For example, the Standard Grading Schedule of the American Insurance Association is used in most U.S. cities (excluding New York) to establish fire insurance rates. As a rule, cities will attempt to meet as many standards in the schedule as possible, so as not to have a lower rating than necessary. But for cities with population over 200,000, the only criteria provided by the Schedule for the number of fire engines and ladders to be located in each part of the city are based exclusively on geography and land use. For certain "high value districts" the Schedule requires every point to be no further than one mile from an engine company and no further than 1.25 miles from a ladder company. Moreover, within 1.5 miles of any point there must be at least 3 engine companies, and within 2 miles at least 2 ladders. These standards may vary slightly from area to area, but for each type of area, the same kind of geographical standard applies.

The main deficiency of geographical standards is that they are meant to be substitutes for more meaningful standards involving the time between receipt of a call for service and the arrival of emergency units. But this response time depends on many factors aside from geographical ones: the delays incurred in dispatching the units, the speed at which the units can travel, and the probability that particular units will be available. (It is little comfort to know that a fire house is within a mile of your home if the units located at that house would very likely be busy at the time you had a fire.)

Thus, as a general rule, it is not possible to determine whether an adequate number of units are located in each geographical area solely by inspecting a map of the city which shows the home location of each unit.

B. Another Traditional Approach: Workload or Hazard Formulas

Instead of relying on a single factor such as geography, so-called workload or hazard formulas combine in a subjective manner virtually *all* factors which might be thought relevant for allocating units. They give an appearance of accuracy because of the large number of factors included.

Perhaps the most well-known such formula was developed for police use by O. W. Wilson in the late 1930's. Wilson combined indicators of activity (such as numbers of arrests, number of calls for service of particular types, number of doors and windows to be checked) with other factors (such as number of street miles, number of licensed premises, and number of crimes) to arrive at a "hazard score" for each area. An area's score is computed by taking a weighted sum of the fractions of each of the factors associated with the area. The weight for each factor is a subjective indicator of its relative importance. In applying the formula, the total number of men (or patrol cars) are to be distributed among the areas in direct proportion to their hazard scores.

This procedure often produces unsatisfactory allocations that may have to be "juggled" by hand computations to arrive at a "reasonable" allocation. For instance, the 5 or 10 percent weighting often given to calls for service may be inadequate to avoid lengthy queue delays in certain areas during periods of high demand, but a higher weighting could affect other properties of the system in undesirable ways. The inherently linear form of a hazard formula precludes description of the highly nonlinear and complex interactions among system components which are often observed in practice. Such a formula also attempts a simple deterministic depiction of a system in which many of the variables are probabilistic. In addition, since some factors (such as arrests) are likely to be highest in sufficiently staffed areas, hazard formulas may have the perverse effect of indicating a need for additional personnel in areas which are already relatively overallocated. But the major difficulty arises in trying to determine how to improve the selection of the subjective weightings, a problem for which there seem to be no underlying principles or guidelines.

C. Modeling Methods

From a modeling viewpoint urban emergency service systems have two distinctive features: (1) probabilistic variations in demands and service requirements over time and (2) distributions of incidents and response units over the space of the city. The first gives rise to congestion when too much service is demanded in too short a time period and may be examined using queueing theory. The second feature gives rise to travel time distributions, patrol patterns, etc., which can be examined using essentially geometrical considerations. The two types of models are then ordinarily combined, perhaps with additional models of other aspects of performance, to produce an allocation algorithm.

Queuing Models   Since it is characteristic of emergency systems that a person's life or well-being may well depend on the immediate dispatch of a unit to an arriving call, *a primary objective of all urban emergency systems is to reduce to a low level the possibility that an urgent call will have to be placed in queue for more than a few seconds.* The probabilistic nature of the arrival times and service times of calls is such that one can never guarantee that *every* call will result in the immediate dispatch of a unit. Thus, the objective of a queuing analysis is to assure that the probability of an important call encountering a queue (or the expected waiting time in queue) is below some specified threshold.

To take a simple hypothetical example, we might imagine a city in which each police patrol car is assigned a geographical response area ("sector" or "beat") in such a way that no other car responds into its area. Then, if a given patrol car were busy, all calls from its sector would have to be placed in queue. Given reasonable assumptions, formulas are available for the queuing characteristics of such a single-server system. One could use these formulas to determine the required number of patrol units as follows: A threshold would be selected for the maximum value of the probability of a queue (or average waiting time) to be permitted in any sector; then the sectors would be selected small enough to assure that the threshold is not exceeded. The total number of sectors designed in this way would then determine how many units are needed.

Applications to real emergency services usually require a multi-server queueing model. The N units may be located at one place (e.g., ambulances at a hospital), or they may be distributed throughout a region. In the simplest model, a call is placed in queue only when all vehicles are busy servicing prior calls, all calls have the same priority and exponential service-time distribution, and the arrival process is Poisson. Stevenson has applied the steady-state version of this model to determining how large the number of ambulances in a region must be to assure that the probability of a queue does not exceed a specified threshold. Given an estimate of the arrival rate and other parameters during each time period, an administrator can select a desired threshold probability and determine how many ambulances to have on duty by time period.

The same model has been used in St. Louis for the allocation of police patrol cars. The city is divided into nine patrol districts, and a call is assumed to enter a queue whenever all the cars in its district are busy. For each four-hour time period, the Police Department estimates, using the multiserver queuing model, how many cars will be needed so that at most 15 percent of each district's calls will experience a queuing delay.

If dispatchers can distinguish the urgent calls from the less important ones, it is useful to introduce priorities into the model. One method, due to Cobham, assumes that higher priority calls are served first, but retains the assumptions that one unit responds to each call and that all service times have the same exponential distribution. Although in most police departments calls are not explicitly assigned priorities according to specified rules, Larson has found this model useful as an approximation to current performance of police dispatchers and as a tool for analysis of the potential benefits of more precise priority schemes. It has the advantage that it places emphasis on reducing the delays which are associated with important calls.

Greater realism could be introduced into this model by (1) permitting each priority level to have different service-time distributions, and (2) allowing the service time to vary with the number of units busy. But the effort required to design such models cannot be justified unless allocation decisions are found to be sensitive to the current model's assumptions and unless a comparable effort is devoted to collecting and analyzing service-time data.

One refinement of the multi-server queuing model has been found practical, and indeed necessary, for predicting the number of units busy at operations of a fire department. Fire dispatchers typically send *several* units to each alarm, while the previous models assume that the one unit is sent to each incident. In addition, fire units do not all complete service at the same time, since each may have distinct duties to perform.

Chaiken has developed a queuing model which allows for these features of fire operations. In particular, in this model

different types of alarms may require different numbers of units of various kinds;

the units may arrive singly, or in groups, and they may depart in similar fashion; and

the length of time the units are busy at the incident depends on the type of incident.

This is an infinite-server model, so that units required in one region of the city are assumed to be dispatched from there or from another region, if necessary. In applying this model, one specifies, for each region, a threshold for the probability of needing to dispatch units from elsewhere. The model gives the probability that more than n units are busy at once, and one chooses the smallest n for which this probability is under the threshold.

Applying the model in New York City, Chaiken found that at low alarm rates (such as occur in the early hours of the morning), the numbers of units needed to meet the requirements of the queuing model are well *below* the numbers needed to meet simple geographical requirements; therefore the geographical factors predominate. However, in some parts of the city the queuing model implies a need for more units at times of high alarm rate than would be suggested by geography alone. The same model could be utilized for analyzing operations of other emergency services which dispatch two or more units to certain types of incidents.

The results from these models, as well as from more complex models which incorporate queuing phenomena (e.g., Savas), have a common property: the number of units needed increases with the call rate, but not in direct proportion to the call rate. This simple observation is an illustration of the unsuitability of entering call rates in a linear fashion in workload formulas.

Travel Time Models Although the typical travel times of four to ten minutes may be dominated by queuing delays during periods of saturation, travel time may comprise the greatest fraction of total response time during normal operating periods. Thus, models are required which relate properties of travel time to the number of units on duty, geographical characteristics, arrival rates of calls, and service times at incidents. If all the units serving a region are located at a single point, changing the number of units has little effect on travel time, as in the "base case" for ambulances studied by Savas. But if the units are spread around the region, travel time models can replace traditional geographical factors in determining the number of units to have on duty during periods of relatively light demand.

Geometrical models can be used for determining the relationship between travel *distance* and the spatial distribution pattern of the units. For a model in which units are *randomly* located, with an average of r units per square mile, Larson has found that the right-angle travel distance for the closest unit has a Rayleigh distribution with mean approximately $0.63/r^{1/2}$. Similar inverse square-root laws are found under other assumptions. If the units are not randomly located, but instead are positioned in such a way as to minimize average travel distance, the mean is found to be $0.47/r^{1/2}$. Kolesar and Blum have found similar results for the travel distance of the nth-arriving unit when more than one are dispatched.

Such geometrical models, used in conjunction with empirical data relating travel time to travel distance, provide an estimate of the travel time distribution, given the spatial distribution of the units. If the units can occupy only a small number of specified locations, the travel time for each possible arrangement can be calculated directly from empirical data and estimated arrival rates in subregions, an approach which has been used by Hogg. The entire travel time distribution should be found, if possible, since ultimately one may be interested in the probability that travel time exceeds a specified limit.

The results from the geometrical models may then be tied with queuing results to determine the travel time properties of the system, given N total units on duty and the other relevant operating characteristics. The queuing model is used to obtain the probability that n units will be available for dispatch, and the geometrical model gives the travel time distribution for this circumstance. Larson follows such an approach to estimate the average travel time to incidents in a region, given homogeneous travel speeds and random positions of the units. Alternatively, one could assume that the available units are moved, if necessary, so as to occupy the locations which give minimum travel time. Similar methods may be used to estimate measures of travel time other than the mean.

In applying these models to determining how many units to have on duty, two constraints are used. The first specifies a limit on some measure of queuing (the probability of a queue, the probability of needing distant units, or the expected waiting time), and the second sets a limit on an appropriate measure of travel time. It is then possible to determine the smallest number of units meeting both constraints. Such a calculation provides sensible allocations for all regions, whether they experience high call rates or not.

In some circumstances it may be possible to find an empirical relationship between an appropriate measure of travel time and the *average* density of available units, in which case one can dispense with the geometrical models. Kolesar and Blum have studied the output from a variety of travel time models, as well as data derived from experiment, and have concluded that if unavailabilities are not too severe the average travel time $\overline{T}$ is inversely proportional to a power of the average density $\overline{r}$ of available units: $\overline{T} = \alpha/\overline{r}^{\beta}$. When this approximation holds, the parameters α and β can be determined from data collected in each region, and the resulting formula can be used directly from a queuing model to select the number of units needed.

Methods Using Other Criteria   The simple models described above may not, by themselves, be sufficient to determine the number of units needed. First, for services which engage in important activities other than response to calls, criteria relating to these activities have to be taken into account. Second, and more fundamentally, easily quantifiable criteria (probability of encountering a queue, average travel time) do not often have a clear relationship to the true performance of the system. For instance, one would like to known the actual benefits which accrue by decreasing response time. Such benefits might be lives saved, stolen goods recovered, property damage averted, etc. Although preliminary research along these lines has been performed the currently available empirical information is not an adequate foundation for an administrator's use in selecting allocation policies. Thus, at present, one is forced to use available performance measures such as response times. A careful and realistic use of such measures can provide reliable proxies for more fundamental measures, as has been discussed by Carter and Ignall.

Given such a necessary reliance on performance measures, an administrator would usually want to employ several simultaneously to arrive at reasonable allocations. In addition to queuing and travel times, he could incorporate factors pertaining to other activities (e.g., preventive patrol) and to administrative matters (e.g., workloads).

Two criteria, travel time and response workload, are analyzed by Carter and Ignall in a queuing model for determining the extent to which an added fire unit provides relief to overworked units in a command. It would be natural to assume that when units are added to a command, the *number* of responses made by each of its units would decrease, and this may be one of the secondary benefits of adding units which is particularly interesting to an agency administrator. However, in instances where two or more units are ordinarily dispatched to each incident, but fewer are sent when some of them are unavailable, adding a new unit may increase the chances of a full (rather than partial) response to such an extent that the average number of responses of units in the area actually increases. Thus their model shows that if it is desired to reduce the workloads of units in addition to improving the response time, a greater number of units may be needed than is suggested by simpler models.

Larson has developed a dynamic programming model for allocating police patrol cars to commands (e.g., precincts) which permits police administrators to specify constraints, which may vary by command, on average travel time, the frequency with which cars pass by an arbitrary point while on preventive patrol, the smallest number of units which can be assigned to any one command, and additional criteria related to crime rates or other activities of patrol cars. The algorithm supplies each command with enough patrol units to satisfy all the constraints and then treats the queuing delay as the objective function to be minimized using whatever additional patrol units are available. With limited police resources, it is possible that a specific set of policy objectives is unobtainable. If so, the algorithm indicates the additional number of patrol units required to meet the stated objectives. To allocate the currently available number of units, the algorithm then requires a more modest set of objectives.

When applied in New York City, the results suggested that during periods of relative congestion (e.g., Friday and Saturday evenings), average queuing delay could be decreased significantly by diminishing resources in residential commands with relatively light demands and increasing resources in commands which are heavily loaded. Such a redeployment of resources does not noticeably degrade performance in residential commands, since sufficiently many patrol units are retained to satisfy all policy constraints. Yet, average queuing delays in heavily loaded areas can often be reduced from thirty minutes to less than two or three minutes.

In general, the quality of the allocations which an administrator can expect from any of the models described above depends on how much effort he is willing to have his staff devote to the application. An analyst who is not a member of the concerned agency cannot make an appropriate determination of what constitutes an "excessively long" delay before the arrival of a unit, or how much preventive patrol will be considered adequate, or what level of workload is "too great."

In the case of fire departments, where the various units dispatched to a single incident may arrive at different times, the analyst is not even in a position to know which arrival patterns are "better" than others. However, the field chiefs, who are completely familiar with their department's operating procedures at a fire, can provide valuable information. Through asking a series of questions such as "Would you prefer two fire engines arriving 1.5 minutes after an alarm, or one arriving at the 1-minute mark and one at the 2-minute mark?", it is possible to derive a chief's utility function for arrival times. Work in progress by Keeney to develop such utility functions should make it possible to select the allocation of units so as to maximize a chief's utility of the resulting patterns of arrival times.

It should also be noted that, in regard to any of these methods for determining how many units to have on duty, there may be some difficulty in assigning individuals to shifts or tours of duty which best "fit" the desired assignments. Legal and administrative constraints can make this problem quite difficult. A heuristic approach is discussed by Edie. A more general approach using a computer algorithm has recently been reported by Heller.

III. DESIGN OF RESPONSE AREAS

A problem commonly shared by all spatially distributed urban systems is the design of response areas (districts or sectors or beats) that indicate where a particular patrol unit, fire engine, or ambulance is to have primary responsibility. In designing these administrative areas, agency administrators have stated several diverse (often conflicting) objectives: minimization of response time, equalization of workload, demographic homogeneity of each area, and administrative convenience. No single mathematical technique for design of districts is likely to take into account all the relevant factors. Yet, some of the recently-developed models have provided useful insights into the problem.

Traditionally, police planners have been instructed to design patrol sectors as squares, circles, or as straight lines along particular streets. The objective of square or circular designs is to keep at a minimum the time required for the patrol unit to travel to the scene of a reported incident in its sector. For instance, O. W. Wilson states that ". . . a square beat (sector) permits a maximum quadrilateral area with a minimum distance between any two possible points within it."

However, travel speeds may depend on direction of travel, in which case it will be desirable to design the sector so that the longer sector dimension corresponds to the direction with higher travel speeds. Using quantitative techniques, it is possible to predict the travel time characteristics of any proposed sector design and thereby determine which designs actually do minimize some indicator of travel time.

As an example, consider an urban region in which the streets form a mutually perpendicular grid (e.g., as occurs in central Manhattan and certain other cities), running, say, east-west and north-south. Then, a shortest route of travel for the assigned patrol unit requires the unit to traverse the total east-west distance, plus the total north-south distance, between the unit's initial position and the position of the incident. Under the simplifying assumption that each patrol unit responds only within its sector, Larson has shown that average intrasector travel time is minimized by designing the sector so that the average time required to travel east-west equals the average north-south travel time. Since it is not unusual to find regions (such as in Manhattan) where the north-south speed is about 4 times as great as the east-west speed, this implies that the sectors should also be 4 times as long in their north-south direction. In this case, such a sector design can be expected to reduce average intrasector travel time by approximately 20% over that obtained by the rule-of-thumb design—square or circular sectors.

Some of these ideas have already been applied by Bottoms, Nilsson, and Olson in the city of Chicago. They have designed a sector plan of the city using rectangular sectors in such a way that the average intrasector travel time does not exceed approximately three minutes.

Certain complications to travel, such as one-way streets and railroad crossings, should be taken into account when estimating travel times. Larson has computed the mean extra distance traveled due to these complications. Although the results indicate that the average travel time does not usually increase very much when such complications are present, a small fraction of responses in a one-way-street grid may require three or more additional minutes for the unit to reach the scene.

Influence of Intersector Cooperation

When an incident is reported from a response area whose units are busy, most emergency service systems will dispatch an available unit nearby in another response area. Such an arrangement is nearly mandated by queuing considerations, but it introduces subtle complications into the design of response districts. In the case of mobile units, it even raises questions about the need for restricting response areas of the units to be nonoverlapping. These consequences of intersector cooperation will be discussed below.

Police Patrol: "Flying"

Police administrators are often heard to argue in favor of assigning patrol units to nonoverlapping sectors in order to establish a "sector identity" on the part of the patrol officer. This identity, which derives from patrolling and from citizen contacts made while responding to calls for service, is supposed to cause the officer to feel responsible for public order in his sector. However, given nonoverlapping sectors, one can show by a simple probabilistic argument that the fraction of dispatches which cause units to travel outside their own sectors is usually equal to or greater than the fraction of time that units in that area are unavailable for dispatch. Thus, if a police department's patrol units are "busy" 40 percent of the time (a typical value), then at least 40 percent of all dispatch assignments cause the assigned patrol unit to leave its own patrol sector. For this reason, at least 40 percent of all citizen contact occurring while responding to calls-for-service takes place in sectors other than the patrol unit's own sector.

The predicted amount of intersector dispatches (called "flying" in some police departments) has been substantially verified both by our own work and by the reports of others. These data showed that the amount of intersector dispatching is never significantly less than the percentage of time unavailable, and it may be significantly more. Intersector dispatches ranged from 37 to 57 percent of the total.

The extent of flying brings into question not only the philosophy behind nonoverlapping sectors but also a widely popular statistical procedure for computing workloads of police patrol cars. Usually a sector is associated with a "workload" which is proportional to the number of calls for service generated from within the sector. However, it is clear that the number of intersector dispatches may be sufficiently large so that one should calculate workloads of units from records of their dispatch assignments and not from the rates of calls for service in individual sectors.

There is one additional property of nonoverlapping sector systems that we should mention. It involves the "randomness" of preventive patrol. With nonoverlapping sectors, preventive patrol coverage in a sector is reduced to zero whenever the corresponding patrol unit is busy. Anyone, including potential criminals, can monitor a patrol unit's activity in some manner (e.g., visual observation, listening to the police radio) and determine when a particular car is not patrolling. Then, since units are assigned to nonoverlapping sectors, a crime can be committed with assurance that the patrol unit will not pass during the commission of the crime.

Given the undesirable features of a nonoverlapping sector system, how can an administrator revise and improve operations? First, if the sector concept is to be retained, the large fraction of calls which are *low priority* (i.e., they do not require rapid response) can be "stacked" and handled by the car assigned to the sector of the call when that car becomes available. This procedure reduces the amount of flying and enhances "sector identity."

Second, the sector concept can be modified so that patrol units are assigned to overlapping areas (or sectors). This procedure enlarges the area with which each patrol officer should develop an "identify." In addition, it increases the "randomness" of patrol, a desirable outcome which is not achieved simply by stacking on nonoverlapping sectors.

Response Districts for Fire Units

A fire unit's primary response district consists of all points to which it would be dispatched if an alarm were generated there, even if all units were available. In the event of unavailabilities, the unit may also respond to alarms elsewhere. Fire departments have traditionally designed districts so that the dispatched units are the ones *closest* to the fire. This means that all points on the dividing line between two districts are equally close to some pair of companies.

With the modification of interpreting "closest" in the sense of "shortest travel time," one might expect this procedure to minimize overall response time. However intuitively reasonable this rule-of-thumb may appear, a recent analytical study by Carter, Chaiken, and Ignall has shown that "equal travel time" dividing lines are usually *not optimal* and that overall average travel time is minimized by following a policy that often requires a unit other than the closest unit to be assigned to a particular fire.

The philosophy underlying this result is one that often appears in systems with unpredictable demands in the near future—it may be preferable to incur an immediate travel time penalty so that the system (e.g., the collection of all fire apparatus) is left in a state which best anticipates future demands. That is, assigning, say, the second closest unit to the most recent fire may result in favorable positioning of units for the *next* reported fire, thus minimizing expected overall response time. Assignment of the closest unit to the first fire might have required an unusually large amount of time to respond to the next reported fire.

Carter, Chaiken and Ignall have also shown that the boundaries which minimize overall average response times will, in many cases, also reduce workload imbalance (where workload imbalance is defined to be the difference in the fraction of time worked by the busiest unit and by the least busy unit). Thus, implementation of their derived procedures can result in reduction in both response time and workload imbalance.

Their boundary structuring procedures have been worked out in detail for the case of two operating units; current research is being directed at extending the results to systems with many cooperating units. The qualitative features of the results have already been used to arrive at preferable dividing lines in New York City Fire Department operations—and these results are currently being implemented.

IV. LOCATING UNITS AND FACILITIES

Closely related to problems of response area design are problems of location, including

which site to select for an additional facility;
when consolidating two or more existing facilities into one new facility, where to place the new facility;
pre-positioning, or where to locate units at the start of a tour;
repositioning, or how, and under what circumstances, to change the locations of units during a tour to correct for unavailabilities as they develop.

Although there is an extensive literature on the subject of "facility location," most of it is presented in economic terms and ignores probabilistic aspects of operations. Revelle, Marks and Liebman recently reviewed a variety of applications of algorithmic location theory to public sector problems without discussing the allocation of urban emergency units. To the extent that analytical methods have been used for locating emergency service units, the application has been limited to a small number of units or to a small number of potential sites for the units. Under these circumstances, one can either determine the travel time properties of all possible combinations of locations or utilize simple algorithms which assist in searching for the optimal locations.

In the work of Savas the implications of dispatching ambulances from two fixed locations in a particular part of Brooklyn (as an alternative to the previous single site) were explored by considering every possible division of n units between the two locations and using a simulation model to estimate average response times. Larson and Stevenson also considered only two sites, but the location of one of them was permitted to be arbitrary. A steepest-descent search routine enabled them to find the location of the second site which minimized average travel time. The results suggest that the optimal location of a new site is rather insensitive to the precise locations of existing units. Hogg considered nineteen potential sites for fire units in Bristol, of which at most 9 were to be occupied. For this purpose, she developed an algorithm for rejecting the least desirable sites.

A considerably larger body of analytical work has been completed, or is underway, concerning the *repositioning* of units during the course of a tour. Two examples are discussed below.

Local Repositioning: Police Patrol Cars

Consider the case of two square patrol sectors which have a north-south boundary in common. We will assume that each unit patrols its sector randomly, and the demands are uniformly distributed over the entire two-sector region. Each unit responds in its own sector, unless it is unavailable, in which case the other one responds. The question of interest is, "At the moment when one of the units becomes busy, is there any advantage to repositioning the remaining available unit? If so, how should this be accomplished?"

Whenever one unit becomes *unavailable,* consider the following three alternatives for the free unit:

Alternative 1. The free unit continues to patrol its original sector (i.e., no repositioning).
Alternative 2. The free unit patrols both sectors uniformly (i.e., "uniform repositioning").
Alternative 3. The free unit is stationed on the boundary line between the two sectors at the north-south halfway point (i.e., "fixed point repositioning").

It is straightforward to show that Alternatives 1 and 2 have the same average travel distance and the distance for Alternative 3 is 75 percent as large. Thus, in an average travel distance sense, uniform patrol repositioning (Alternative 2) offers no advantage over no repositioning (Alternative 1). On the other hand, fixed point repositioning offers a 25 percent reduction in average travel distance when compared to Alternatives 1 and 2. Similar results hold for regions of four cooperating sectors and for more complicated examples.

The results suggest that only local repositioning (among nearby sectors) is advantageous in a travel distance sense only if patrol is concentrated near the boundaries of the appropriate sectors. In practice, strict fixed point repositioning may not be advisable because of lost preventive patrol coverage; still, if the free unit must remain patrolling, a large part of the travel distance reduction can be retained provided the patrol occurs near the appropriate sector boundaries. In fact, we have heard patrolmen remark that on an informal basis two units will occasionally agree to "cover" both sectors when the other unit is unavailable; this "covering" usually takes the form of concentrated patrol near the common sector boundary. To gain travel distance reductions when such covering occurs, it is necessary that the dispatcher be aware of the identity of the cooperating units so that he can assign the covering unit to a call in the busy unit's sector.

Global Repositioning: Relocation of Fire Units

By "global repositioning," we mean the reassignment of one or more available units to areas which may be at some distance from the areas to which they are currently assigned. For many years urban fire departments have relocated available units when a number of units in one area become busy fighting a large fire. Indeed, these relocations are pre-planned, so that when a second alarm (or higher) is sounded, specified units respond to the fire while other units simultaneously move into certain fire houses which have been vacated by units at the fire. Such large-scale repositioning is not as widely used in other urban service systems, although situations continually arise (e.g., police precinct-level congestion) in which repositioning of forces would reduce travel times and dispatch delays or provide some preventive patrol.

The following factors are important in determining whether to make a relocation:

How long is the expected duration of the existing unavailability?
How many units will have to relocate to accomplish the desired final locations?
How long will it take for the units to travel to their new locations?
Is the magnitude of the expected improvement in performance large enough to warrant the effort required to move units?
Is there a good reason to believe that the units to be moved will be needed at their present locations in the near future?

Work still in progress at the New York City-Rand Institute is designed to produce an algorithm which will operate in a computer-assisted dispatch system and will recommend relocations both for large fires and for unavailabilities which occur through an accumulation of smaller fires.

Several approaches have been tried. Swersey developed an integer programming model to determine which fire houses should be empty and which full. His objective was basically to minimize the average travel time to incidents, taking into account the average time that busy units would remain busy. In addition his procedure provided a penalty for each unit relocated. Once a solution to this model has been found, a standard assignment problem can be solved to recommend which units should move to which empty houses. Unfortunately, it was not possible to solve Swersey's model rapidly enough, using either branch-and-bound or approximate heuristic techniques to make it an appropriate tool for real-time applications.

The relocation method which is now planned for implementation has been developed by Kolesar and Walker based on a suggestion of Chaiken. Instead of focusing on average travel time, which can be close to minimum for a variety of states of the system, this method utilizes ideas of "coverage." A point is said to be "covered" if at least one available engine company (or ladder company) is within T minutes of the location. If one or more neighborhoods are expected to be uncovered for an undesirable amount of time, a heuristic algorithm first determines which vacant houses to fill, then which available units to relocate to the vacant houses. While this algorithm is not "optimal" in any sense, it appears to compute very reasonable relocations using a comparatively small amount of computer time.

V. CRIME PREVENTIVE PATROL

Although other urban service agencies have certain patrolling activities (e.g., fire departments "patrol" areas looking for fire hazards), the patrolling function is most important in urban police operations. A patrol unit is said to be performing "crime preventive patrol" when passing through an assigned area, with the officers checking for crime hazards (e.g., open doors and windows) and attempting to intercept any crimes in progress. By removing opportunities for crime, preventive patrol activity is supposed to *prevent* crime. By posing the threat of apprehension, preventive patrol is supposed to *deter* individuals from committing crimes.

Most mathematical studies of police preventive patrol have occurred in the past several years, although some earlier work in "search theory" is also relevant to the problem. The term "random patrol" was introduced into police literature in 1960 by Smith who stressed the need for unpredictable patrol patterns. Blumstein and Larson developed a simple analytical model for spatially homogeneous random patrol in order to estimate the probability that police would pass a crime in progress. Elliott quoted one of Koopman's search theory results and attempted also to compute probabilities of space-time coincidence of crime and patrol. Bottoms, Nilsson and Olson have applied some of these ideas to operational problems in the Chicago Police Department.

To illustrate one simple model, consider the problem of estimating the probability that a patrolling unit will intercept a crime while in progress. For a crime of short duration T_c which occurs on street segment i one can argue that a reasonable upper bound estimate of the probability of space-time coincidence of crime and patrol is

$$P_c = \frac{se_i P_p T_c}{L}$$

where s = speed of patrolling vehicle
  e_i = a number between 0 and 1 reflecting the relative patrol coverage of segment i
  P_p = fraction of total time spent patrolling
  L = a weighted sum of the segment lengths in the patrol sector, the weights being the e_i's

Even this simple formula provides certain insights. It illustrates that crime-intercept probability is directly proportional to coverage (e_i), fraction of time spent patrolling (P_p), duration of the crime (T_c), and patrol speed (s) and is inversely proportional to a weighted sum of segment lengths (L). The interaction of the response and patrol activities is also apparent: during periods of heavy call-for-service demand (i.e., when P_p is small), crime intercept probability is small. A potential trade-off exists between the amount of screening and/or delaying of calls for service, reflected in the value of P_p, and the likelihood of intercepting a crime in progress.

In applications of this formula one typically finds intercept probabilities below 1 in 100. Such low detection probabilities bring to question whether the threat of apprehension provided by preventive patrol is actually great enough to deter crime.

Given such scarcity of preventive patrol effort, any effective allocation of effort must reflect the relative likelihoods of crimes occurring at various times and places. Even raising intercept probabilities from 0.01 to 0.02, say, could result in a doubling of on-scene apprehensions. By structuring a model of preventive patrol operation one finds that the allocation of preventive patrol effort is mathematically similar to an allocation of search effort problem studied by Koopman and reformulated by Charnes and Cooper. Application of search theory ideas to allocating relative patrol effort (e_i's) to maximize detection probabilities yields the following properties:

On street segments with sufficiently low crime rates, no preventive patrol should be allocated.

On segments which should receive preventive patrol effort, the effort should grow slower than linearly with crime rate.

This behavior again illustrates the inadequacy of linear hazard formulas which imply that preventive patrol coverage should be *directly proportional* to frequency of crime occurrence. Although much more refinement of these techniques is required before they can be implemented by police, we would expect the qualitative features of the solution to hold.

VI. SIMULATION MODELS FOR EVALUATING PROPOSED CHANGES IN ALLOCATION PROCEDURES

An agency administrator is typically faced with a number of proposed changes in his allocation policy at one time. For example, the results of the models described in previous sections may suggest that he should adopt priority queuing schemes, add units at certain times of day, select new locations for some units, change response areas or patrol patterns, and modify the procedures for relocating units. In addition, certain technological innovations such as automatic car locator systems may have been proposed to accomplish some of the same objectives. Before making a choice among the alternatives, the administrator will want to have a realistic comparison of the benefits which can be expected from each approach.

For a thorough evaluation of such a comprehensive change in allocation procedures, one generally has to turn to much more complex and detailed models than the ones already discussed, for example simulation models. These models can provide information about the effect of a proposed policy change on a wide range of variables: response times to particular types of calls, workload of units, queuing delays, availability of units, etc. Such simulation studies have been undertaken by Savas to investigate the reduction of travel times which could be achieved by spatially repositioning ambulances, by Swersey to analyze the operations of the dispatch centers of the New York City Fire Department, by Carter and Ignall to compare a wide range of combinations of fire department allocation policies, by Larson in a study of the allocation of police patrol and the potential benefits of utilizing a car locator system, and by Adams and Barnard to study the value of an automated dispatch system for the San Jose Police Department. Recent work on efficient computer coding of geographical data has been of some assistance in designing such simulation models of urban emergency service systems.

A common feature of these studies has been the finding that rather simple and inexpensive administrative innovations can often make a contribution to system performance which is equivalent to that of much more expensive hardware or increases in manpower. Swersey's study provides such an example. In this case, the fire dispatching office in Brooklyn was experiencing an increase in alarm rates and consequent delays prior to dispatch of units which were beginning to be of some concern to the Department. The simulation showed that computerized methods for recording, storing, or retrieving location information about alarms would not solve the essential difficulty, which had to do with the fact that a single man had final responsibility for every dispatch decision. Swersey's suggestions for dividing this responsibility, a basically administrative change which has been implemented, provided substantially decreased delays during peak-alarm hours.

Similarly, Larson's simulation has demonstrated that the absence of an explicit priority structure for calls to police departments produces unnecessary delays for urgent and moderately important calls. Most departments have been reluctant to implement such a structure, stating that their policy is to provide rapid service to all citizens. However, some departments have begun to reject some calls and implement priority structures for dispatching cars to the remainder.

In addition, the Larson simulation was used to study the best use of automatic vehicle locator systems in police departments. The technology of such systems is well developed and recently field tests and operational installations have been reported. Each system provides a central dispatcher with estimates of the positions of all service units (e.g., buses, patrol cars, taxicabs) and with other status information (e.g., current speed and direction, current type of activity).

In the Larson study analysis showed that superimposing an automatic vehicle locator system on a patrol force assigned to nonoverlapping sectors causes an average travel time reduction in the order of 10 to 20 percent, the exact value depending on the fraction of time each car is busy, number of sectors in a command, spatial distribution of calls, etc. Analysis also showed that a system with fully overlapping sectors utilizing car position information has approximately the same travel time characteristics as current nonoverlapping sector systems without car position information. Thus, if there are reasons to want overlapping sectors, even to the extent of not assigning sectors to cars, there would be little or no degradation in travel time characteristics of the overlapping sector system, compared to current systems, provided high resolution car position information is available. Apparently, the pre-positioning advantages gained by assigning cars to mutually exclusive sectors are recovered by knowing exact car positions in a system with no deliberate spatial pre-positioning. (As mentioned in §II, arguments based on "regional identity" and "randomness of patrol" seem to favor some type of overlapping sector plan.)

This analysis is an example of an instance in which applying technology to a system "operating as usual" may not fully utilize the new technological capabilities.

SUMMARY AND CONCLUSIONS

Many of the allocation problems experienced by emergency service agencies, whether police departments, fire departments, or ambulance services, are not unique to any particular agency, nor are they confined only to the largest cities. Typically, an increasing rate of calls for service over a period of years creates a situation in which a substantial fraction of callers with urgent emergencies must wait "too long" for the arrival of a unit. Moreover, the agency's field personnel may be spending so much of their time responding to calls that they are unable to pay adequate attention to other important duties, or they may feel overworked.

Hiring a large number of additional personnel and providing them with new units, stations, and other equipment can almost always resolve the problem, but this solution may be neither feasible nor necessary. To determine the best course of action, an agency administrator should consider a variety of alternatives, using quantitative methods to estimate the potential benefits of any combination of them.

As a first step, the arrival patterns of calls by time and place should be analyzed, with attention paid to the number of calls which do not require immediate service. This may lead to plans for queuing or rejecting the calls of lowest priority, in which case callers should be informed as to whether a unit will be dispatched. Next, the number of units actually needed to serve the remaining calls and perform other essential functions should be determined in accordance with the methods of §II. This analysis may reveal opportunities for improving service by moving units from one region to other or changing the hours of day at which units operate. On the other hand, the results may show that the desired level of performance cannot be achieved without adding units, in which case the administrator will have a good estimate of the actual number of men needed.

Next, the locations (or patrol areas) and the dispatching rules for the units should be scrutinized to isolate opportunities to reduce delays further and balance workloads, using the methods described in §§III and IV. In addition, plans should be made to reposition units as the need becomes apparent, since no static plan for locating the units can produce the best allocations under all circumstances. Police departments should also consider the extent to which preventive patrol can be enhanced by designing overlapping sectors and by providing the patrolmen with guidelines for their patrol patterns.

Although in many instances we do not yet know how to make the link between true measures of performance of emergency systems and the quantities which can be studied using analytical models, it is now apparent that models can indicate clearly which aspects of performance are likely to improve or to be degraded as a result of a specific policy change. Many of the research goals for allocation of police patrol forces proposed in 1968 in a study for the Department of Justice are now being approached, if not achieved. Wide interest is now apparent, as illustrated by reported applications of quantitative techniques in police departments in Boston, New York, St. Louis, Chicago, Cleveland, Tucson, Arizona, Phoenix, Arizona, and Great Britain. The whole subject of the allocation of fire units has been developed in the past two or three years and has given an entirely new complexion to fire research. In the next few years we expect that the models will improve in their sophistication and utility and agency administrators will make increasing use of quantitative models as their virtues become apparent.

In the reading above, reference was made to a study performed by Kolesar and Walker (17). This paper was awarded the Lanchester Prize by an awards committee of the Operations Research Society of America as "the best English-language published contribution to operations research" in 1974. Because of its length and mathematical level, at least in some segments of the paper, it is not reprinted here, but it is highly recommended for those who wish to gain insight into systems studies at a higher technical level than is presented in this text.

In the early days of operations research much was said about optimizing. In this author's opinion, far too much attention was given to perfecting the means of optimizing in the years that followed. High-level systems to this day are essentially unoptimizable, as is indicated in the following quotation from Hitch (15),[26a] from his address as the retiring president of the Operations Research Society in 1960:

> The future is uncertain. Nature is unpredictable, and the enemies and allies are even more so. He has no good general-purpose technique, neither maximizing expected somethings, nor max-mining, nor gaming it to reveal the preferred strategy. How can he find the optimal course of action to recommend to his decisionmaker?
>
> The simple answer is that he probably cannot. The same answer is also the beginning of wisdom in this business. There has been altogether too much obsession with optimizing on the part of operations researchers, and I include both grand optimizing and sub-optimizing. Most of our relations are so unpredictable that we do well to get the right sign and order of magnitude of first differentials. In most of our attempted optimizations we are kidding our customers or ourselves or both. If we can show our customer how to make a better decision than he would otherwise have made, we are doing well, and all that can reasonably be expected of us.

The following example from Quade and Boucher (25, pp. 55-59) is presented early in their book to attempt to show some of the complexities that exist in the

[26a] Reprinted by permission from "Uncertainties in Operations Research," *Operations Research*, vol. 8, pp. 443-444, Copyright 1960, ORSA. No further reproduction permitted without consent of the copyright owner.

relationships between the major system elements: objectives, alternatives, costs, effectiveness scales, effectiveness, and criteria. This rather short example does effectively highlight some of these underlying complexities.

An Example:
Selection of a New Aircraft Engine

E. S. Quade
W. I. Boucher

As an example of how these elements of analysis figure in a relatively narrow decision problem, let us consider the selection of a new aircraft engine, and assume that the *objective* is simply to increase engine performance. Then the *alternatives* are obviously the various possible engine types that achieve this objective by such means as exotic fuels or novel design. The *costs* would be of two general kinds: the total capital resources (such as manpower and research facilities) that must be allocated to the research, and the time required to achieve a successful prototype. In this simple case, the *effectiveness scale* relates directly to the objective, and might be taken as the difference between the specific fuel consumption typical today and that achieved by further research, for fixed engine weight. The *effectiveness* of a particular alternative engine type would then be its estimated reading on this scale. The greater the difference, the better the engine, since we desire to decrease the specific fuel consumption by research. In general, the amount of improvement will depend upon the amount of effort expended upon research, so that estimated costs and effectiveness might be related as shown in Fig. 1.

Such different levels of performance might result from a situation that H. Rosenzweig will discuss more fully later, in which alternative 1 corresponds to a very

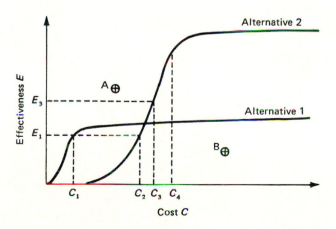

Figure 1 Cost and effectiveness.

conservative improvement over operational engines, and alternative 2, to a larger state-of-the-art advance.

Note, however, that even if we assume that both these alternative research programs can be completed on time and are subject to essentially the same amount of uncertainty, we still could not decide between them. What is missing is some knowledge of why the improved performance is needed. Thus, although alternative 1 achieves only a modest level of effectiveness (E_1), it does so at one-third the cost of alternative 2. If the level E_1 is adequate, why not select alternative 1 and thereby minimize cost? Indeed, quite often cost will be limited by decree to some level such as C_2, in which case alternative 1 is the obvious choice. On the other hand, the goal of the research may be to achieve some minimal new level of effectiveness, such as E_3, no matter what the cost. Then alternative 2 is obviously the choice.

The point to be made is that, in general, it is not possible to choose between two alternatives just on the basis of the cost and effectiveness data shown in Fig. 1. Usually, either a required effectiveness must be specified and the cost minimized for that effectiveness, or a required cost must be specified and the effectiveness maximized. Clearly, the results of the analysis of effectiveness should influence the selection of the final criterion. For example, if C_3 is truly a reasonable cost to pay, then the case for C_4 is much stronger, in view of the great gains to be made for a relatively small additional investment. As a matter of fact, this approach of setting *maximum* cost so that it corresponds to the knee of the cost-effectiveness curve is a very useful and prevalent one, since very little additional effectiveness is gained by further investment.

Overspecification of Criteria

On the other hand, both required cost *and* effectiveness should not be specified; this overspecifies the criterion, and can result in asking for alternatives that are either unobtainable (point A in Fig. 1) or underdesigned (point B in the same Figure). An extreme case of criterion overspecification is to require maximum effectiveness for minimum cost. These two requirements cannot be met simultaneously, as is clear from Fig. 1, where minimum cost corresponds to zero effectiveness, and maximum effectiveness corresponds to a very large cost.

Maximizing Effectiveness/Cost

Somewhere in the middle are criteria that apparently specify neither required cost nor effectiveness. One which is widely used calls for maximizing the ratio of effectiveness to cost. This seems to be a workable criterion, since, in general, we want to increase effectiveness and decrease cost. Nevertheless, as we can see by examining Fig. 2, it has a serious defect. Since the effectiveness-cost ratio for either alternative is simply the slope of a line drawn from the origin to a given point on the curve for that alternative, and since, in this example, the ratio obviously takes on a maximum at the knee of the curve, our choice between the two alternatives seems to be settled at once. Thus, alternative 1 is clearly preferred with this criterion. However, if E_3 is the minimum level of effectiveness acceptable from a research program, then alternative 2 is the obvious choice. The point to be made here is that unless the decisionmaker is

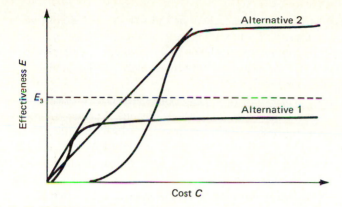

Figure 2   Effectiveness/cost ratio.

completely unconcerned about *absolute* levels of effectiveness and cost, then a criterion such as this, which suppresses them, must be avoided.

Theoretically, it is possible to escape this need for specifying either the required cost or effectiveness by expressing cost and effectiveness in the same units, such as dollars or equivalent lives saved. For if this can be done, then it is possible to subtract cost from effectiveness, and take as the criterion the maximization of this difference. But seldom, if ever, can cost and effectiveness be expressed in similar units, and we may assume that the earlier description of a criterion applies.

Dominance

Infrequently it happens that selection between alternatives is easy. An extreme case of this is shown in Fig. 3, and occurs when an alternative such as 3 is more effective than any other at every cost. In such a case it is clearly advantageous to select alternative 3, which is said to *dominate* alternatives 1 and 2 at all levels of investment and effectiveness. Note that it is still not permissible to overspecify the criterion and

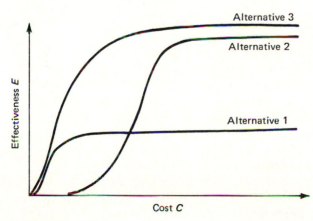

Figure 3   Dominance.

require maximum effectiveness for minimum cost. For the situation of dominance only permits us to select alternative 3; minimum cost still corresponds to zero effectiveness for alternative 3, and so forth. Even though dominance designates alternative 3 as preferred, the required level of effectiveness must be specified before the preferred level of investment can be selected.

In this example of propulsion research, as in many others in advanced research or specific component design, the goal has been simple and obvious. Further, in such cases an appropriate scale of effectiveness is usually obvious and related directly to the goal. Finally, the measurement of effectiveness (that is, the location of an alternative upon that effectiveness scale) is straightforward in such cases. Since the example at hand will be discussed by H. Rosenzweig in some detail, we can conclude our discussion of it here, and pass on to the more difficult, but perhaps more interesting, questions the analyst must face in identifying goals, selecting scales of effectiveness, and performing the measurement of effectiveness for some of the other problems which were mentioned earlier.

SIMULATION

By far, the modeling technique most frequently used in IE/OR/MS is simulation. Simulations do not optimize, but they are a means for estimating outcomes. A good solution, if not an optimum one, is determined with simulation by modifying a representation of system structure or input parameters and observing the output until desirable output is discovered. It is hoped that the results of the simulation lead one to policies with respect to the real system under consideration that will produce the outcomes desired.

There are two basic types of digital computer simulations; one is called "discrete" and the other "continuous." Discrete means that the identity of individual units is maintained in the simulation. A discrete simulation of the workforce in a plant, for example, would maintain the identity of individuals throughout the simulation. A continuous simulation of the workforce, however, might show the number of direct employees in a plant to be 326.5 at the end of the month of March. The 0.5 employee may be rationalized a number of ways: (1) as a part-time employee, (2) as an employee considered to be only 50 percent productive, or (3) as a reflection of employee absences due to illness, etc.

Simulation is so common today in applied IE/OR/MS that the following two examples are offered to provide some insight into how computer simulation works. The first example is a GPSS simulation of children going to a convenience store after school to purchase either a package of gum, a candy bar, or an ice cream cone. The average time to do this is assumed to be of interest. A number of simplifying assumptions have been made to limit the size of the problem. The fourth computer card in the program of Table 7-2, the simulate card, simply "tells" the GPSS compiler program to check this GPSS program for syntactical errors (i.e., for misspellings or for placing the characters in inappropriate columns) and, if it is correct, to perform the

Table 7-2 Child-to-Store GPSS Simulation Program

BLOCK NUMBER	*LOC	OPERATION	A,B,C,D,E,F,G	CARD NUMBER
	*	CHILD–TO–STORE SIMULATION		1
	1	FUNCTION	RN1,D3	2
	0.55,15/0.85,22/1.0,18			3
		SIMULATE		4
1		GENERATE	60,,,120	5
2		TRANSFER	0.2,,ABC	6
3		ADVANCE	300	7
4		TRANSFER	,BCD	8
5	ABC	ADVANCE	200	9
6	BCD	ADVANCE	FN1	10
7		QUEUE	WLINE	11
8		SEIZE	CLERK	12
9		DEPART	WLINE	13
10		ADVANCE	50	14
11		RELEASE	CLERK	15
12		TABULATE	STRT	16
13		TERMINATE	1	17
	STRT	TABLE	M1,260,10,42	18
		START	120	19
		END		20

simulation. Card 5, the generate card, following GPSS format rules, will create a total of 120 transactions (i.e., students) coming out of the school one at a time 60 seconds apart with the intent of purchasing and consuming one of the three items mentioned above. For the purpose of this example, it is assumed that students going to the store randomly take a short cut 20 percent of the time. The short cut takes 200 seconds, whereas the long way takes 300 seconds but does not involve the likelihood of meeting Ralph, a very unpopular dog known to bite children if it is in a bad mood. Card 6 causes 20 percent of the transactions (children) to proceed to card ABC (card 9), where a transaction is delayed 200 seconds. The term ADVANCE in GPSS language produces the effect of a delay. The 80 percent of the students who did not proceed to ABC go to card 7, where they are delayed for 300 seconds, after which they proceed to BCD from card 8. The advance (delay) at BCD (card 10) is the time it takes at the convenience store to pick out a package of gum (15 seconds), a candy bar (22 seconds), or a frozen ice cream cone from the freezer chest (18 seconds). The delay time used at this step is defined by FN1 (function 1). Cards 2 and 3 summarize the data from Fig. 7-2, which indicates, as is assumed, that gum is selected 55 percent of the time by students, candy bars 30 percent of the time ($0.85 - 0.55 = 0.30$), and ice cream cones the remaining 15 percent of the time. RN1, designated in card 2, specifies that a random number, drawn from a random number seed, will determine the independent value of X in Fig. 7-2, resulting in one of the three items being selected and the student being delayed for the time appropriate to this choice.

The student, at card 11, has made a choice and enters a line (queue), called

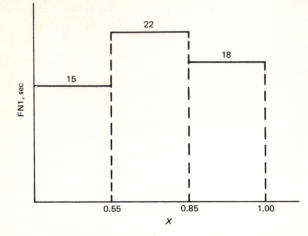

Figure 7-2 Shopping time function.

WLINE, of students waiting to pay the clerk, if a line exists. When the clerk is ready to wait on a particular student in the line, that transaction (student) enters card 12, "seizing" (i.e., occupying) the clerk. At this point, in effect instantaneously, the student is no longer waiting to reach the clerk and, therefore, passes through card 13, departing the queue.

The student occupies the clerk for 50 seconds, and is delayed doing so, and then releases the clerk. The release of the clerk permits the next student in line to seize the

Table 7-3 Store Timetable

TABLE ENTRIES	STRT IN TABLE 120	MEAN ARGUMENT 358.741	
	UPPER LIMIT	OBSERVED FREQUENCY	PERCENT OF TOTAL
	260	0	0.00
	270	2	1.66
	280	3	2.49
	290	1	0.83
	300	13	10.83
	310	1	0.83
	320	0	0.00
	330	1	0.83
	340	2	1.66
	350	0	0.00
	360	0	0.00
	370	68	56.66
	380	11	9.16
	390	3	2.49
	400	1	0.83
	410	13	10.83
	420	1	0.83

REMAINING FREQUENCIES ARE ALL ZERO

clerk. After releasing the clerk in card 15, the transaction passes through card 16, which specifies that certain information, called STRT (table of store time), be retained to be tabulated later. In this case the total time elapsed since generation (designated as M1 in card 18) is entered in a table that begins with a lower limit of 260, has class intervals of width 10, and contains 42 intervals.

After the tabulation function in card 16, the transaction is eliminated in card 17. Every transaction entering card 17 decrements the "start" count (card 19), and when it reaches zero the simulation ends. Table 7-3 is a table of elapsed times tabulated by card 16 for one run of this model. Note that average elapsed time for the 120 students is 358.741 seconds.

This example demonstrates how essential information, even in probabilistic form, can be represented in a simulation model, and that model run to determine outcomes (e.g., average elapsed time from the schoolhouse door until the convenience store clerk is paid and consumption can begin).

The next simulation example, "A Tutorial Example of Industrial Dynamics Simulation," by Hicks (14), is continuous. The modeling methodology is called industrial dynamics (9), and the computer compiler developed for it was named DYNAMO (24).

A Tutorial Example of Industrial Dynamics Simulation

Philip E. Hicks

INTRODUCTION

This article is a tutorial example of industrial dynamics simulation, which is a technique or methodology specifically developed for analysis of large-scale socio-economic systems. The concepts represented herein can be employed effectively as an introduction to this type of simulation at the junior level or as a basis, along with more detailed description, for an overview introduction of technique in a graduate course in industrial dynamics. The program can be run on any model 7090, 360/50, or 360/65 IBM computer with standard peripheral equipment, as well as on many other computers for which DYNAMO compilers have been developed. The original DYNAMO compiler, developed at the Sloan School of Industrial Management, Massachusetts Institute of Technology, required six man-years of programming effort. The DYNAMO compiler was specifically developed for implementing the industrial dynamics methodology.

This example is a modification of the newsboy problem employed in an introductory text in operations research. The problem has been modified to possess the following desired characteristics:

1 It is a relatively small problem.

2 It contains a management decision rule imbedded within the system structure.

3 It permits demonstration of the typical strengths, weaknesses, and general capabilities of industrial dynamics, thereby suggesting the appropriate class of problems to which this technology may be effectively applied.

4 It describes a practical industrial dynamics simulation technique.

PROBLEM STATEMENT

Assume that a newspaper publisher has the logistics problem of employing either too few or too many newsboys. The newsboys are on salary; therefore, the above condition causes either a restriction of newspaper sales due to insufficient newspaper selling capacity or idle newsboys on salary. Assume further that paper demand for the next thirty days is known. Figure 1 specifies the thirty-day forecast for newspapers in thousands of newspapers.

Newsboys brought into the system learn from other newsboys during the first week and the learning newsboys consequently are not paid during this period. Newsboys leaving the system depart according to a constant attrition percentage of the number of trained newsboys in the system. To permit specification of a nonlinear function in the model, assume that the publisher's distribution cost per newsboy is a nonlinear function of the total number of trained newsboys in the system. Figure 2 indicates the relationship for unit newsboy cost. Total cost is assumed to be the sum of the distribution cost and a constant fixed cost. One final assumption represents the management decision rule for hiring additional newsboys. This rule compares the present newspaper selling capacity, as a linear function of the number of trained newsboys in the system, with an exponentially smoothed demand for newspapers for all the previous days. If newspaper selling capacity is less than the averaged demand, an

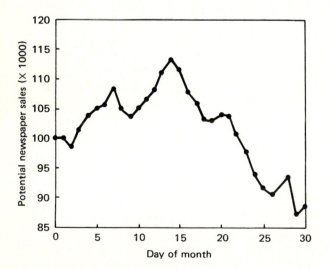

Figure 1 Potential newspaper sales by days.

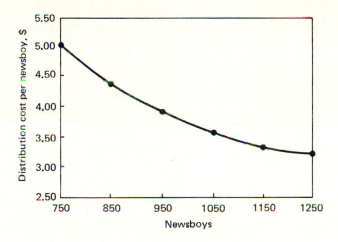

Figure 2 Distribution cost per newsboy.

additional fifty newsboys are brought into the system, otherwise no additional trainees are introduced. This latter rule represents the management decision rule imbedded within the model.

Model formulation of the above described problem would begin with industrial dynamics flow-charting. Figure 3 is a standard industrial dynamics flow chart for the newsboy problem described above.

The circles in Figure 3 identify auxiliary equations which represent new information formed from other information flows represented by the dotted lines. The new information developed then serves as an input to some other part of the model or is part of the output desired. The valve-shaped symbol, representing a rate equation (e.g., labeled Newsboy hiring rate), regulates the flow of a particular resource. The combination of information acting on a rate equation often represents the controllable management decision being employed. The rectangular symbol, which represents a level equation (e.g., labeled *Newsboys in training*), is a resource accumulation point in the system.

Starting with the source of untrained newsboys in Figure 3, it is clear that the input rate to the first accumulation (i.e., newsboys in training) is the newsboy hiring rate. The flow rate from the newsboys-in-training level to the trained-newsboy level reduces the level of newsboys in training and increases the trained-newsboys level. The newsboy leaving rate, however, causes a reduction in the trained-newsboys level. This flow is quite analogous to the cascade water flow shown in Figure 4. Valves 1, 2, and 3 in the figure correspond to the newsboy hiring rate equation, the newsboy training rate equation, and the newsboy leaving rate equation, respectively. Levels 1 and 2 correspond to the newsboys-in-training level and trained newsboys level, respectively. The only other resource accumulation in this model is profit to date. The profit rate equation determines the profit accumulation over time, as a function of other variables in the system.

The newspaper selling capacity is a function of the number of trained newsboys

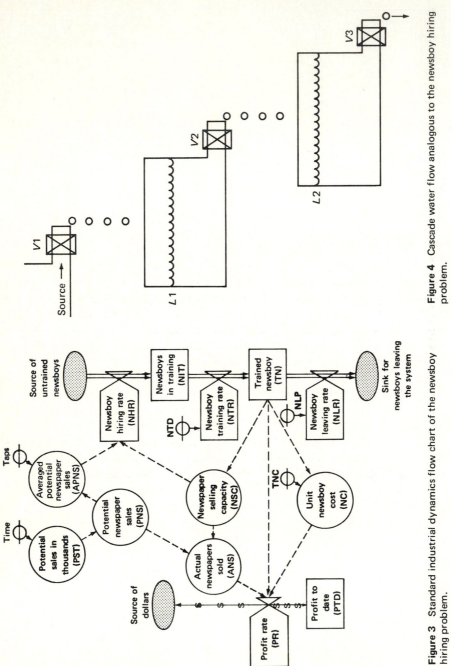

Figure 3 Standard industrial dynamics flow chart of the newsboy hiring problem.

Figure 4 Cascade water flow analogous to the newsboy hiring problem.

in the system, which is determined solely by this variable and a pertinent parameter. Unit newsboy cost is indicated on the flow diagram as a function of the trained newsboys level previously specified in Figure 2. Potential newspaper sales in thousands are shown as a function of time, and averaged potential newspaper sales are shown to be a function of the potential newspaper sales. The equation for averaged potential newspaper sales (APNS) is a first-order exponential smoothing equation. The alpha fraction of the difference between the previous period's actual and forecasted sales is determined by the time-constant TAPS. It may be noted that the circled variables in Figure 3 represent information which serves as input to rate equations. As an example, actual newspapers sold, the number of trained newsboys, and unit newsboy cost determine the income and direct costs which establish the profit rate. In the newsboy hiring rate equation, averaged potential newspapers sales and newspapers selling capacity represent information inputs to the hiring decision.

Before we leave the flow diagram, it is instructive to consider the following negative feedback loop, which is of interest in this problem. Clearly, if potential newspaper sales increase, averaged potential newspaper sales will also increase, and, if sales are greater than selling capacity, additional newsboys will be hired. The hiring of additional newsboys will, after a training delay, increase the newspaper selling capacity and nullify the difference between the desired and actual selling capacity. Figure 5 is a representation of a typical management negative-feedback-loop control process in which an apparent condition is compared with a goal; if the difference is sufficient, action results from the decision process which attempts to change the apparent condition in such a way as to nullify the difference between the goal and this condition. In a typical negative feedback loop there are two characteristics of particular interest—the period and the gain. The period is determined by the number and duration of sequential delays in the loop, and the gain represents the degree of reaction.

To illustrate period and gain consideration, the following example is offered. Figure 6 is intended to illustrate a one-way road containing a sharp curve. Assume that a sports car can enter the curve on three different trials employing a different steering mechanism on each trial. Path "A" for the sports car is representative of a steering mechanism possessing an insufficient gain characteristic, which results in a smooth but

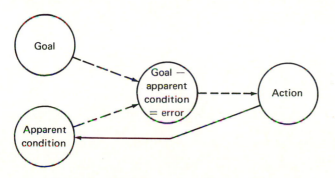

Figure 5 Typical management negative feedback loop.

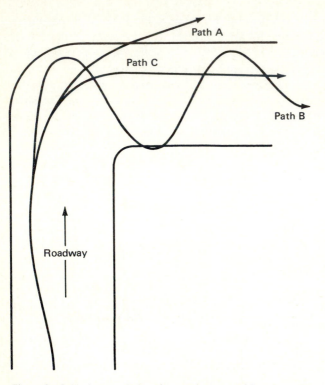

Figure 6   Gain characteristics of an analogous problem.

insufficient change in direction. Path "B" is representative of a high-gain system which changes direction too rapidly, causing oscillatory behavior which can be equally undesirable. There is some intermediate gain characteristic, represented by path "C," which produces sufficient change in direction, yet does not produce oscillatory behavior characteristic of over-control. In this problem, the fifty newsboys hiring increment represents the gain in the system. If too little gain is employed, the adjustment period may be too long; yet, if the gain is too high, erratic and typically oscillatory behavior results from excessive adjustment and consequent overshoot.

In the newsboy hiring problem there is assumed to be a most desirable adjustment process in regulating the supply of newsboys. The supply must track an unknown demand. If a system such as the newsboy hiring problem exhibits undesirable characteristics, two types of changes can be made in the system. Parameter changes are typically the easiest to make, but rarely result in high-order change in output. System restructure is usually necessary if a major change in output is desired.

In an attempt to improve the behavior of the system, the following changes were made in a revised version of the model:

1   In the hiring decision, compare averaged potential newspaper sales (APNS) with 95 percent of newspaper selling capacity (NSC).

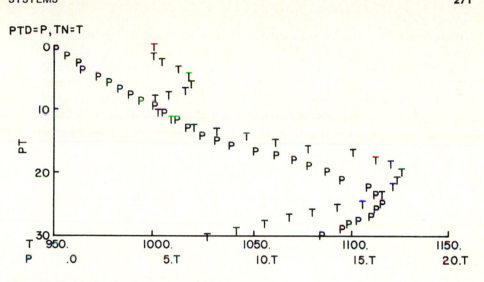

Figure 7 Model plots for the present model.

2 Increase the newsboy leaving rate (NLR) from 2 percent to 5 percent if newspaper selling capacity (NSC) is equal to or greater than averaged potential newspaper sales (APNS).

The above two steps are conceivable control decisions open to management.

In the interest of brevity, detailed computer programs for the newsboy hiring problem are not included in this article. The programs can be obtained, however, directly from the author.

Figures 7 and 8 are plotted outputs for the original and revised newsboy hiring

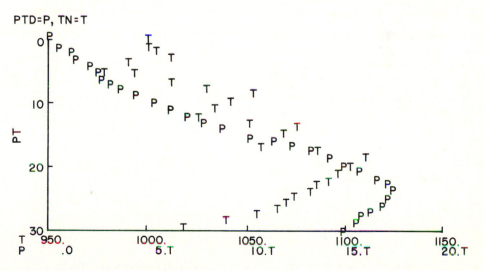

Figure 8 Model plots for the revised model.

problem, respectively. Profit to date is plotted as P, and the number of trained newsboys is plotted as T. Profit to date for the thirty-day period was $13,789.00 and $14,772.00, respectively, for the original and revised models. For the demand presented in this problem the revised model appears to offer improvement as compared to the original model. Comparison of both models for a variety of different demand situations would improve the evaluation of the two models.

CONCLUSIONS

The newsboy hiring problem in one respect is not typical of industrial dynamics problems in general, because the problem considered here was a microlevel problem chosen to permit detailed understanding of the assumed relationships. Industrial dynamics problems, however, are typically macrolevel in nature, for if they are microlevel, an analytical solution might well be possible and would be far superior. At the macrolevel, simulation can be the only available means for analyzing the model in question. Another reason for handling macrolevel problems is that industrial dynamics is a continuous-system modeling technique. At the microlevel, model representation at the lower levels of aggregation must consider more discrete relationships, in which case a simulation program such as GPSS typically has more capability than DYNAMO for handling such modeling requirements.

At this point in time, industrial dynamics is a controversial simulation methodology. However, the works of Hamilton et al. and Forrester offer examples of attempts to solve significant problems for which alternate viable solution methodologies have not been offered thus far. The success or failure of industrial dynamics in solving real-world problems will determine its ultimate utility to the model builder of the future.

The computer program, not included in the original article, is contained in Table 7-4. Note that the newsboy problem goes through three distinct phases in this study. First, a detailed written description of the situation is prepared. In military circles, this description would be referred to as a scenario. It should be obvious that an accurate description of the relationships that exist between all relevant factors is an essential first step. Probably the most common reason for producing poor solutions to problems can be traced to weaknesses in data or inadequate understanding or description of the fundamental relationships of the relevant factors.

The second and third phases to the above problem involved first the development of a flow chart relating variables, and then the translation to the mathematical form of Table 7-4. This three-phase translation is typical in simulation models. The quality of initial description and subsequent translations determines the quality of the model in accurately determining outcomes. There is an obvious analogy in a Frenchperson observing an accident, and then describing the accident to an English constable, who then reports to a Hungarian judge, who must make a judgment in the interest of justice. If the original description and two subsequent translations are good, there may

Table 7-4 Newsboy Simulation Program

```
RUN 1
NOTE
NOTE NEWSBOY HIRING PROBLEM — PRESENT MODEL
NOTE
A PST.K=TABLE(TPST,TIME.K,0,30,1)                    (POTENTIAL SALES IN THOUSANDS)
C TPST*=100/100/98/102/103/105/106/108/105/104/105/107/108/112/113/112/1
X1 08/106/103/103/104/104/100/98/94/92/90/92/93/87/89
A PNS.K=(PST.K)(1000)                                (POTENTIAL NEWSPAPER SALES)
L NIT.K=NIT.J+(DT)(NHR.JK—NTR.JK)                    (NEWSBOYS IN TRAINING)
N NIT=(NTD)(NLR)
R NTR.KL=NIT.K/NTD                                   (NEWSBOY TRAINING RATE)
C NTD=7                                              (NEWSBOY TRAINING DELAY)
L TN.K=TN.J+(DT)(NTR.JK—NLR,JK)                      (TRAINED NEWSBOYS)
N TN=1000
R NLR.KL=(TN.K)(NLP)                                 (NEWSBOY LEAVING RATE)
N NLR=(TN)(NLP)
C NLP=0.02                                           (NEWSBOY LEAVING PERCENTAGE)
A NC.K=TABLE(TNC,TN.K,750,1250,100)                  (UNIT NEWSBOY COST)
C TNC*=5.00/4.35/3.90/3.60/3.30/3.15                 (TABLE OF UNIT NEWSBOY COST)
R PR.KL=(0.1)(ANS.K)—5750—(NC.K)(TN.K)               (PROFIT RATE)
L PTD.K=PTD.J+(DT)(PR.JK)                            (PROFIT TO DATE)
N PTD=0
A NSC.K=(TN.K)(100)                                  (NEWSBOY SELLING CAPACITY)
A ANS.K=CLIP(PNS.K,NSC.K,NSC.K,PNS.K)                (ACTUAL NEWSPAPERS SOLD)
L APNS.K=APNS.J+(DT)(1/TAPS)(PNS.J—APNS.J)           (AVERAGED POTENTIAL NEWSPAPER SALES)
N APNS=PNS
C TAPS=7                                             (TIME TO AVERAGE POTENTIAL SALES)
R NHR.KL=CLIP(50,0,APNS.K,NSC.K)                     (NEWSBOY HIRING RATE)
PRINT 1)PNS, APNS/2)NSC,ANS/3)NIT,TN/4)NC,PR/5)PTD
PLOT PTD=P/TN=T
SPEC DT=1/LENGTH=30/PRTPER=1/PLTPER=1
```

be reason for hope that the results will be good also. In communication theory, the equivalent of the translation effects above is referred to as noise. It is easy to demonstrate the effect of noise by relating some simple incident sequentially to a dozen people and then comparing the final message with the original statement.

The following final example of a systems study (8)[27] employs simulation, as is often the case in applied systems solutions. This article has also been edited to remove the abstract and references in the interest of brevity.

[27] Reprinted from "Analysis of Solid Waste Management Operations in Cleveland, Ohio: A Case Study," *Interfaces,* by R. M. Clark and J. I. Gillean, vol. 6, no. 1, November, 1975, pp. 35–42, published by The Institute of Management Sciences.

Analysis of Solid Waste Management
Operations in Cleveland, Ohio:
A Case Study

Robert M. Clark
James I. Gillean

INTRODUCTION

Solid waste management can be divided into two major areas: collection, including storage, transfer, and transport; and disposal, including any accompanying treatment. Publicly and privately, about $6.4 billion per year is spent in the United States on solid waste management. Of the $3.0 billion spent publicly for municipal solid waste management, 80 percent is attributed to the collection activity.

The collection operation can be subdivided into two unit operations, collection and haul. The collection unit operation consists of removing solid waste from the storage point at the place of generation. This operation begins when the collection vehicle leaves the garage and includes all time spent on the route, whether or not it is productive. The haul unit operation starts when the collection vehicle departs for the disposal site from the point where the last container of solid waste is loaded, and includes the time spent at the disposal site and the time after leaving the site to return to the first container on the next collection route, or to the motor pool when the day's work is complete. The haul unit operation includes, therefore, the total round trip travel time from the collection route to the disposal site.

Three alternatives are normally considered for solid waste disposal: direct shipment from municipalities to a sanitary landfill; direct shipment from municipalities to a transfer station where solid waste is transferred to larger vehicles and then shipped for ultimate disposal; and direct shipment from municipalities to an incinerator where the solid waste is burned and the residue is shipped for ultimate disposal.

Waste collection and disposal systems are sometimes planned on the premise that they are independent operations when, in fact, the interdependencies that exist between the two are numerous and often very significant. Disposal methods can influence collection methods, and conversely, collection methods can exert a strong influence on disposal practices. Because of these interactions and interdependencies, much attention has been given to systems analysis and mathematical modeling techniques for sorting out various available alternatives.

Although many investigators have explored the application of deterministic, analytical, and stochastic simulation models to urban solid waste management problems, their applications were based on either assumed data or data collected on a one-time basis. These approaches have been useful for demonstrating the potential of "systems" or operations research techniques in helping the solid waste manager to make important operational decisions. However, none of these studies have considered the problem of obtaining data continuously and utilizing it, coupled with mathematical models, to make management decisions. This paper reports on the successful utilization of a data acquisition and analysis system in combination with a simulation model which were used to implement significant changes in the operations and management of the Cleveland Division of Solid Waste Collection and Disposal. In addition to providing the basis for making part changes in the Division's operations and management, these techniques are being used to plan for additional capital investments which will make Cleveland's solid waste operations even more efficient in the future.

In order to understand the problems and successes associated with this effort, it is necessary to know something of the history and political climate that influenced the developments described in this paper.

POLITICAL SITUATION

In most large American cities, the population is reluctant to vote for additional taxes, and yet either demands more services or refuses to relinquish the services they already have. Faced with demands for higher wages as well as increases in the purchase price of equipment, facilities, and other nonlabor-related items, many cities have an eroding tax base. As middle and upper-income families move to the suburbs taking with them needed tax revenue, lower-income families who produce less tax revenue, but who require just as many services, take their place. The condemnation of property for highways and other nontaxable uses also reduces potential income for the city.

In the study area, a situation similar to the one described, coupled with the defeat of a much needed tax levy, created a financial impact felt in all city departments, but nowhere more acutely than in the department responsible for the collection and disposal of solid waste. The City's Division of Waste Collection and Disposal was moving waste from the point of generation, transporting it to the disposal point, and disposing of it with no effort required by the general citizenry. Because of rising costs and limited revenues, a decision was made to sharply curtail some of the division services.

Prior to this financial crisis, the City had initiated a cooperative program with the U.S. Environmental Protection Agency (EPA) in which its solid waste routes were used as data sources for a pilot national solid waste data network. Working with the City's Division of Waste Collection and Disposal, EPA began collecting data on a regular basis in October, 1970. Two, and later four, routes were selected for continuous evaluation. Data was obtained from the collection vehicle operator on each route in the form of daily reports.

Shortly after the national data system was initiated, the aforementioned tax levy, designed to provide funds for city services, was defeated. The Commissioner of the Waste Collection and Disposal Division was faced with a number of difficult decisions regarding possible reductions in service levels. Having several months' worth of data available from the routes being monitored within the city, the solid waste managers were able to compare their six-man crews giving backyard, once-per-week service with other routes being evaluated within the pilot network.

After considering various approaches to reducing costs, backyard service was eliminated, and the collection crews was reduced by two men, leaving one driver and three laborers on each route. Several months later, the collection crew was reduced from three to two laborers. Data collected on both routes from October, 1970 through May, 1971 are shown in Table 1. The cost per ton for waste collected for an average day dropped from a value close to $30.00 per ton to approximately $13.00 per ton, with an estimated annual savings of over $4,000,000 per year.

Data from the national network proved to be such a valuable tool for the city's solid waste managers that a project was initiated to collect similar data from all of the city's routes. The city, in cooperation with ACT Systems of Winter Park, Florida, and EPA's Office of Solid Waste Management Programs, initiated the development of a Data Acquisition and Analysis System (DAAS) and a simulation model.

DATA COLLECTION SYSTEM DEVELOPMENT

The Data Acquisition and Analysis System was designed to utilize a daily collection route information form (trip tickets) as the basis for the required information flow. Table 2 shows the types of information that are generated by the system, and which can be grouped in three general data information categories—route, collection, and

Table 1 Collection Route Data

Month 1970 to 1971	Crew size	Equipment cost per day in dollars	Manpower cost per day in dollars	Total cost per day in dollars	Cost per ton	Cost per residence served per week in dollars
			Route no. 1			
October	6	18.64	194.88	213.52	30.30	0.499
November	6	16.69	194.88	211.57	36.50	0.494
December	6	20.72	194.88	215.60	34.81	0.501
January	4	19.72	130.56	150.28	22.50	0.351
February	4	22.16	130.56	152.72	20.40	0.356
March	4	19.13	130.56	149.69	20.19	0.350
April	3	24.85	98.48	115.33	14.50	0.270
May	3	23.85	98.88	123.73	14.14	0.289
			Route no. 2			
October	6	27.32	194.88	222.20	26.10	1.011
November	6	25.15	194.88	220.03	26.19	1.005
December	6	34.07	194.88	228.95	27.00	1.041
January	4	37.57	130.56	168.13	16.60	0.765
February	4	37.42	130.56	167.98	17.00	0.763
March	4	39.12	130.56	169.68	15.21	0.771
April	3	37.62	98.48	128.10	11.28	0.583
May	3	29.12	98.88	128.00	12.64	0.581

TABLE 2 Management Information Report Outputs

Route information report	Collection information report	Collection cost information report
—Route number	—Route number	—Route number
—Days of data used	—Homes served per collection	—Cost to route per day
—Average vehicle size	day	—Cost to collect per day
(cu. yards) and type	—Weight per home-collection	—Cost to transport per day
—Average crew size	day	—Equipment cost per day
—Motor pool to route	—Persons served per collection	—Manpower cost per day
(hours & miles per day)	day	—Total cost per day
—Collection operation	—Waste generated (lbs. per	—Total breakdown cost (man-
(hours & miles per day)	person-day)	hours)
—Transportation operation	—Collection time per home	—Incentive cost per day
(hours & miles per day)	—Collection time per 100 lbs.	—Overtime cost per day
—Total down time	—Collection time to total	—Cost per ton
—Lunch time	time worked	—Weekly cost per home
—Weight (tons per day)	—Total time worked to	
	standard time	
	—Loads per day (to	
	incinerator landfill and	
	transfer station sites)	
	—Weight per cubic yard of	
	first load	

cost. Information can be provided for every level of management, from the route to the Commissioner of Waste Collection and Disposal.

A weekly commissioner's report is provided in which the critical data from each station's report are reprinted. These data are consolidated to provide sums and daily averages for the overall daily collection. In addition to the tabulated information, certain exceptional information for the five highest routes and the five lowest routes is printed out for the following categories: average weight collected per day; average time spent collecting per day; homes served per collection crew; collection time per residence unit; collection time per 100 lb.; actual time to paid time; incentive cost per day; cost per ton; and cost per home per week.

Although the Data Acquisition and Analysis System provided extremely useful information which could be used to evaluate current operations and to plan for future changes, it was also obvious that some type of model would be useful for examining various alternatives for making the City's solid waste operations more efficient. The development of a Solid Waste Management Simulation Model was therefore initiated.

SIMULATION MODEL DEVELOPMENT

A Solid Waste Management Simulation Model was developed and implemented in December 1971. Using data from the Data Acquisition and Analysis System, the following types of information were provided for the model:

• Generation data, which were composed of population, dwelling units, densities, weight, and cost information.
• Collection data, including distances, volume, pickup time, route identification, vehicle type, and crew size and costs.
• Transport data, related to distance, time and speeds.
• Disposal data, such as distance traveled, offload time, disposal site, and related costs.

The simulation model itself consists of five master routines, each of which corresponds to the basic operations in the solid waste collection and disposal activity. These routines are as follows: input data logic; truck generation logic; collection logic; transportation logic; and disposal logic. It should be emphasized that the model is purely deterministic in nature and does not specifically consider random variations in the input data. However, a range of values for each class of input data was assumed in order to obtain some ideas as to the effect that random variations might have on the various policies being considered. A general flow diagram which demonstrates the way in which the model simulates the system is shown in Figure 1.

The input data consists of the basic statistics that describe the system. For example, solid waste generated, truck size, crew size, number of homes served, collection time, and transport miles are all input data to the model and are, therefore, fixed for each simulation run. The values for each of these categories may change depending on the type of collection system being modeled.

The model generates the number of trucks required for the collection process at the beginning of each day and is based on the number of routes being considered. This, in turn, is based on the assumed solid waste generated per day. An 8-hour working day was used as the base line for comparison of various types of collection systems. However, when the complete cycle of collection and disposal was considered, a week's effort was used for evaluating the various alternatives.

The collection logic assumes a set of work rules and contains the supporting data to be used for a specified collection system. For example, when a two-man collection system, which consists of a truck, a driver, and a helper, is being considered, it is assumed that the driver will assist the helper for approximately 60 percent of the time spent collecting on the route. For this assistance, the driver is assumed to receive a wage adjustment. The collection rate in the model was assumed to vary from 1.1 lb./sec-man to 1.5 lb./sec-man, and the walking rate for the collectors between pickup point was assumed as 2 ft./sec.

When a truck has made its trip to the disposal or offload point, the vehicle enters the disposal logic routine. At the offload point, the truck is assigned to an unloading lane where it may enter a queue or unload immediately depending on the conditions at the offload point. After the vehicles are empty, the truck and crew may return to the motor pool or to the collection route, depending on their having completed their assigned service areas and on the time of day.

The routines were utilized to model the performance of an individual truck and crew as well as the performance of the entire system. The first step in the modeling process was to assume a three-man crew and rear-loading packer collecting over a

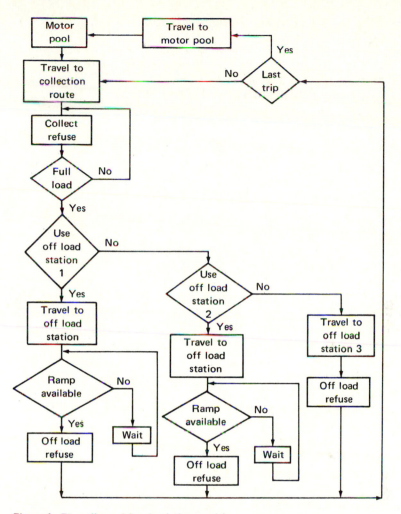

Figure 1 Flow diagram for simulation model.

typical route, using the collection logic. Various dwelling densities were assumed, and a long and short-haul option were considered. Next, various collection systems were tested using the collection logic until several good systems were found. These systems were tested, using the disposal logic to test various transfer station configurations. Once these configurations were determined, the truck generation logic was used to estimate the required fleet size for the entire city. This procedure was followed, using numerous variations in the input data.

After the model had been conceptualized, programmed, and validated, it was used to evaluate various collection alternatives which might make the current system a more efficient operation. Figure 2 shows an example of some of the alternatives considered in this evaluation including various service conditions, generation rates, collection system, and crew sizes. Many other conditions were also examined.

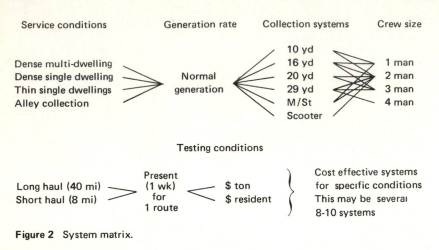

| Service conditions | Generation rate | Collection systems | Crew size |

Figure 2   System matrix.

MANAGEMENT AND OPERATIONAL CHANGES

Prior to the initiation of the study, the City of Cleveland collected approximately 350,000 tons of solid waste at a cost of $14,800,000.00 per year. The total work force consisted of 1,640 men and approximately 300 trucks. The city was divided into six collection districts with the waste being shipped to either a landfill site or an incinerator site.

Operational Changes

Several operational changes have taken place in the Cleveland system over the time period being considered. The initial system involved 224 routes and 6-man crews giving backyard collection service, and utilizing 16 and 18 cubic-yard vehicles. This evolved into 4-man crews giving curb side service during the period January to April 1971. During the period, the total work force was reduced from 1640 men to approximately 1200 men. In April of 1971, a transition was made to 3-man crews and approximately 165 routes. From April of 1971 to May of 1972, a gradual reduction in the number of routes from 165 to 140 was made, resulting in a total work force of 850 men. Between April, 1971 and March, 1974, an investment program in new, more efficient collection trucks was initiated, and the fleet was phased from 16 and 18-yard vehicles into a fleet composed of 20 and 25-yard vehicles. This fleet change and other management/organizational changes allowed for increased time in productive collection and reduced transportation time, thereby allowing a reduction in the number of routes from 140 to 102 in March of 1974. The total number of solid waste employees on duty with the City of Cleveland remained at 850. Each of these changes was based on the data from the Solid Waste Data Collection System and predictions from the simulation model. They could, however, not have been implemented without a favorable attitude on the part of city management.

   During this period the city's population declined from 740,000 in 1969 to 710,000 in 1974, but the Solid Waste Generation in thousands of tons has gradually leveled out. Because of the operational changes made, the annual budget which was

$14.8 million in 1970 dropped to a low of $8.8 million in 1972. It has gradually increased since then due to a series of pay raises awarded to the solid waste workers. The current operational budget is $12.2 million, and will be discussed in more detail later.

Management Changes

As these major operational changes were implemented, accompanying changes were made in the management of the organization. The organization was changed so that the residential collection was accomplished under 3 rather than 6 superintendents. Their responsibilities were expanded to include maintaining their areas in a clean condition, and being responsible for the public relations in their area between the citizens of Cleveland and the Division of Waste Collection and Disposal. In the old organization, 29 foremen reported to the 6 superintendents, and the foremen were responsible for the operation of 3 to 6 collection crews. Each foremen operated independently. In the new operation, the foreman position was changed to that of a Community Operations Supervisor (COS). Either 5 or 6 COS's, depending upon the size of an area, reported to each superintendent. These COS's are responsible for the operation of either 6 or 7 crews, and are also responsible for maintaining public relations in the community to which they are assigned. Standardized formats were prepared to insure uniformity of operations in all collection areas. Two-way radio communication equipment was installed in each collection truck, and a control center was set up in each area headquarters. The pre-planning of this organization, the training of the people, and clear-cut lines of responsibility resulted in a significant improvement of the collection operations themselves. Evidence supporting this includes the fact that within two weeks of the major implementation change, operations were being conducted on a normal basis. Another indicator is the fact that after a major strike that occurred approximately one month after the new system was implemented, clean-up was accomplished in approximately one and one-half weeks, rather than the estimated 4 weeks. Certainly, other factors influenced these good performances, including the attitude of the workers. It is difficult to measure the effect of an organization other than by the results of the overall program, and the results have been good.

PLANNING STUDIES

The improvement in solid waste operations in Cleveland is directly attributable to the application of Management Science techniques to the city's solid waste operations. Perhaps an even more important contribution which these techniques have made is that the solid waste managers can now explicitly evaluate the cost of every alternative considered in addition to evaluating many more alternatives. A couple of examples here are intended to highlight how the model has been used to assist in planning decisions.

Modifying Current Operations

The current solid waste budget is approaching a level, once again, where it becomes a candidate for careful scrutiny and possibly some cost reduction measures. Analysis of current operations, through the use of the model, show that the current number of routes could be reduced to 90, and the number of personnel in the division could be reduced to approximately 500. This would result in an estimated savings of $3.0 to $4.0 million per year, or an annual budget of close to $9.5 million per year. However, it has been the city's position that the 850 employees currently employed in solid waste collection and disposal represent those with the most seniority. It has been the city's position to date that these employees will be retained as a reward for their long service and dedication. With this analysis, the cost of this policy is now explicit and the city managers can now make their choice in an environment where the costs are known.

Facility Planning

It is possible that the Cleveland solid waste disposal system could be made even more efficient by installing transfer stations at critical points in the city. These stations would allow the collection crews to spend more productive collection time by reducing the driving time of the collection trucks to the existing off-loading sites. By using the model, a couple of sites were selected as those which provide the optimal solid waste collection and off-loading configuration. The Cleveland City Council is currently considering legislation which will enable the construction of these facilities. If these facilities are built, it is anticipated that the number of routes and collection crewmen will be reduced below 90.

SOCIO-ECONOMIC EFFECTS

In order to attain a more efficient solid waste collection system, nearly 800 employees have been laid off, creating a considerable social cost in terms of unemployment and welfare benefits paid to the unemployed solid waste workers. After the layoffs of approximately 500 men, a telephone survey was initiated to determine how many of the men were still unemployed one year after the layoffs started. Out of 300 phone calls, only 100 ex-employees were still able to be contacted, and only 3 of those were able to find employment. The authors attempted to convince the City of Cleveland to invest some of the funds saved in the layoffs in a job training program for the unemployed workers. The city agreed to explore the idea if a grant could be obtained from the Federal Government to help cover some of the training costs. However, efforts to work out the details with EPA met with failure. Therefore, the attainment of this increased efficiency by the city also incurred a social cost of undetermined magnitude.

SPIN-OFF EFFECTS

As might be imagined, the spin-off effects of this study have been considerable. EPA's Office of Solid Waste Management redesigned its entire technical assistance program to utilize the techniques demonstrated in the study. It also initiated close to forty community case studies, using the information system described herein, in order to establish productivity measures for solid waste collection. Studies, much like the one described for Cleveland, are being conducted for Columbus, Ohio, Richmond, Virginia, and St. Petersburg, Florida. Results to date have been similar to those achieved in Cleveland.

SUMMARY AND CONCLUSIONS

The changes in the Cleveland Solid Waste Management system have been of considerable magnitude. Based on a 1970 annual budget of $14.8 million, over $14.6 million has been saved from 1971 through 1974. The collection staff has been reduced from 1640 employees in 1970 to 850 employees now. The collection system has been completely redesigned, and a totally new collection fleet purchased. Management for the solid waste program has been reorganized, and a capital investment program for transfer stations is being considered. In addition to all of these changes, the decision process within the solid waste program itself has changed. From an out-of-control operation, solid waste management in Cleveland is now guided by sound decision-making principles. All of these changes have resulted from the application of management science techniques.

Work is currently underway which will hopefully improve the methodology discussed in this paper. It is hoped that this new research will increase the transferability of these techniques to other solid waste management operations.

DECISION THEORY AND UTILITY

In the real world it is rare when decisions are made with certainty; most decisions involve risk. Managers are paid to make decisions under risk. To remain managers they must, however, make enough good decisions to more than outweigh the bad ones, as compared to the capability of other available potential managers.

A commonly discussed criterion for valuing risky decisions is called the "maximum expected value criterion." Under this criterion it is assumed that value can be estimated as the product of the value of an outcome and its probability of occurrence. For example, if the probability of receiving $100 in a letter you have just received from a parent is 20 percent, the expected monetary value of the letter is $100 × 0.20 = $20. Deferring "utility" aspects of the problem until later, this criterion has been used extensively in what are called "decision tree" problems.

Assume, for example, that a contractor is installing wood flooring in a building this week and moving in certain pieces of equipment. One major piece of equipment will not arrive until next week. According to weather information, however, a change in the weather is expected this coming weekend. All available sources of information indicate a 15 percent chance of high humidity and a 5 percent chance of rain.

One choice available to the contractor is to install the permanent roof during this week and suffer an additional cost of $3,400 for removing a wall to install the piece of equipment coming next week. A second choice available is to construct a temporary roof of polyethylene at a cost of $1,600. Damage to the flooring and equipment is estimated to be $7,600 if it is humid and $22,600 if it rains. Another choice, and the most risky, is to install the permanent roof after the equipment is received next week. If humidity occurs, damage is estimated to be $12,000, and if it rains $30,000. What should the contractor do? The choices are: (1) install the permanent roof now, (2) install a temporary polyethylene roof now, and (3) do nothing and hope the weather prediction is wrong about the weather changing before the equipment arrives next week.

Figure 7-3 is a decision problem tree for the above contractor problem. To summarize, if the permanent roof is installed now, the contractor accepts a loss of $3,400 and need not be concerned further with the problem. If the contractor does not build the permanent roof and instead builds the temporary roof, there is an 80 percent, 15 percent, and 5 percent chance of losing $1,600, $7,600, and $22,600, respectively. If no roof is built now, there is an 80 percent, 15 percent, and 5 percent chance of losing $0, $12,000, and $30,000, respectively.

The expected losses for installing the temporary roof or not are as follows:

Install temporary roof: 0.8($1,600) + 0.15($7,600) + 0.05($22,600) = $3,550
No roof: 0.8($0) + 0.15($12,000) + 0.05($30,000) = $3,300

The "X" on the "install temporary roof" path of Fig. 7-3 indicates that it should be dropped from further consideration. The expected value of $3,300 is less than the $3,400 for installing the permanent roof now, so the best choice using the expected value approach is to construct no roof and wait and see what happens.

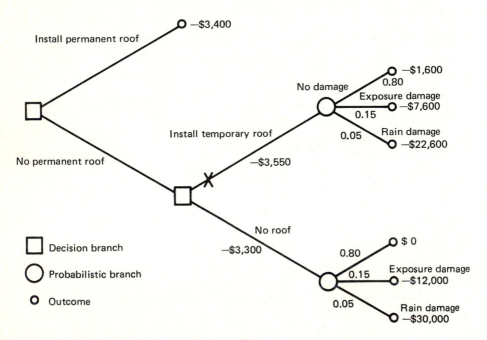

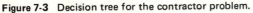

Figure 7-3 Decision tree for the contractor problem.

No definitive meaning has been given to the word "value" thus far. Hall gives some insight into the complexities associated with determining value by stating (11, p. 283):

> A fundamental problem in a general theory of value is to define value so as to include all its forms. One answer to this problem is to say that value resides in *any sort of interest or appreciation* of an object, event or state of affairs. Such appreciation involves feeling and ultimately desires, tendencies, or needs underlying the feeling. Therefore value *is* the feeling; *value and feeling of value are the same thing.* This is the psychological concept of value, and the propositions developed on this concept comprise the *psychological theory of value.*
>
> According to this theory, the measure of value is found in such conceptions as *intensity of feeling,* or the *strength* of the desire; such concepts are meant ultimately to reflect the importance of the psychological and biological needs and desires presupposed for establishing and maintaining the "good life."

A value system then is specific to individuals. It depends on a person's relative needs for security, pleasure, peer approval, aspirations, etc. In the previous contractor problem it was assumed that value is a linear function of dollars. This is typically not the case.

Assume that you and I have agreed to cut cards as a game of chance. Assume first that if I lose I pay you \$1, otherwise you pay me \$1. No difficulty exists as yet. Assume, however, that having lost \$1 to you I indicate that I have a train to catch and suggest that we cut one more time for \$20,000. Your expected gain is \$20,000(0.5) − \$20,000(0.5) = \$0. Employing the expected value criterion, it would seem that you would be as willing to cut cards for \$20,000 as \$1. What has slowed you down is consideration of the relative values of winning \$20,000 versus losing \$20,000. The pain of losing \$20,000 may be valued to be greater than the potential pleasure of winning \$20,000. Figure 7-4 is a typical utility function for an individual. Not only do the psychological values of gains versus losses influence this nonlinearity, but the regressiveness of federal taxation adds another measure of nonlinearity.

Assume that Fig. 7-5 is a derived utility function for the contractor. Figure 7-6 is the decision tree for the contractor with the added assumption of maximizing the utility. Note that the dollar value of each outcome in Fig. 7-3 has been converted to a utility value by employing the utility function of Fig. 7-5. In terms of the contractor's utility, Fig. 7-6 indicates that it now appears more advantageous to install the permanent roof, accepting the loss of \$3,400.

Of course, this method has utility only if we are able to derive a utility function for an individual. Note in Fig. 7-5 that the scaling of utility is such that −\$30,000 is equal to −45,000 *U*'s. Using this as a starting point, it would be possible to ask what value the contractor would place on alternative B (i.e., the purchase of insurance) to equate it to alternative A in Fig. 7-7*a*. If the answer is \$21,000, we can establish a point on the utility curve of Fig. 7-5 equating \$21,000 to −27,000 *U*'s. By a similar procedure, this point can be used as a basis for establishing a third point, as indicated in Fig. 7-7*b*.

Not only are utility functions specific to individuals, but individuals possess different utility functions for different resources. Added to these complexities is the

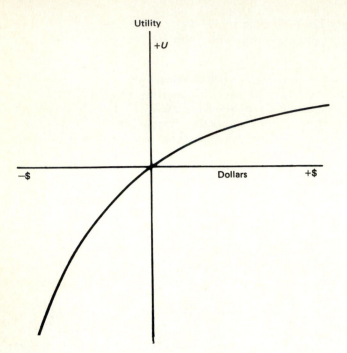

Figure 7-4 A utility function for an individual.

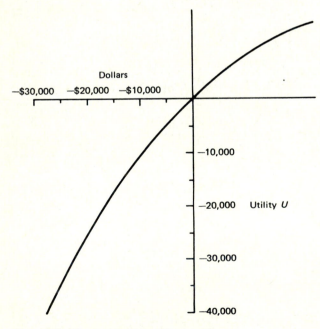

Figure 7-5 The contractor's utility function.

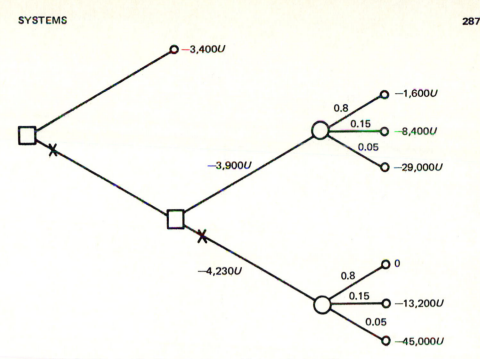

Figure 7-6 Decision tree for the contractor problem employing utility.

fact that utility can vary significantly over a relatively short period of time, as is suggested by the utility curves for an individual named George in Fig. 7-8. It should be apparent that the practical consideration of utility in IE/OR/MS problems is complex. There has been very little development in this area in a practical sense, but it is likely that considerable development will occur to meet the needs in modeling sociotechnical systems in the future.

In the analysis of human systems it is not uncommon to need to specify a value or a functional relationship for a variable where data simply do not exist. In engineering circles, this is called making a "guesstimate." A common way to provide an air of authenticity to a guesstimate is to have a group of people (i.e., a committee) come up with it. Unfortunately, a group not as well informed as one individual in the group will typically produce a poorer estimate than the individual would have provided. In most committee environments, sociological and psychological factors are greater determinants of the outcome of the committee than logic or facts.

The need for expert opinion is often essential in systems studies, as indicated by Helmer (13, p. 11)[28]:

> While model-building is an extremely systematic expedient to promote the understanding and control of our environment, reliance on the use of expert judgment, though often unsystematic, is more than an expedient; it is an absolute

[28] From page 11, Part I, "Social Technology," in *Social Technology* by Olaf Helmer, with contributions by Bernice Brown and Theodore Gordon, © 1966 by Olaf Helmer, Basic Books, Inc., Publishers, New York.

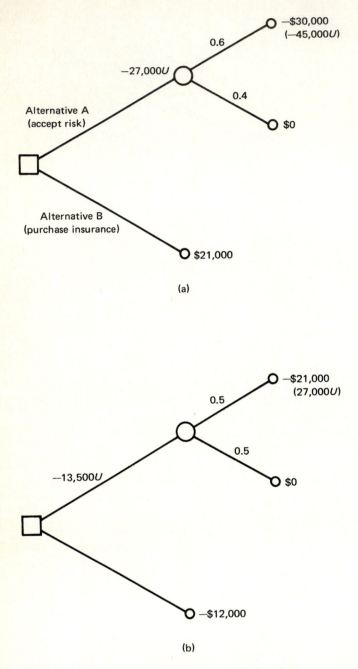

Figure 7-7 Decision tree derivations for a utility function.

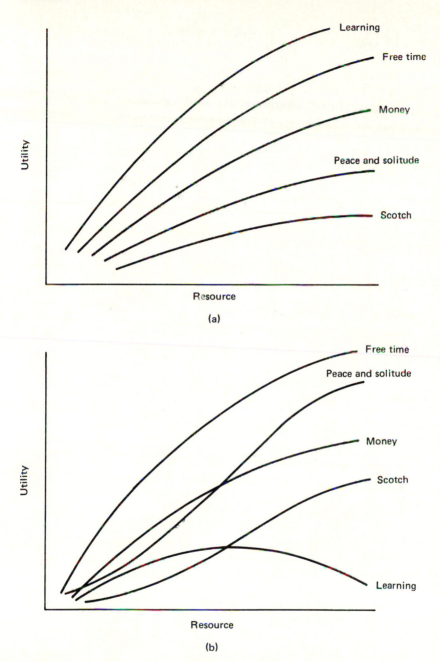

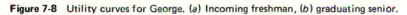

Figure 7-8 Utility curves for George. (*a*) Incoming freshman, (*b*) graduating senior.

necessity. Expert opinion must be called on whenever it becomes necessary to choose among several alternative courses of action in the absence of an accepted body of theoretical knowledge that would clearly single out one course as the preferred alternative. [This can happen if there is either] a factual uncertainty as to the real consequences of the proposed courses of action, or, even if the consequences are relatively predictable, there [is] a moral uncertainty as to which of the consequent states of the world would be preferable.

An approach referred to as the Delphi technique is under development at present to assist in perfecting these estimates. The following article by Milkovich et al. (21)[29] is reprinted here as an example of research in this area of decision-making. The article is also suggestive of the type of systems applications for which such an IE/OR/MS technique may be useful. The abstract and references have been removed in the interest of brevity.

The Use of the Delphi Procedures in Manpower Forcasting

George T. Milkovich
Anthony J. Annoni
Thomas A. Mahoney

INTRODUCTION

Manpower planning defined simply as the process of determining how the organization should move from its current manpower position to its future desired manpower encompasses most elements of manpower management. Forecasting manpower requirements and supplies, as an integral subprocess in the manpower planning system, functions to provide a framework within which the other manpower processes, such as staffing, rewarding, developing, operate to achieve the organization's goals. Forecasting generally involves systematic procedures for forming expectations in which decision makers have greater confidence than an unguided or less supported guess. Most manpower forecasting models currently in use rely upon historical relationships among parameters and upon the expectation that these historical relationships and changes in them will persist into the future. While manpower specialists in organizations recognize that future conditions may be unique, few systematic procedures for incorporating future uncertain states of nature have been developed, implemented or tested regarding manpower resource decisions. We report a case study of the application and evaluation of the delphi technique which systematically uses expert judgment in generating a manpower forecast.

The delphi technique, a set of procedures originally developed by The Rand Corporation in the late 1940's, is designed to obtain the most reliable consensus of

[29] Reprinted from "The Use of the Delphi Procedures in Manpower Forecasting," by George T. Milkovich, Anthony J. Annoni, and Thomas A. Mahoney, *Management Science*, vol. 19, no. 4, December, 1972, pp. 381–387, published by The Institute of Management Sciences.

opinion of a group of experts. Essentially, the delphi is a series of intensive interrogations of each individual expert (by a series of questionnaires) concerning some primary question interspersed with controlled feedback. The procedures are designed to avoid direct confrontation of the experts with one another.

The interaction among the experts is accomplished through an intermediary who gathers the data requests of the experts and summarizes them along with the experts' answers to the primary question. This mode of controlled interaction among the experts is a deliberate attempt to avoid the disadvantages associated with more conventional uses of experts such as in round table discussions or direct confrontation of opposing views. The developers of the delphi argue the procedures are more conducive to independent thought and allow more gradual formulation to a considered opinion. In addition to an answer to the problem, the interrogation of the experts is designed to call out the parameters each expert considers relevant to the problem, and the kinds of information he feels would enable him to arrive at a confident answer to the question.

Typically, the answer to the primary question is a numerical quantity (in this study the number of employees required). It is expected that the individual expert's estimates will tend to converge as the experiment continues even if the estimates expressed initially are widely divergent.

The literature on the delphi reports either the answers to specific problems generated by its use or results of laboratory experiments investigating the effects that varying its procedures had on the estimates generated. Studies concerned with its usefulness relative to other techniques or its use for manpower planning problems are rare.

The purpose of this study was to investigate the usefulness of the delphi procedures in projecting manpower requirements. Usefulness of the procedures is considered on two dimensions. One is the accuracy of the results generated by the delphi versus conventional regression models compared to the actual employment decisions made by the organization. The other is to investigate the information elements and implicit models used by the experts in an attempt to formulate an improved forecasting model.

A low profit margin national retail firm agreed to furnish both data and managerial employees who comprised the panel of experts for this study. The firm's expected demand for buyers was selected as the primary question. This decision was based on the firm's contention that a buyer represents the single most critical skill in their organization and the fact that the firm had been unable to generate reliable forecasts of this crucial skill using conventional methods.

A relatively informal method of generating forecasts of employment of buyers had been employed. The manager in charge of all buying activities prepared an annual forecast of employment after consultation with buyers and managers of functions affecting buying activity, e.g. store expansion. This annual forecast was supplied to the personnel function for recruiting purposes and was adjusted and revised during the year as necessary. Past experience had indicated that frequent revision was necessary for recruiting to meet desired staffing levels.

Procedurally, this study used both regression based models and the delphi procedure to estimate the number of buyers the firm will need "one year from now."

The number of buyers actually employed by the firm "one year later" was subsequently recorded without the results of either approach influencing the firm's decisions. Comparisons of the projections generated by the delphi and conventional models with the firm's actual employment decisions served as the criterion for analyzing the relative accuracy of the delphi method.

While it is recognized that the actual employment decision by the firm for a one-year period may not represent optimal behavior, it is considered the best available proxy of the firm's "true demand" for buyers.

DESCRIPTION OF THE EXPERIMENT

The delphi forecasting approach was conducted with a panel of seven experts who were company managers. They were selected on three criteria: their direct or indirect involvement in the firm's informal method of generating forecasts, their willingness to participate, and their availability for the duration of the study. Altogether, five questionnaires submitted at approximately eight-day intervals were used to interrogate key personnel involved in determining the firm's employment behavior regarding buyers.

In Questionnaire #1, panel members were presented with a statement of a problem and a brief description of the company. Each panel member was asked to indicate what specific information he felt would be needed to solve the problem (to accurately forecast company demand for buyers one year hence), and to indicate how that information would be used, once made available. Individual information sheets were prepared for each panel member, including only that information which the panel member had specifically requested. These information sheets were returned as part of Questionnaire #2.

In Questionnaire #2, each panel member was asked to formulate the best estimate possible based on the information he had received, to indicate how he had "combined" the information to yield that estimate, and was invited to request additional information which would enable him to "refine" his estimate. Once again, individual information sheets were prepared for each panel member consisting of the information requested via Questionnaire #2. These information sheets were returned, along with the interquartile range of initial estimates, as part of Questionnaire #3.

In Questionnaire #3, panel members were asked to formulate an estimate based on all information in their possession, and to indicate how the additional information received had either confirmed the previous estimate or, alternatively, indicated the need for a revised estimate.

For Questionnaire #4, summary information sheets were compiled, which included all information requested by panel members in previous rounds. These sheets were distributed, along with the interquartile range of previous estimates, to all panel members. In Questionnaire #4, panel members were once again asked to formulate estimates, and to indicate how additional information either confirmed the previous estimate or indicated the need for a revised estimate.

In Questionnaire #5, the only information disseminated was the interquartile range of previous round estimates. Panel members were asked, based on all information at their disposal, to formulate a final estimate, and to indicate how that estimate had been made.

RESULTS

Information Requested

The experts were asked to utilize only that information which was furnished to them and their "general experience" in merchandising. Any additional information not requested but used to generate the estimated demand was to be reported. Information requested was furnished, if available, on subsequent rounds. Individual panel members received only that information which they explicitly requested during the first three rounds. In the fourth round, a summary of all information requested by all experts was disseminated to each expert.

A total of 39 different elements of information was requested with only the ten items in Table 1 requested by two or more experts. Not surprisingly, the data concerning product demand, sales outlets, and buyer productivity were requested by all seven experts. Table 1 reveals that the requests were highest for historical data on the number of retail units, the number of buyers employed, the average sales volume per buyer and the projected sales volume and store expansion plans. While there was considerable commonality over the basic parameters, the majority of the information items requested were unique to a single expert. Under interrogation, the experts revealed that this unique information entered into their judgments about the anticipated rates of change in sales volume and buyer productivity. The sheer volume of requests was the greatest in the first two rounds, and no new information was requested after the third round.

Convergence

The anticipated covergence of the experts' projections is shown in Table 2 and Figure 1. The range of estimates decreased from 23 to 11, while the median number of buyers

Table 1   Information Requests

Elements	Round 1	2	3	4	5	Total
Projected gross sales volume	7					7
No. of buyers over past periods	7					7
Automation plans for buyers decisions	4	1				5
Buyers productivity index	4	3				7
No. of retail units over past periods	4	2	1			7
No. of retail units planned	2	3	1			6
Average retail unit volume	2		1			3
Gross sales over past periods	2	1	1			4
Average turnover in past periods	1	1				2
Total	33	11	4	0	0	

Table 2   Projected Demand for Buyers by Round

Round	A	B	C	D	E	F	G	Mdn	No. of change	Range
				Experts						
2	55	35	33	35	55	33	32	35	—	23
3	45	35	41	35	41	34	32	35	4	13
4	45	38	41	35	41	34	34	38	2	11
5	45	38	41	35	45	34	34	38	1	11
No. of changes	1	1	1	0	2	1	1		7	

projected increased from 35 in Round 2 to 38 in Round 5. The greatest incidence of change in the projection occurred in Round 3 when 4 out of 7 experts changed their estimates. The interquartile range of the experts' estimates was first fed back to all the experts at the beginning of Round 3. The summary of all the total information elements requested by all experts was included for the first time in Round 4 with little impact apparent on the projections (only 2 out of 7 experts changed their projections).

Each expert's reasons for any change in a projection are recorded in Table 3. The principal reasons given are long-range expansion and growth plans of the firm, and the anticipated effects of automating the reordering system upon buyer productivity.

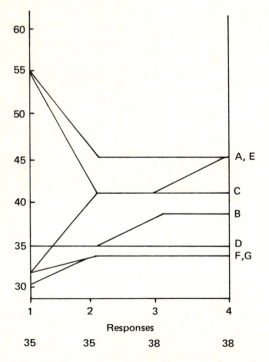

Figure 1   Successive estimate of buyer requirements.

Table 3   Reasons for Projection Changes

	Projection in rounds		
Expert	2	3	Reason given for change
A	55	45	• Automating reorder decisions deviated too much from interquartile range
C	33	41	• Expansion data
E	55	41	• Automating reorder decisions
F	33	34	• Projected new outlet construction
	Rounds		
	3	4	
B	35	38	• Projected number of new stores
G	32	34	• Expansion data
	Rounds		
	4	5	
E	41	45	• Long-range growth greater than originally anticipated

Accuracy

Prior to this study, the firm had been investigating the use of linear regression equations and other models to forecast their demand for buyers. The results of these models had never been incorporated into the organization's planning systems due primarily to the lack of confidence in the projection generated.

Table 4 compares the results of these regression equation results with the

Table 4   Forecasts Yielded by Alternative Methods

Method	Point	Range
Actual firm behavior[1]	37	—
Delphi	38	34–35
Regression[2]		
Projected number of retail outlets $(t + 1)$	43.11	38.08–48.13
Current gross retail sales	45.99	41.64–50.34
Current number of retail outlets	49.40	50.20–53.60

Note:

[1] Actual firm behavior includes one unfilled vacancy.

[2] Using projected number of retail outlets yielded the closest forecast when compared to actual firm behavior. Current gross retail sales explained the greatest proportion of variance in the employment level of buyers, and current number of retail outlets yielded results with the smallest standard error of estimation.

projection generated through the delphi procedures and compared them to the firm's actual experience for the period under consideration. The three regression equations chosen had the smallest standard errors of estimation, the largest coefficients of determination and the closest point and interval estimate when compared to the firm's actual behavior. The forecast generated by the systematical albeit clinical delphi procedures is closer to the firm's "true demand for buyers" than any of the more conventionally generated projections. In fact, none of the three regression equation interval estimates even includes the firm's actual decision of 37 buyers.

In addition to the accuracy of the procedure, the usefulness of the delphi was also investigated by further analyzing the experts' information requests and judgments in an attempt to formulate a forecasting model that incorporates elements used by the experts.

Under interrogation by questionnaire, all the experts indicated they made use of current sales volume and a simple index of buyer productivity. The differences in the experts' actual projections resulted from their judgments concerning the anticipated rates of growth of both sales and productivity.

The experts who considered that automating buyers reorder decisions would increase productivity tended to be most accurate when compared to the firm's actual behavior.

The intuitive model incorporating the elements used by the experts seems to be a general growth model which can take the following form:

$$D_m(t+n) = \frac{\text{sales volume index } (t) (1 + \alpha)^n}{\text{productivity index } (t) (1 + b)^n}$$

where D_m represents the demand for manpower, α and b represent the rate of change in sales volume and productivity respectively. It would seem, based on the interrogation of the experts, that their judgments entered the process most heavily in assessing the rates of change, α and b.

An attempt was made to estimate these rates of change using historical data; however, the results were less accurate than the regression generated results. Historical data underestimated both the rates of change in productivity and sales volume, while the systematic use of the experts' opinions anticipated the organization's manpower needs relatively accurately.

The clinical judgments of experts systematically collected generated results closer to the firm's actual behavior than conventional models in this case. Admittedly, the regression equations were very naïve; however, given the state of the art reflected in various surveys of the literature and of company practices, they are representative. The greatest utility of the delphi procedure seems to lie in its hypothesis or model-generating powers. While models and hypotheses can be developed from several alternative sources, the delphi procedure does represent an established method of soliciting the decision processes and implicit models of experts, managers and administrators. Further, the information needs revealed by the process can provide a useful source of elements for a manpower information system necessary for effective manpower planning.

Obviously, there are a number of shortcomings in the procedures. While they are discussed extensively elsewhere, a few of the crucial ones are:

(1) *Role of the Intermediary.* Standard feedback takes the form of answers to an expert's inquiry for data, summaries of all inquiries and interquartile ranges of the estimates. The summaries of all inquires are brief and do not include the richness of interpretation each expert brings to bear on the problem. This is the price paid for not allowing the experts to interact directly.

(2) *Independent Expert Responses.* Experts are initially instructed not to discuss the experiments with others; however, in practice, it is difficult to prevent discussion of these issues.

(3) *Number of Rounds.* Five rounds seemed to be the typical number used in reported experiments. However, in our case, most of the convergence and most of the data requests occurred in the early rounds leaving the usefulness of later rounds open to question.

(4) *Changes of Estimates.* Five out of the seven experts changed their estimate only once, while one did not change his initial estimate at all. From the reported experiments in nonlaboratory settings, this is a low frequency of change. It may be attributed to the short range (one year) of the forecast, and more changes in successively approximating the "true" answer might occur in a long-range problem with greater uncertainty.

At the minimum the delphi appears to be highly useful in generating preliminary insights into highly unstructured or underdeveloped subject areas such as manpower planning. Further, a carefully developed consensus of managers' opinions may be acceptable when direct empirical data are unreliable or unavailable.

One of the problems associated with such techniques is the identification of experts. One of the best measures of an expert is "reliability," (i.e., past record). Research (3) has indicated that another rather interesting basis for estimating the ability of experts is to ask them to appraise their competence in carrying out a specific estimating assignment.

In recent extension of the Delphi technique described in the above article expert opinion is weighted on the basis of the relative reliability of the experts involved. There is likely to be much development in this area in the near future.

There are potential hazards in attempting to combine the preferences of individuals in the absence of adequate theory. One example, and a rather convincing one, is the voting paradox contained in Arrow's text (1, p. 3), as follows:

A natural way of arriving at the collective preference scale would be to say that one alternative is preferred to another if a majority of the community prefer the first alternative to the second, i.e., would choose the first over the second if those were the only two alternatives. Let A, B and C be the three alternatives, and 1, 2 and 3 the three individuals. Suppose individual 1 prefers A to B and B to C (and

AT THE AGE OF
18, PRINCE
LADISLAUS of
Pomerania
LED A FORCE
OF ONLY 1800
MEN AGAINST
THE COMBINED
ARMIES OF
FRANCE AND
AUSTRIA,
WHICH NUMBERED
OVER
208,000 TROOPS!

He was soundly defeated.

Figure 7-9   A princely failure. [*From Quade and Boucher (25, p. 278)*.]

therefore A to C); individual 2 prefers B to C and C to A (and therefore B to A); and individual 3 prefers C to A and A to B (and therefore C to B). Then a majority prefer A to B, and a majority prefer B to C. We may therefore say that the community prefers A to B and B to C. If the community is to be regarded as behaving rationally we are forced to say that A is preferred to C. But, in fact, a majority of the community prefer C to A. So the method just outlined fails to satisfy the condition of rationality as we ordinarily understand it.

This would suggest that in systems study, as in life, things have a way of getting complicated.

SUMMARY

This chapter has indicated a much broader role for IE/OR/MS than was typical twenty years ago. The gradual broadening of systems analysis, initially technical systems, to human systems greatly increased the potential for failure, frustration, and confusion, but also significant contribution. There is no telling what can be accomplished with true dedication to purpose, as is indicated in Fig. 7-9. Trying is important however.

REFERENCES

1   Arrow, K. J.: *Social Choice and Individual Values,* John Wiley & Sons, Inc., New York, 1951.
2   Brach, R. M., and James M. Daschbach: "State Criminal Court Systems," *AIIE Technical Papers 1972,* American Institute of Industrial Engineers Conference and Convention, Anaheim, Calif., 1972.
3   Brown, B., and O. Helmer: "Improving the Reliability of Estimates Obtained from a Consensus of Experts," p. 2986, The Rand Corp., September, 1964.
4   Chaiken, J. M., and R. C. Larson: "Methods for Allocating Urban Emergency Units: A Survey," *Manage. Sci.,* vol. 19, no. 4, December, 1972.
5   Chinal, J. P.: "The Systems Approach: A French Experience," *Interfaces,* vol. 5, no. 2, February, 1975.

6 Churchman, C. W.: *The Systems Approach*, Dell Publishing Co., Inc., New York, 1968.

7 Churchman, C. W., R. L. Ackoff, and E. L. Arnoff: *Introduction to Operations Research*, John Wiley & Sons, Inc., New York, 1957.

8 Clark, R. M., and J. I. Gillean: "Analysis of Solid Waste Management Operations in Cleveland, Ohio: A Case Study," *Interfaces*, vol. 6, no. 1, November, 1975.

9 Forrester, J.: *Industrial Dynamics*, M.I.T. Press, Cambridge, Mass., 1961.

10 Halbrecht, H.: "So Your Mother Wanted You To Be a Doctor But a Ph.D. in OR?," *Interfaces*, vol. 2, no. 2, February, 1972.

11 Hall, A. D.: *A Methodology for Systems Engineering*, D. Van Nostrand Company, Inc., Princeton, N.J., 1962.

12 Hamilton, H. R., et al.: "A Dynamic Model of the Economy of the Susquehanna River Basin," Battelle Memorial Institute, Columbus, Ohio, 1966.

13 Helmer, Olaf: *Social Technology*, Basic Books, Inc., New York, 1966.

14 Hicks, P. E.: "A Tutorial Example of Industrial Dynamics Simulation," *Simulation*, January, 1972.

15 Hitch, C. J.: "Uncertainties in Operations Research," *Oper. Res.*, vol. 8, July–August, 1960.

16 Jennings, J. B.: "Blood Bank Inventory Control," *Manage. Sci.*, vol. 19, no. 6, February, 1973.

17 Kolesar, P., and W. E. Walker: "An Algorithm for the Dynamic Relocation of Fire Companies," *Oper. Res.*, vol. 22, no. 2, March–April, 1974.

18 Krick, E. V.: *Methods Engineering*, John Wiley & Sons, Inc., New York, 1965.

19 Lewis, H. A.: "A Perspective of the Industrial Engineer's Role in Urban Mass Public Transportation," *Proceedings 1974 Spring Annual Conference*, American Institute of Industrial Engineers Conference and Convention, New Orleans, La., 1974.

20 Likert, R.: *New Patterns of Management*, p. 103, McGraw-Hill Book Company, New York, 1961.

21 Milkovich, G. T., A. J. Annoni, and T. A. Mahoney: "The Use of the Delphi Procedures in Manpower Forecasting," *Manage. Sci.*, vol. 19, no. 4, December, 1972.

22 Nadler, G.: *Work Design*, 4th ed., Richard D. Irwin, Inc., Homewood, Ill., 1967.

23 "Orders of Magnitude," 16-millimeter film, Charles Eames Studios, Venice, Calif.

24 Pugh, A.: *DYNAMO User's Manual*, M.I.T. Press, Cambridge, Mass., 1963.

25 Quade, E. S., and W. I. Boucher: *Systems Analysis and Policy Planning*, American Elsevier Publishing Company, Inc., New York, 1968.

REVIEW QUESTIONS AND PROBLEMS

1 Give an example of a system within a system within a system.

2 Give one example each of a system of things and a human system.

3 Why is choosing the right objective more important than choosing the right system?

4 Give an example of a system problem for which there are at least five objectives, and state the objectives.

5 Assume that there are four objectives, A, B, C, and D, with respect to a systems problem. The lengths of the lines below indicate the relative value of obtaining each of these four objectives. Employing Churchman's Procedure 1, by visual inspection only, determine normalized relative values for the four objectives.

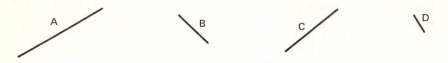

6 Determine relative normalized values for the line segments above by measuring them and comparing these values with your results from Prob. 5.

7 Differentiate between a discrete and a continuous digital computer simulation.

8 Analyze the GPSS child-to-store simulation program of Table 7-2 and determine the minimum time between a student's leaving the school and being served by the convenience store clerk.

9 There is a negative feedback loop in the newsboy industrial dynamics problem in this chapter. Determine the extent of action (i.e., gain) in the negative feedback loop in this problem that results if a sufficient error is detected. Hint: carefully examine the written description of the newsboy problem (i.e., phase 1).

10 Mr. Johnson is considering selling ice cream and popcorn at five summer camping areas for the summer. One choice is to use his company trucks to do so. The expenses associated with the use of these trucks are estimated to be $100. Another choice available is to rent trucks for $3,000 for the summer. A third choice, unappealing because Mr. Johnson does not particularly like his brother-in-law, is to use his brother-in-law's trucks. He estimates that his brother-in-law would expect various rewards for the use of his trucks that amount to approximately $1,000. Mr. Johnson has estimated the income he would expect to receive from employing each of the three types of trucks, as indicated below:

Weather	Probability	Company trucks	Brother-in-law's trucks[*]	Rental trucks
Excellent	.3	$24,000	$22,000	$28,000
Good	.5	20,500	17,000	23,000
Bad	.2	12,200	8,000	12,000

[*]The incomes shown for using his brother-in-law's trucks assume that Mr. Johnson's wife will not help him, for which he estimates there is a 70 percent chance. If she does help, he will save $4,000 net regardless of the weather because of her free labor for the summer.

Which choice should he make?

11 Ms. Wall intends to sell stock in a company within a week because she expects the company to go into bankruptcy within a year. If she sells today she will suffer a loss of $23,000. If she waints until the end of the week there is a 25 percent, 35 percent, and 40 percent chance that her loss will be $10,000, $18,000, and $30,000, respectively. After this week, Ms. Wall expects the value of the stock to decline until bankruptcy occurs. If Ms. Wall's utility function is identical to that of the contractor in Fig. 7-5, which choice should she make to maximize her utility?

12 What is a "guesstimate"?

13 If the most qualified individual in a group comes closer more often than anyone else in the group in judging a value using the Delphi method, would it be better after the Delphi process has been completed to select the estimated value offered by this most qualified individual or a weighted estimate based on the "reliability" of each member of the group?

Part Three

The Future

Chapter 8

Trends and Opportunities

Progress, for the most part, only substitutes one partial and incomplete truth for another; the new fragment of truth being more wanted, more adapted to the needs of the time, than that which it displaces.

John Stuart Mill

Industrial engineering has been around for half a century and OR/MS a quarter of a century. What have they accomplished? What are their prospects?

In the mid-1970s, with all this help, the United States finds itself in an economic recession. Why?

PRODUCTIVITY

There are those who believe that inflation and international noncompetitiveness are major factors in the present recession. In the last fifty years, and especially in the last twenty, living on credit has become a way of life. Credit cards have become so popular that wallets have been redesigned to hold them. The most colossal offender in this "shortcut to Valhalla" is the United States. It seems evident now that the "Great Society" simply costs a great deal more than was anticipated.

In addition, one out of six workers in the United States works for the government. The U.S. Navy has eight times as many captains as it has ships to captain. Parkinson's (10) "growth of the admiralty" is a present-day reality in the United States. These are random samples, but they are indicative. Those who live on credit beyond their means must ultimately answer the "knock at the door." An individual finds that on the other side of the door is a bill collector; for an overexpended government, on the other side of the door is "inflation."

Is the United States noncompetitive? Wilson of North American Rockwell says (14, p. 9), "Today we are rapidly losing our competitive advantage. This is our national industrial economic disease—our inability to produce goods which can be sold at a competitive price in the world market."

One simple answer is that the United States has taught a lot of other people to be productive. Through massive foreign aid, we have given them the capital equipment to initiate modern industrialization. An achieving people is much less likely to want to go to war to meet its national objectives. We had our reasons; whether they were humanitarian or simply practical is a moot question at this time.

Through foreign aid and the exportation of U.S. technology we have so helped our defeated enemies of World War II, Germany and Japan, that we are now struggling competitively in international markets. One obvious advantage of countries with traditionally lower standards of living is lower labor costs. One turn of events, most disturbing to this author, is that we taught our competitors to be productive and then forgot what we taught them. Much of what Taylor (12) espoused at the turn of the century we have taught others, and since then we have moved away from our own teachings.

Methods study, work measurement, and materials handling continue to deal with the greatest controllable cost in manufacturing. Yet, so little attention has been given to these areas in recent years that it borders on embarrassing for someone trained in these areas to observe the "state of the art" in typical plants throughout the United States. Unfortunately, Americans are so fad-oriented that "if it's not new, we're not interested in it."

The present recession has, fortunately, caused some people to reflect. For years now, numerous companies have moved away from time standards. In this author's opinion, this is to fail to recognize two basic requisites of labor cost control: (1) that management knows how long it should take to perform a task, and (2) that the employee knows that management knows.

The humanistic (or behavioral) approaches of Herzberg (5) and others have a lot to offer. They are not *the answer,* however. What is needed is the combined potential of the "knowing where we are" attributes of time standards and the motivational attributes of humanistic approaches. At the high levels of present-day investment per worker, it is essential that workers be interested in maximizing whatever throughput is possible from the equipment they control, directly or indirectly. One recent encouraging development is the intention of the U.S. Air Force to require work measurement on large-scale procurements [i.e., MIL–STD–1567 (USAF)].

There is a generally held belief that unions came into existence because they were the only means whereby workers could make a rather thoughtless management give serious consideration to their grievances. In the early days of the labor movement there were valid grievances that begged for solution. Management's insensitivity to the issue of safety alone offered more than enough justification for the initiation of unions. At present, some corporations work hard at being thoughtful, in the hope of not having to deal with unions on a day-to-day basis. In a typical unionized company, by an imaginative combination of union rules, it is possible for a union to drive productivity to essentially zero. Unions sometimes do this to punish the management; unfortunately, they are punishing themselves and everyone else at the same time.

Possibly the most frustrating aspect of unionism to a sincere management is the "manufactured" union incident designed to convince union membership that their monthly dues really are buying something.

There is reason to believe that unions, generally speaking, are not champions of productivity, as suggested by Senator John Tower in the following quotation (13):[1]

The major cause of the inflationary spiral that has plagued the nation in recent years is the concentration of economic power in the hands of organized labor.

At one time, I believed government fiscal and monetary policies were the principal causes of inflation. I no longer think so. They are factors, but they are less significant than the inordinate might of the unions.

This imbalance came about largely because the basic labor laws under which we operate today are outmoded products of the 1920s and 1930s, enacted to deal with circumstances of that era.

These laws have put up a statutory wall of protection for organized labor behind which it can operate with little or no restraint.

We have heard much about the combines and trusts of the Nineteenth Century, but the power exercised by unions today is something the old robber barons would never have dreamed of.

The chief problem this causes, as far as the average citizen is concerned, is that organized labor can virtually bring an industry to its knees, forcing it to accept wage contracts that go beyond any rational relationship to productivity. These settlements drive up production costs—and prices.

Organized labor is able to bring our whole economy almost to a standstill—or at least to affect large segments of it through dock strikes, railroad strikes and so forth—until its inflationary demands are met.

Unions shouldn't have the power to arbitrarily punish everybody else in the United States because they are not getting precisely what they want.

Our position in international trade also is involved. In addition to causing inflation at home, high labor costs weaken us in competing with other industrial nations.

The leaders of organized labor are really acting against the long-range interests of their constituency: They may obtain some short-term gains in wages, benefits, etc., but in the final analysis they are going to shrink the number of jobs available to workers in this country.

High labor costs mean that jobs are going to be exported to countries with lower wage levels.

[1] Reprinted by permission from *Nation's Business*, December, 1971. Copyright 1971 by *Nation's Business*, Chamber of Commerce of the United States.

This is already being done. We're just not labor-competitive any longer. The only thing that is saving us is our technology and market know-how.

The most common basis of paying for a worker's labor has been and still is the purchase of the worker's time. The worker suffers the inconvenience of being at work rather than elsewhere, and is rewarded for the inconvenience. There need not be any direct relationship, however, between the purchase of the worker's time and productivity.

High on the list of interests of the worker is job security. A common realization of adult life is that one should have a job; a job provides one with the means to acquire the necessities of life.

Behavioralists such as Herzberg (5), McGregor (9), and Maslow (8) believe the answer to productivity is essentially motivation (i.e., of the nonremunerative variety). In dealing with professional people they are probably at least partially right. In dealing with worker A putting nut B on bolt C they may well be very wrong. Fein takes the following view (3, p. 2):

> We know from first-hand experience that job security is not enough; witness the low productivity in the civil service jobs. I believe that job security is an essential *precondition* to increasing the will to work. Without it employees resist management's attempts to raise productivity. But adding job security by itself will probably not appreciably change the productivity picture.
>
> In my opinion, the 85% of the non-involved employees need job security *plus* the incentive of a piece of the action to raise their productivity. These workers come to work to eat, to exchange their efforts and skills for what they can buy back outside of the work place. Their work has too little meaning to them for it to motivate them. Management must develop financial incentive plans which will appeal to these employees and encourage them to apply themselves more diligently to their work.

It has been well documented in numerous instances over the past fifty years that if a plant on day work (i.e., paying for time) is converted to an incentive plant, output and employee pay (assuming a 100 percent participation plan) increase in the range of 30 to 35 percent. As fixed costs become a greater and greater percentage of total manufacturing cost, such increases in total facility output represent significant decreases in unit manufactured cost. It is one of those rare "everybody wins" type of situations. The employee brings home more pay, and the owners and consumers share in the advantages of a lower unit cost. Why is it that plants in the United States have moved away from incentive plans in the last thirty years? The answer is not obvious.

To press the point a bit further, what is particularly unfortunate is that the scheme suggested above is not new and untried. Over forty years ago, James F. Lincoln (6) initiated what he called "incentive management." The abnormally high productivity of the Lincoln Electric Company has been well known, discussed, and rediscussed for the past forty years. Why does it stand alone? Why have other companies failed to adopt similar approaches?

Fein states (3, p. 57), "These [Lincoln Electric] employees participate in their work to a far greater extent than visualized by psychologists in their writings. The key to this participation is that the employees *want* to do it; there is no holding back. A manager from the outside would drool on witnessing Lincoln employees at work."

For some, this topic may seem a bit removed from OR, MS, systems, and so forth. With a little reflection, however, it should be obvious that productivity is supposed to be what IE/OR/MS is all about. Surely, if IE/OR/MS is important as a means of productivity, then the study *of productivity* is important.

SYSTEMS

Magee (7), in his TIMS presidential address in April, 1972, described three phases in the development of management science to date. The first phase, called the "primitive" phase, occurred during the 1950s, and had as characteristics "tactical problems, emphasis on quantitative and optimum solutions, and professionals transferred from other sciences." The second phase, occurring in the 1960s, was called the "academic" phase. Its characteristics were "emphasis on technique, largely quantitative, more 'management scientists' trained as such, and greater interest in theory than in observation." The final phase, the "maturing" phase, occurring in the 1970s, was described as "balance between theory and observation, attention to qualitative aspects, and interest in 'process' as much as in 'solution'." This suggests that MS has gone through considerable change in the last thirty-five plus years, and so have IE and OR. The change fundamentally has been from a "reductionist" point of view to a "systems" point of view. This has been a revolutionary change in scientific thinking and to a large extent explains the greatly changed scope of IE/OR/MS during this period. This revolution in scientific thought was very articulately expressed in the following invited address by Professor Russell L. Ackoff (1)[2] at a joint meeting of IE/OR/MS societies in 1972. References have been deleted in the interest of brevity.

Science in the Systems Age: Beyond IE, OR, and MS

Russell L. Ackoff

I believe we are leaving one cultural and technological age and entering another; that we are in the early stages of a change in our conception of the world, a change in our way of thinking about it, and a change in the technology with which we try to make it serve our purposes. These changes, I believe, are as fundamental and pervasive as were those associated with the Renaissance, the Age of the Machine that it introduced, and the Industrial Revolution that was its principal product. The socio-technical revolution we have entered may well come to be known as the *Resurrection*.

THE MACHINE AGE

The intellectual foundations of the Machine Age consist of two ideas about the nature of the world and a way of seeking understanding of it.

The first idea is called *reductionism.* It consists of the belief that everything in the world and every experience of it can be reduced, decomposed, or disassembled to ultimately simple elements, indivisible parts. These were taken to be atoms in physics; simple substances in chemistry; cells in biology; monads, directly observables, and basic instincts, drives, motives, and needs in psychology; and psychological individuals in sociology.

Reductionism gave rise to an *analytical* way of thinking about the world, a way of seeking explanations and, hence, of gaining understanding of it. For many, "analysis" was synonymous with "thought." Analysis consists, first, of taking what is to be explained apart—disassembling it, if possible, down to the independent and indivisible parts of which it is composed; secondly, of explaining the behavior of these parts; and, finally, aggregating these partial explanations into an explanation of the whole. For example, the analysis of a problem consists of breaking it down into a set of as simple problems as possible, solving each, and assembling their solutions into a solution of the whole. If the analyst succeeds in decomposing the problem he faces into simpler problems that are independent of each other, aggregating the partial solutions is not required, because the solution to the whole is the sum of the solutions to its independent parts.

It should be noted—even if with unjustified brevity—that the concepts "division of labor" and "organizational structure" are manifestations of analytical thinking.

In the Machine Age, understanding the world was taken to be the sum, or resultant, of understandings of its parts that were conceptualized as independently of each other as was possible. This, in turn, made it possible to divide the labor of seeking to understand the world into a number of virtually independent disciplines.

The second basic idea was that of *mechanism.* All phenomena were believed to be explainable by using only one ultimately simple relation, *cause-effect.* One thing or event was taken to be the cause of another, its effect, if it was both necessary and sufficient for the other.

Because a cause was taken to be sufficient for its effect, nothing was required to explain the effect other than the cause. Consequently, the quest for causes was environment-free. It employed what we now call "closed-system" thinking. Laws—like that of freely falling bodies—were formulated so as to exclude environmental effects. Specially designed environments, called "laboratories," were used so as to exclude environmental effects on phenomena under study.

Environment-free causal laws permit no exceptions. Effects are completely determined by causes. Hence, the prevailing view of the world was *deterministic.* It was also *mechanistic,* because science found no need for teleological concepts—such as functions, goals, purposes, choice, and free will—in explaining any natural phenomenon; they were considered to be either unnecessary, illusory, or meaningless.

The commitment to causal thinking yielded a conception of the world as a machine; it was taken to be like a hermetically sealed clock—a self-contained mechanism whose behavior was completely determined by its own structure. The major question raised by this conception was: Is the world a self-winding clock, or does it require a winder? Most took the world to be a machine created by God to serve His purposes, a machine for doing God's work. Additionally, man was believed to have been created in God's image. Hence, it was quite natural for man to attempt to develop machines that would serve His purposes, that would do His work.

The conception of work that was used derived from the conception of the world as consisting ultimately of particles of matter with two intrinsic properties, mass and energy, and an extrinsic property, location in a space-time coordinate system. Work was taken to be the movement of mass through space or the application of energy to matter so as to transform either matter or energy, or matter into energy. Work that was to be mechanized was analyzed. Such analysis came to be called "work study." It was thus decomposed into work elements, indivisible tasks. To these, elementary machines—the wheel and axle, the inclined plane, and the lever—energized by other machines, were applied separately or in combination. Separate machines were developed to perform as many elementary tasks as possible. Men and machines, each performing one or a small number of elementary tasks repetitively, were organized into processing networks that became mass-production and assembly lines.

The Industrial Revolution brought about mechanization, the substitution of machines for man as a source of physical work. This process affected the nature of work left for men to do. They no longer did all the things necessary to make a product; they repeatedly performed a simple operation in the production process. Consequently, the more machines were used as a substitute for men at work, the more working men were made to behave like machines. The dehumanization of men's work was the irony of the Industrial Revolution.

THE SYSTEMS AGE

Although eras do not have precise beginnings and ends, the 1940s can be said to have contained the beginning of the end of the Machine Age and the beginning of the Systems Age. This new age is the product of a new intellectual framework in which the doctrines of reductionism and mechanism and the analytical mode of thought are being supplemented by the doctrines of *expansionism, teleology,* and a new *synthetic* (or systems) mode of thought.

Expansionism is a doctrine maintaining that all objects and events, and all experiences of them, are parts of larger wholes. It does not deny that they have parts, but it focuses on the wholes of which they are part. It provides another way of viewing things, a way that is different from, but compatible with, reductionism. It turns attention from ultimate elements to a whole with interrelated parts, to *systems.* Preoccupation with systems emerged during the 1940s. Only a few of the highlights of this process are noted here.

In 1941 the American philosopher(ess) Suzanne Langer argued that, over the preceding two decades, philosophy had shifted its attention from elementary particles, events, and their properties to a different kind of element, the *symbol.* A symbol is an element whose physical properties have essentially no importance. Charles W. Morris, another American philosopher, built on Langer's work a framework for the scientific study of symbols and the *wholes* of which they were a part, *languages.* By so doing he expanded the center of attention. In 1949 Claude Shannon, a mathematician at Bell Laboratories, developed a mathematical theory that turned attention to a still larger phenomenon, *communication.* Another famous mathematician at the Massachusetts Institute of Technology, Norbert Wiener, in his book *Cybernetics,* put communication into a still larger context, that of *control.* By the early 1950s, it became apparent that interest in control and communication were only aspects of an interest in a still larger phenomenon, *systems,* to which the biologist Ludwig von Bertalanffy drew attention with his work. "Systems" has since been widely recognized as the new organizing concept of science. The concept is not new, but its organizing role in science is.

A system is a set of interrelated elements of any kind; for example, concepts (as in the number system), objects (as in a telephone system or human body), or people (as in a society). The set of elements has the following three properties.

1   The properties or behavior of each part of the set has an effect on the properties or behavior of the set as a whole. For example, every organ in an animal's body affects the performance of the body.

2   The properties and behavior of each part and the way they affect the whole depend on the properties and behavior of at least one other part in the set. Therefore, no part has an independent effect on the whole. For example, the effect that the heart has on the body depends on the behavior of the lungs.

3   Every possible subgroup of elements in the set has the first two properties. Each has an effect, and none can have an independent effect, on the whole. Therefore, the elements cannot be organized into independent subgroups. For example, all the subsystems in an animal's body—such as the nervous, respiratory, digestive, and motor subsystems—interact, and each affects the performance of the whole.

Because of these three properties, a set of elements that forms a system always has some characteristics, or can display some behavior, that none of its elements or subgroups can. Furthermore, membership in the set either increases or decreases the capabilities of each element, but it does not leave them unaffected. For example, parts of a living body cannot live apart from that body or a substitute. The power of a member of a group is always increased or decreased by such membership.

A system is more than the sum of its parts; it is an *indivisible whole.* It loses its essential properties when it is taken apart. The elements of a system may themselves be systems, and every system may be a part of a larger system.

Preoccupation with systems brings with it the *synthetic* mode of thought. In the analytic mode, it will be recalled, an explanation of the whole was derived from explanations of its parts. In synthetic thinking, something to be explained is viewed as part of a larger system and is explained in terms of its role in that larger system. For example, universities are explained by their role in the educational system, rather than by the behavior of their colleges and departments. The Systems Age is more interested in putting things together than in taking them apart.

Analytic thinking is, so to speak, outside-in thinking; synthetic thinking is inside-out. Neither negates the value of the other, but by synthetic thinking we can gain understanding that we cannot obtain through analysis, particularly of collective phenomena.

The synthetic mode of thought, when applied to systems problems, is called the *systems approach.* This way of thinking is based on the observation that, when each part of a system performs as well as possible, the system as a whole seldom performs as well as possible. This follows from the fact that the sum of the functioning of the parts is seldom equal to the functioning of the whole. This can be shown as follows.

Suppose we collect one each of every model of available automobile. Suppose further that we then ask some top-flight automotive engineers to determine which of these cars has the best carburetor. When they have done so, we note the result. Then we ask them to do the same for transmissions, fuel pumps, distributors, and so on through each part required to make an automobile. When this is completed, we ask them to remove the parts noted and assemble them into an automobile, each of whose parts is the best available. They will not be able to do so, because the parts will not fit together. Even if they could be assembled, in all likelihood they would not work well together. System performance depends critically on how the parts fit and work together, not merely on how well each performs independently.

Furthermore, a system's performance depends on how it relates to its environment, the larger system of which it is a part, and to other systems in that environment. For example, an automobile's performance depends on the weather, the road on which it is driven, and how well it and other cars are driven. Therefore, in systems thinking we try to evaluate the performance of a system by evaluating its functioning as a part of the larger system that contains it.

It will be recalled that in the Machine Age cause-effect was the central relation in terms of which all explanations were sought. At the turn of this century the distinguished American philosopher of science, E. A. Singer, Jr., noted that cause-effect was used in two different senses. First, it was used in the sense already discussed: a cause is a necessary and sufficient condition for its effect. Secondly, it was also used when one thing was taken as necessary but *not* sufficient for the other. He used as an example an acorn, which is necessary but insufficient for an oak; various soil and weather conditions are also necessary. Similarly, a mother—despite women's liberation—is only necessary, not sufficient, for her child. Singer chose to refer to this latter sense of cause-effect as *producer-product.* It can also be thought of as a probabilistic or nondeterministic cause-effect.

Singer went on to show that studies of phenomena that use the producer-product relation were compatible with, but richer than, studies restricted to the use of deterministic causality. Furthermore, he showed that a theory of explanation based on producer-product permitted functional, goal-seeking, and purposeful behavior to be studied objectively and scientifically. These concepts no longer needed to be taken as meaningless or inappropriate for scientific study.

Later, biologist G. Sommerhoff came independently to the same conclusions as Singer had. In the meantime, in a series of papers that laid the groundwork for cybernetics, Arturo Rosenblueth, Norbert Wiener, and J. H. Bigelow showed the great value of conceptualizing machines and man/machine systems as functioning, goal-seeking, and purposeful entities. In effect, they showed that, whereas it had been fruitful in the past to study men as though they were machines, it was now at least equally fruitful to study machines, man/machine systems, and, of course, men as though they were goal-seeking or purposeful. Thus, in the 1950s *teleology*—the study of goal-seeking and purposeful behavior—was brought into science and began to dominate our conceptualization of the world.

For example, in mechanistic thinking behavior is explained by identifying what caused it, never its effect. In teleological thinking, behavior can be explained either by what produced it or by what it is intended to produce. For example, a boy's going to the store can be explained either by the fact his mother sent him, or by the fact that he intends to buy ice cream for supper. Study of the functions, goals, and purposes of individuals and groups has yielded a greater ability to evaluate and improve their performance than mechanistically oriented research did.

THE POST-INDUSTRIAL REVOLUTION

The doctrines of expansionism and teleology, and the synthetic mode of thought are both the producers and the products of the Post-Industrial Revolution. But this revolution is also the product of three technological developments, two of which occurred during the (First) Industrial Revolution. One of these emerged with the telegraph in the first half of the nineteenth century, followed by the invention of the telephone by Alexander Graham Bell in 1876, and of the wireless by Marconi in 1895. Radio and television followed in this century. Such devices mechanized communication, the transmission of symbols. Since symbols are not made of matter, their movement through space does not constitute physical work. The significance of this fact was not appreciated at the time of the invention of communication machines.

The second technology emerged with the development of devices that can observe and record the properties of objects and events. Such machines generate symbols that we call *data*. The thermometer, odometer, speedometer, and voltmeter are familiar examples of such instruments. In 1937 there was a major advance in the technology of observation when it "went electronic" with the invention of radar and sonar.

Instruments can observe what we cannot observe without mechanical aids, or magnitudes and differences too large or small for our senses. Note that such instruments, like communication machines, do not perform physical work.

The third and key technology emerged in the 1940s with the development of the electronic digital computer. This machine could manipulate symbols logically. For this reason, it is frequently referred to as a thinking machine.

These three technologies made it possible to observe, communicate, and manipulate symbols. By organizing them into a system, it became possible to *mechanize mental work,* to *automate.* This is what the Post-Industrial Revolution is all about.

Development and utilization of automation technology requires an understanding of the mental processes that are involved in observing, recording, and processing data, communicating them, and using them to make decisions and control our affairs. Since 1940 a number of interdisciplines have been developed to generate and apply knowledge and understanding of mental processes. These include communication and information sciences, cybernetics, systems engineering, operations research, and the management and behavioral sciences. Such fields provide the software of the Post-Industrial Revolution.

THE ORGANIZING PROBLEMS OF THE SYSTEMS AGE

Because the Systems Age is teleologically oriented, it is preoccupied with systems that are goal-seeking or purposeful; that is, systems that can display *choices,* of either means or ends, or both. It is interested in purely mechanical systems only insofar as they can be used as instruments of purposeful systems. Furthermore, the Systems Age is most concerned with purposeful systems, some of whose parts are purposeful; these are called *social groups.* The most important class of social groups is the one containing systems whose parts perform different functions, that have a division of functional labor; these are called *organizations.* Systems-Age man is most interested in groups and organizations that are themselves parts of larger purposeful systems. All the groups and organizations, including institutions, that are part of society can be conceptualized as such three-level purposeful systems.

There are three ways in which such systems can be studied. We can try to increase the effectiveness with which they serve their own purposes, the *self-control* problem; the effectiveness with which they serve the purposes of their parts, the *humanization* problem; and the effectiveness with which they serve the purposes of the systems of which they are a part, the *environmentalization* problem. These are the three strongly interdependent organizing problems of the Systems Age.

SCIENCE IN THE SYSTEMS AGE

Up to this point I have tried to deal with the question: What in the world is going on in the world? My response to this question provides a vantage point from which I would now like to look at science in general and at the management sciences in particular.

Since its inception, science has not only been taking the world apart, but it has also been taking itself apart, although not without reason or benefit. The decomposition of science could not have been avoided. The reason is revealed in the statement with which Colin Cherry opened his book, *On Human Communication*:

> Leibnitz, it has sometimes been said, was the last man to know everything. Though this is most certainly a gross exaggeration, it is an epigram with considerable point. For it is true that up to the last years of the eighteenth century our greatest mentors were able not only to compass the whole science of their day, perhaps with mastery of several languages, but to absorb a broad culture as well.

The continuous accumulation of scientific knowledge that occurred during and after the eighteenth century made it necessary to divide and classify this knowledge.

Scientific disciplines were the product of this effort. Science formally separated itself from philosophy only a little more than a century ago. It then divided itself into physics and chemistry. Biology emerged out of chemistry, psychology out of biology, and the social sciences out of psychology. This much was completed at the beginning of this century. But scientific fission continued. Disciplines proliferated. The National Research Council now lists more than one hundred and fifty of them.

Disciplines are categories that facilitate filing the content of science. They are nothing more than filing categories. Nature is not organized the way our knowledge of it is. Furthermore, the body of scientific knowledge can, and has been, organized in different ways. No one way has ontological priority. The order in which the disciplines developed was dictated to a large extent by what society permitted scientists to investigate, not by any logical ordering of subject matter. Scientists started to investigate the areas that least challenged deeply held social, cultural, religious, and moral beliefs of the time. The subject matter of science was chosen—and not always successfully—so as to maximize the probability of survival of scientists and science. As science gained prestige, it pressed against the social barriers that obstructed its development; one by one they were breached.

But scientists and philosophers wanted to invest the history of science with more logic than history itself provided. Therefore, they sought to rationalize the order of disciplinary development by invoking the concept of a hierarchy of the sciences. They argued that physics deals with objects, events, and properties of both that were ultimately simple, hence irreducible and directly observable. Each successive discipline, it was argued, dealt with increasingly complex functions and aggregations of these objects, events, and properties. Hence, each discipline except physics was taken to rise out of, and to be reducible to, the one that preceded it. Physics was taken to be basic and fundamental. Dependence between sciences was taken to flow in only one direction.

This hierarchical myth is still widely accepted in and out of science. It is the basis of a caste system in the community of science that is as severe and irrational as any that has existed in society.

It is still widely believed that the physical sciences alone deal with ultimate reality and that they have no need of the other disciplines in their effort to do so. This belief is maintained despite the fact that we can demonstrate that no concept used in any one discipline is ultimately fundamental and incapable of being illuminated by work in other disciplines.

Consider, for example, a concept used in physics that, perhaps more than any other, is thought to be its exclusive property: *time.* Physicists have dealt with time in one of two ways. They have either taken it to be a primitive concept that cannot be defined, and hence a concept whose meaning can only be grasped by direct experience of it; or they have dealt with it operationally, defining it by the operations used to measure it. Techniques of measuring time in physics all derive from use of the rotation of the earth around the sun as a basic unit. Clocks, sun dials, water clocks, sand clocks, and so on are instruments to divide this unit into equal parts. Thus, in physics time is dealt with as an ateleological astronomical concept. It is generally assumed that contributions to understanding it cannot be made by any other discipline.

This is not the case. Time can be considered teleologically, not as a property of the universe that is out there for us to take, but as a concept deliberately constructed by man to serve his purposes.

People develop alternative ways of individuating events. For example, a person may differentiate between breakfasts by their content, location, or by those with whom he had the meal. Some of these individuating properties may be adequate only in special circumstances. He may have the same breakfast with the same people at the same place on different days. Two events that occur to the same individual may be the same with respect to every property except one, time. Two events that occur to the same individual at the same time cannot be otherwise identical: they must differ in some respect; otherwise, they would not be two events.

Therefore, from a functional point of view, time is a property of events that is sufficient to enable a person to individuate any two changes in the same property of the same thing. Because we measure time using physical phenomena, we erroneously conclude it is a physical concept. The error becomes apparent in situations in which astronomical measures do not serve our purposes well. In measuring the rate of growth of plants, for example, C. W. Thornthwaite found astronomical time inadequate. He sought a biological clock and found one in the pea plant; he used the time between appearances of successive nodes on the pea plant as units of time for his work. These units were of different duration when measured astronomically, but they made possible more accurate prediction and control of harvests than did hours and days. We measure time by using events that are identical in all respects save time; and, in principle, these can be of any type—which type we use is determined by our purposes.

As the application of science increased it became useful to organize its findings functionally around areas of application, into professions, as well as into disciplines. Old professions that preceded science borrowed from a number of scientific disciplines and new ones did so as well.

The disciplinary and professional classifications of scientific knowledge are orthogonal to one another, and hence can be represented by a matrix in which the disciplines form the rows and the professions the columns. New rows and columns can be expected to be added in the future.

As the problems to which science was addressed became more complex—and particularly as it began to address itself to problem complexes, systems of problems that I like to call *messes*—a new organization of scientific and technological effort was required. The first response to this need occurred between the two World Wars and took the form of *multidisciplinary* research. In such research the problem complex investigated was decomposed into unidisciplinary and uniprofessional problems that were taken to be solvable independently of each other. Hence, they were assigned to different disciplines or professions, separately solved, and the solutions were either aggregated or allowed to aggregate themselves. With the emergence of systems thinking, however, it was realized that the effect of multidisciplinary research on the treatment of the whole was frequently far from the best that could be obtained.

This realization gave rise to *interdisciplinary* research, in which the problem complex was not disassembled into disciplinary parts, but was treated as a whole by representatives of different disciplines working collaboratively. Operations research and the management sciences were among the interdisciplines born of this effort. So were cybernetics, the organizational sciences, the policy sciences, planning science, general systems research, and the communication sciences, among others.

Universities began to educate the young for such work. Those so educated were not of any one discipline but of the intersection of several. Hence, their loyalty was not directed to one discipline but to an interdisciplinary concept. But this did not last long. The interdisciplines sought recognition and status by emulating the disciplines and professions. Academic departments and professional societies were formed along conventional lines. The interdisciplines began to identify themselves with the instruments that they developed and used—that is structurally—rather than with what these instruments were used for—that is, functionally. They began to introvert, to look inward and contemplate their own methods and accomplishments, rather than the messes that had given rise to their activities. Jurisdictional disputes and efforts to individuate interdisciplines arose between activities created to eliminate just such disputes and individuation.

As the problem complexes with which we concern ourselves increase in complexity, the need for bringing the interdisciplines together increases. What we need may be called *metadisciplines,* and what they are needed for may be called *systemology.*

The formation of interdisciplines in the last three decades can now be understood as a transitional development, a beginning to an evolutionary synthesis of human knowledge, not only within science, but between science and technology, and most importantly, between science and the humanities. Consider the distinction between science and the humanities. I believe that in the Systems Age science will come to be understood as the search for similarities among things that are apparently different; and the humanities will come to be understood as the search for differences among things that are apparently the same. The former seeks generality; the latter uniqueness. This makes science and the humanities like the head and tail of a coin; they can be looked at and discussed separately, but they cannot be separated. Consider why.

In the conduct of any inquiry, we must determine the ways in which the subject under study is similar to other subjects previously studied. Doing so enables us to bring to bear what we have already learned. But, in addition, it is also necessary to determine how the subject at hand differs from any we have previously studied: in what ways it is unique. Its uniqueness defines what we have yet to learn before we can solve the problems it presents. Thus, the humanities define the aspects of messes that we still have to learn how to handle, and science provides ways both of handling or researching the aspects that have previously been dealt with, and of finding ways of approaching the aspects that have not been studied previously.

The effective study of large-scale social systems requires the synthesis of science and the humanities, of science and technology, and of the disciplines within science and the professions that use them.

Despite the need for integration, universities and professional and scientific societies preserve the autonomy of the parts of science and their application. What is needed is not a temporary association of autonomous interdisciplines such as we have here at Atlantic City, but a permanent integration of interdisciplines that yields a broader synthesis of methods and knowledge than any yet attained.

We need a fusion of interdisciplines that extends well beyond those represented here. Nevertheless, a fusion of these three would be a significant step in the right direction. But as far as I can tell—after considerable effort to merge two of them—they would rather die separately than live together. And they are dying despite their growth. Death is *not* a function of the number of cells in a body, but of their vitality: the membership of even a cemetery can expand continuously.

None of what I have said denies the usefulness of either disciplinary science or the professions. They will remain useful in dealing with problem areas that can be decomposed into problems that are independent of each other. But the major organizational and social messes of our time do not lend themselves to such decomposition. They must be attacked holistically, with a comprehensive systems approach.

Nor are my remarks intended to diminish the past accomplishments of IE, MS, and OR—they have been significant and I share with you a pride in them. But their accomplishments are becoming less significant because their development has not kept pace with the growing complexity of the situations with which managers and administrators are faced.

As currently conceived, taught, and organized, industrial engineering is not broad enough to engineer industry effectively by itself. This is obvious. Look at the wide variety of other types of engineers crawling all over industry. The management sciences are not broad enough to make management scientific. What percentage of the decisions of even the managers most dedicated to these sciences are based on science? Operations research is not broad enough to research effectively the operating characteristics of our social system that most urgently need research: discrimination, inequality within and between nations, the bankruptcy of education, the inefficiency of health services, increasing criminality, deterioration of the environment, war, and so on. This is not to say that IE, MS, and OR are no good, but it does say that they are *not good enough.* Each of them suffers not only from the lack of competencies that are required to deal with the messes that preoccupy those who manage most public and private systems, but also because they use Machine-Age concepts and methods in attempts to deal with Systems-Age problems.

Meetings such as this one should be dedicated to the marriage of movements, and to the conception and birth of ways of coping with complexity. But, instead, they are wakes at which interdisciplines are laid out and put on display in their best attire. Eulogies are delivered in which accounts are given about how messes were murdered by reducing them to problems, how problems were murdered by reducing them to models, and how models were murdered by excessive exposure to the elements of mathematics.

But those who attend a wake are not dead. They can still raise hell. And, if they do, even a corpse—like that of James Joyce's Finnegan—may respond and rise with a shout.

It may be interesting to note that nowhere in the titles of IE, OR, and MS is the word "system," and yet that is the basis for their commonality. It seems reasonable to assume that, in particular, IE/OR/MS is concerned with the design and operation of *productive* systems. Almost without exception, all three fields treat systems of things (i.e., technical systems) as "black boxes" within human systems—hence the title of this text at one time was *Productive Systems in a Complex Society: The IE/OR/MS Approach.*

Recognition of the "systems approach" has been enhanced recently with the creation of the IIASA (International Institute for Applied Systems Analysis). Under the direction of Professor Howard Raiffa of Harvard University, twelve countries have formed the institute. The institute proposes "to use a broad research strategy that involves the use of techniques, concepts, and scientific, systematic approach in order to help decision makers choose a desirable course of action in problems of international concern" (2, p. 3).

It would be inappropriate to leave the reader with the belief that all is well in IE/OR/MS. This is not the case.

One of the things Americans do best is complain. Freedom of speech is a valued tradition in the United States, and there is a considerable volume of writing, particularly in recent issues of the OR/MS journals *Interfaces* and *OR/MS Today,* suggesting potential improvements in OR/MS and its application. It is all very healthy; it is the kind of feedback that helps cause corrections along the path. The following is a randomly ordered list of twenty-one identifiable alleged problem areas found in a partial review of recent OR/MS literature by the author:

1   Inadequate data specification and collection
2   Insufficient humility of the practitioner
3   Inappropriate assumptions
4   Insufficient consideration of values and ethics
5   The "publish or perish" influence
6   Inadequate attention to human relations aspects of the study
7   Failure of the practitioners to adequately "sell" themselves or aspects of the study
8   Failure to adequately recognize the translator role of OR/MS between managers and others
9   Infatuation with "optimization"
10   Inadequate attention to implementation aspects of a study
11   Tendency of OR/MS practitioners, rejected by traditional managers, to introvert their OR/MS interests
12   Naïveté with respect to details of the work environment under study
13   Inexperience of the practitioners
14   Failure of the practitioners to adapt their "study language" to suit the limitations of vocabulary of their listeners
15   Inadequate attention to study objectives
16   Failure to gain additional valuable knowledge of the system by discussing it with those who possess intimate knowledge of it (e.g., the workers)
17   Tendency to overcomplicate, resulting in contradictions of common sense
18   Inherent difficulties in multidisciplinary and interdisciplinary research
19   Self-gratification of the practitioners (i.e., the mathematical elegance problem)
20   Inadequate appreciation of overriding profit motive aspects
21   Inadequacies with respect to "human factors" aspects of the problem

If all OR/MS studies possessed all of the deficiencies above there would be cause for alarm. That is not the case. As the person wakened at the end of a concert and asked about the music said, "It wasn't as bad as it sounded." The twenty-one problems are problems that have come to light in reviewing past studies, and they are not as bad as they sound.

In the author's opinion, the twenty-one problems can be grouped in three main categories, as follows: (1) practitioner attributes indicative of educational short-comings (problems 1, 2, 3, 6, 7, 8, 9, 10, 11, 12, 14, 16, 19, 20, and 21), (2) methodological shortcomings (problems 4, 15, 17, and 18), and (3) problems beyond immediate control (problems 5 and 13). If this categorization is correct, or even close to correct, it is apparent that a redesign of the educational process that generates OR/MS practitioners is in order. One of the unfortunate fundamental difficulties in affecting this change is that those in the best position to implement the necessary changes in the curriculum (i.e., the professors) are the greatest violators of problems 3, 5, 9, 11, 19, and 21 because of their perception of what best serves their individual interests. This leads one to the conclusion that our educational system has problems, and indeed it does.

The following quotations from Grayson's article (4) in the *Harvard Business Review* speak to this issue:

Management science [and OR] has grown so remote from and unmindful of the conditions of "live" management that it has abdicated its usability. Managers, for their part, have become disillusioned by management science, and are now frequently unwilling to consider it seriously as a working tool for important problems. The author believes that management science *can* make a contribution to management; and in this article suggests how a bridge between the two groups can be constructed. He also makes it clear that the scientists must be the ones to start construction.

First, both management science [and OR] faculty and students have to get out of the isolated, insulated world of academe. They must go beyond structured cases and lectures and become directly involved in real-world, real-time, live projects, not as a way of applying what they know, but as a way of learning.

Tackle the *real* problem. This will be frustrating, but the frustration of trying to reach a *workable* solution can be used to teach the management scientist or student in a way that is useful both to him and to the business or government unit.

Faculty members should plan to get out of the university, physically and completely, for meaningful periods of time. They should plan their careers so that they schedule at least a year, periodically, in which they get direct, real-world experience in business, non-profit organizations, or the government.

What is particularly unfortunate in the situation described above is that many faculty members, particularly those who are young and inexperienced in the real world, would genuinely welcome an opportunity to spend periods of time in industry if the opportunity were presented. The blame for this condition must lie, at least partially, with industry and government.

The reader has probably noticed that industrial engineers were excluded from the above discussion of twenty-one problems in OR/MS. It is the opinion of the author, coincidently an industrial engineer, that industrial engineers have ameliorated some of the problems contained in this list. The following are possible reasons for this alleged position:

   1   Industrial engineering has been practiced for a longer period than OR/MS and has resulted in greater accommodation with some of these realities of industrial life, in practice.
   2   Industrial engineering professors have a great body of prior successful application from which to draw.
   3   Engineering accreditation guidelines generally insure that aspects of the curriculum fundamental to future successful practice must be included in the curriculum.
   4   Industrial engineers typically work from "within the system" as compared to the consultative type of work arrangement somewhat more common to OR/MS practice.

In an attempt to demonstrate this different point of view of the IE practitioner, the following article is offered, written by Smith (11), a former student of mine. It is at least partially revealing to compare the role of the IE, as suggested by this writer, with the list of twenty-one problems associated with OR/MS practice.

A Student IE's Answer to "What's an IE?"

Ronald Smith

"**Industrial Engineer**." A very impressive title. Slides easily over the tongue. But what is one? Must be an engineer, but what type?

"Industrial" brings to mind bellowing smokestacks, mighty gears turning, the screaming of a steam whistle, forges flaring, millions of fellow working men building a stronger nation, the American way of life! . . . all very soul stirring, but what does an IE do?

First, let's see if we can find a niche for the IE in the vast engineering disciplines. There are four main fields: (In alphabetical order, for some sensitive types)

   1   **Chemical**: Bubble-bubble, toil & trouble: If it is sticky, gooey, gummy, grubby, bubbly, they're in it.
   2   **Civil**: Bridge builders, architects-to-be, road makers, stress and strain types.
   3   **Electrical**: Zap! Crackle! Pop! AC/DC! Tiny black boxes with tiny black boxes within.
   4   **Mechanical**: Clink-clank-clunk! What makes the world go round: gears, spindles, sprockets, & spanners.

(I'm sorry, Ag.E's, but I just could not bring myself to do it . . .)

Now, then, where were the IE's? We mixed some chemical together to make something, wired it up, put wheels & gears on it, and built a shed for it. What is left?

Some people consider IE similar to caulking compound—it fills the gaps between the other engineering fields—a catch-all for what the others don't care to use. Things like quality control, safety, & reliability. After the noble EE's, ME's, etc. have finished, the IE scuttles around tidying-up the process. "If you can't make it as a *real* engineer, you can always become an IE."

Others seem to consider the IE as a sadist whose biggest kick in life is making people work faster-for-lesser. "Efficiency Experts." The management's stool pigeons. "Young wet-nosed punks to tell me, after 10 years with the company, that I'm working too slow!"

These people are entitled to their opinions, but for those persons who have no idea at all about the difference between an IE and the other engineers, probably 90 percent of the people in the world (89 percent of whom have never heard of IE's), the following material may help. It is an attempt to describe some of the many different parts of being an IE, stressing the differences between IE's and those other engineers.

The IE brings a sense of **business reality** to the other engineers. The IE straddles the vast gap between the practical managers and the development engineers. He is the guy who tells the EE that his solid gold, platinum plated, lithium-lined relay contacts are being replaced by tin ones. "Is that third brace really needed?" This is where the IE's training in the various engineering fields comes in handy. He is a translator. The engineer can communicate with an IE when he may be unable to show the non-engineering boss what he is trying to do. And visa-versa, the IE can bring the holy word down from above to the engineers, reminding them that they are here to make money. Also, the IE has a sporting chance if the engineers try to snow him with technicalities; they do not know *what* he may know.

The IE is a **step backer**. He takes the proverbial one step back. He gets the big picture. Often people working intensively on the details of a project or design fail to see where they are going overall. While the ME is wondering if the right-torque framastan should have 5 or 6 spokes for the wombat model, the IE has decided that the fritler doesn't even need wombats.

An IE is a **people buffer**. Not only does he work with the bright shining, purring machines, but also with their hairy, ham-fisted operators. Is Joe Blow working as fast as possible with the thingie cutter? How much should he be paid? Per piece or not? Call in the stop watch people, the IE's. They will help you to set up work standards for comparison, they will suggest ways of job evaluation, possible incentive plans to use, etc.

Joe Blow may complain at first, but the IE's help him, too (if, admittedly, he's not part of the dead wood trimmed away), (that is life, Joe), as more efficient work is usually easier work.

Most IE's really don't go around pulling wings off flies, kicking old ladies' crutches, etc. They are more or less human. They don't want Joe's job. They want to find faster, cheaper ways to do it. Also, safer ways, which brings up the next topic:

An IE is a **people protector**—safety. Keeping Joe Blow's paws out of the cutter's "field of authority." If not for Joe's sake, then it is at least more efficient. Many people consider safety as a joke—never take it seriously, like security. "Hah, the rope will hold; whatch'a worrin 'bout?" IE's make Mr. Blow safe if he wants to be or not. The IE's ears perk up at the sound of "famous last words . . ."

An IE can be a **plant planner**. Building a factory? Should the tinker-tuners go near the fudge-forgers? Should there be a moving belt or a passageway? How much how soon which way where when? Don't ask the CE—he's trying to keep the roof up; the ME's unsticking the door lift; the EE's in the John working on the lights; the ChemE is making smoke in the basement; the AgE is shoveling . . . well, anyway, it is the IE who's had the training, he is the dude who coordinates and plans. It is his neck if the feather fluffer winds up next to the molasses mixer. His training includes human factors as well as straight engineering knowledge.

The ME may see no reason why he shouldn't have the walls painted black to absorb heat, but the IE will also balance in the effect it will have on the employees. Also, where will you put the plant? Once the plant is located and built, it is very expensive to correct any major errors. An IE would be a valuable asset in assembling the assembly plant.

He can be a **sampler**. Who decides, out of 100 wombats, how many samples you must take to be 95 percent certain that 90 percent of the goods are good? A good IE can. He has been trained in statistics, probability, and reliability. Practical, applicable statistics, too, as opposed to what a math major would get. The other engineers? They may have had at most 3 hours of statistical theory. Not much of a confidence level there.

An IE is a **betterwayer**. He has "Is this the best way?" constantly rattling around in his head. Here, perhaps, the difference between an IE and the other engineers is less noticeable. All engineers try to come up with better methods of making something. The ME's, EE's, etc., have the advantage of technical knowledge in their own fields, but the IE may be able to pull something from, say, the ChemE's and apply it to an ME's problem. The IE's big advantage comes in after the advanced model is made. He knows the assembly line better than the other engineers, and can see a great use for the new No. 1 zoot sticker in the No. 3 wombat assembly line. The IE has ideas on what to do with the engineers' new ideas.

An IE is good at **maximizing**. If you want to ship your fresh wombats, and route A takes 3 hours at $4 per wombat, route B takes 6 hours at $2, and route C takes 4 hours at $3.50 each, and you lose $1 for each hour a wombat is outside, what is the best route? This example may be simple, but when you consider a plant's normal production of wombats per day, and all the various methods of transportation available, it gets very sticky indeed.

Many people would be surprised at the elaborate formulas IE's use to solve these types of problems (when they are not pulling wings off flies or winding their stop watches). You just can not pull the answers out of your hat. The other engineers could possibly "plug and shove" to get the answer, but an IE is trained for it.

One main difference between IE's and the other engineers is **ladder climbability**. An EE can look forward to being head of the EE department someday, but that is usually about it. The IE finds it a comparatively easy transfer to management; he usually has a foot half-way into management to begin with. His background knowledge will be an asset. Also, he may be a bit more open minded than the EE, who would favor his old department. An IE's "old department" is the whole plant.

Another important difference is **survivability**. An IE is like a cat—he can always (well, more often than not) land on his metaphysical feet. His knowledge can be used in almost any field, from industry to hospitals, from the military to research projects. If the rubber market drops out, he can jump to, say, aeronautics. The ChemE, however, will be caught in a fairly specialized field, and he may go down with the rubber duckies.

Now then, many IE's are not going to like the fact that we may have brushed over lightly or even skipped their pet IE subject, but it is hard to cover IE'ism completely, which brings up the next characteristic of Industrial Engineering:

Indefinabilityness. One of the most frustrating characteristics of Industrial Engineering to some, but appealing to others, is that IE cannot be defined in a little capsule summary. There are no definite boundaries to IE. It is so wide, from management to time study, from quality control to design analysis, almost anyone can find his niche sooner or later.

An EE is an EE within fairly predictable limits, but when you say "I am an IE," people still haven't got you pinned down, pegged away in a mental cubby hole. You are an unknown factor, they don't known what to expect.

The world needs non-superspecialized engineers—people who can get an overall view, bring together the specialists, and handle new, presently unforseen and unpretrainable-for, events.

I hope this has given you at least some vague idea of what an IE is and how he differs from "them other slip-stick jockies." I admit it is biased, if not a little aggressive, but then many an IE walks around with a large chip on his shoulder, because when he says "I'm an IE!", people still say "Whatzat?".

SUMMARY

What, then, are the prospects for IE/OR/MS? In the author's opinion—considerable. The prior discussion concentrated on problems, and every discipline has them. What was not discussed was the successes. Of course, there has not been as much success in the application of IE/OR/MS as some would like. What is more important, though, is that IE/OR/MS is making a greater and greater impact on the solution of problems of human systems. It seems likely that in the future, an almost inexhaustible supply of difficult "human systems" problems will be available to challenge imaginative IE/OR/MS practitioners for a long time to come. Figure 8-1 is meant to indicate that there is a considerable commonality of interest between the industrial engineer, the operations researcher, and the management scientist. In fact, the commonality of interest has grown to the point today that the technical literature of the three disciplines is essentially one common body of knowledge serving all three disciplines. The fields are distinguished from one another more by what is *not* common to all three fields.

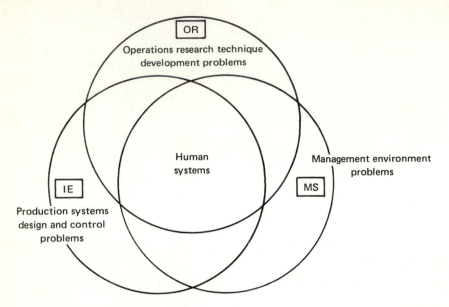

Figure 8-1   Interest areas in IE/OR/MS.

Industrial engineers are trained as engineers; they are more comfortable dealing with engineering details of a problem than are the OR/MS practitioners; their primary environment, at least educationally, is manufacturing; and they are quantitatively educated. Their greatest weakness is likely to be an insufficient knowledge of the management environment, because so much of their curriculum is limited to engineering topics.

Operations researchers are typically very quantitatively trained, particularly in a purely mathematical sense. Their greatest weakness may well be that their curriculum may be highly restricted to mathematical topics, giving inadequate attention to the acquisition of less technical, but essential, knowledge and skills specific to successful practice.

Management scientists typically have a much greater depth and breadth of knowledge concerning the management environment. Far too many management science programs to date, however, have devoted insufficient attention to the development of prerequisite mathematical skills and applicable quantitative techniques.

Practitioners of each of the three disciplines can identify with specific problem areas for which their training best qualifies them for attacking problems at hand. There is also a very large area of common interest (i.e., human systems) in which all three disciplines can make equally valuable contributions for a long time to come. In the author's opinion, there is an essentially unlimited future in IE/OR/MS for capable and imaginative practitioners.

REFERENCES

1   Ackoff, Russell L.: "Science in the Systems Age: Beyond IE, OR, and MS," *Oper. Res.*, vol. 21, no. 3, pp. 661–671, May–June, 1973.

2 Edie, Leslie C.: "Message from the Outgoing President," *OR/SA Today,* vol. 3, no. 4, p. 3, July, 1973.

3 Fein, Mitchell: "Motivation for Work," Monograph no. 4, Work Measurement and Methods Engineering Division, American Institute of Industrial Engineers, Norcross, Ga., 1971.

4 Grayson, C. Jackson, Jr.: "Management Science and Business Practice," *Harvard Business Review,* vol. 51, no. 4, pp. 41–48, July–August, 1973.

5 Herzberg, Frederick: *Work and the Nature of Man,* World Publishing Company, Cleveland, Ohio, 1966.

6 Lincoln, James F.: *A New Approach to Industrial Economics,* p. 36, Devin Adair, 1961.

7 Magee, John L.: "Progress in the Management Sciences," *Interfaces,* vol. 3, no. 2, pp. 35–41, February, 1973.

8 Maslow, Abraham H.: *Motivation and Personality,* Harper and Row, New York, 1954.

9 McGregor, Douglas: *The Professional Manager,* McGraw-Hill, New York, 1960.

10 Parkinson, C. Northcote: *Parkinson's Law,* Houghton Mifflin Publishing Co., Boston, 1957.

11 Smith, Ronald: "A Student IE's Answer to 'What's an IE?," *Ind. Eng.,* vol. 5, no. 7, pp. 26–27, July, 1973.

12 Taylor, Frederick W.: *Scientific Management,* Harper and Row, New York, 1947.

13 Tower, John G.: "Labor Law Reform: Essential to Fight Inflation," *Nation's Business,* December, 1971.

14 Wilson, Richard K.: "Effectiveness in Industrial Engineering," *Ind. Eng.,* vol. 4, no. 4, pp. 8–11, April, 1972.

REVIEW QUESTIONS AND PROBLEMS

1 What categories of cost represent the greatest controllable cost in industry today?
2 What are the two main requisites to labor cost control?
3 How does the basis for pay of workers differ in incentive plants as compared to day-work plants?
4 Prepare a description of an incentive pay system found in the literature.
5 On what main point do Herzberg (5) and Fein (3) seem to disagree concerning the motivation of direct-labor workers?
6 To what extent to you believe MS has "matured," as suggested by Magee (7)?
7 Give one example each of a problem you believe could be best solved by an industrial engineer, an operations researcher, and a management scientist.
8 On the basis of Smith's article (11), which of the twenty-one problem areas seem to be less applicable to industrial engineers?
9 What do you believe to be the future prospects for IE/OR/MS?

Name Index

Ackoff, R., 178, 309
Alexander the Great, 40
Annoni, A., 290
Apple, J., 34
Aquinas, St. Thomas, 12
Archimedes, 8, 9, 42
Arkwright, Richard, 24
Aristotle, 7, 8
Arnold, H. H., 43
Arrow, K., 297
Athenaios, 12

Babbage, C., 24
Barnes, R., 34, 67, 73
Barth, C., 31
Battista della Porte, G., 17
Bell, A. G., 314
Bellman, R., 173
Bentley, W., 214
Bertalanffy, L., 47, 48, 312
Bigelow, J. H., 313
Biringuccio, 17
Blackett, P., 42
Blair, R., 52
Boodman, D., 116
Boucher, W., 46, 52, 258, 259
Boulding, K., 47–48
Boulton, M., 23

Boyle, R., 17, 18
Brach, R., 214, 216, 220
Brandeis, F., 28
Bromilow, L., 51
Brunelleschi, L., 14

Caesar, J., 4, 10
Chaiken, J., 239, 245, 251
Cherry, C., 315
Chinal, J., 193
Christ, 40
Churchman, C., 191, 235, 236, 238
Clark, R., 274
Claudius, Emperor, 11
Clemenceau, G., 36
Colbert, J., 18
Conway, R., 52
Cooke, M., 32, 35

Dantzig, G., 137, 159
Daschbach, J., 214
da Vinci, L., 15–16, 21, 24
de Camp, S., 3, 4
de Forest, L., 20
Demetrius, 9
Descartes, R., 17, 18
Diocletian, 12
Dionysius, 8

Subject Index

Academy of the Lynx, 17
Accounting and cost control, 130–133
Accuracy of Delphi technique, in manpower
    forecasting, 295–297
Activity relationship chart, 85–89
Agricultural revolution, 4
Aircraft engine, selection of, 259–265
Algorithm:
    assignment, 137–146
    transportation, 146–157
American Institute of Chemical Engineers, 20
American Institute of Industrial Engineers, 20
American Road Builders Association, 205
American Society of Electrical Engineers, 20
American Society of Mechanical Engineers, 20,
    25
Analog simulation, 48
Aqueduct, Roman, 10–11
Assembly line balancing, 116–117
Assignable cause, 127
Assignment algorithm, 137–146
Automatic transit systems, 207–208
Acceptable quantity level (AQL), 130

Battelle Institute, 236
Behavioral approach to productivity, 306, 308
Blood bank inventory control, 226–228
    individual hospital, 228–230
    regional, 230–234

Bridge building, China, 13
Bus systems for UMPT, 206–207

Canals, Chinese, 13
Central limit theorem, 75–76
Charting methods, 62
Computer science, 53
Computerized plant layout, 92
Control:
    cost, 130–133
    inventory, 106–112, 226–235
    limits, 127
    production, 112–125
    quality, 125–130
Convergence, in Delphi technique, 293–294
Conveyors, and mass production, 24
CORELAP area layout, 92, 95
Cost control, 130–133
Cost-effectiveness study, 47
Courts, state criminal, 214–226
CRAFT area layout, 92, 93–94
Crime prevention patrol, allocation of, 254–
    255
Criminal courts, state, 214–226
Critical path method (CPM), 120–125
Cybernetics, 45, 47, 312

Data collection in solid waste management
    system analysis, 276–277

333